WITHDRAWN
UTSA LIBRARIES

James C. Gifford

THE UNIVERSE OF THE MIND

SEMINARS IN THE HISTORY OF IDEAS

Editorial Board
George E. Owen *Neville Dyson-Hudson*
George Boas *Earl R. Wasserman*

Covenant: The History of a Biblical Idea
Delbert R. Hillers

Ideas About Substance
Albert L. Hammond

Vox Populi: Essays in the History of an Idea
George Boas

The Universe of the Mind
George E. Owen

THE UNIVERSE OF THE MIND

GEORGE E. OWEN

THE JOHNS HOPKINS PRESS
Baltimore and London

Copyright © 1971 by The Johns Hopkins Press
All rights reserved
Manufactured in the United States of America

The Johns Hopkins Press, Baltimore, Maryland 21218
The Johns Hopkins Press Ltd., London

Library of Congress Catalog Card Number 76-125674

International Standard Book Number 0-8018-1131-7 (clothbound edition)
International Standard Book Number 0-8018-1179-1 (paperback edition)

Originally published, 1971
Johns Hopkins Paperbacks edition, 1971

LIBRARY
University of Texas
At San Antonio

To My

Mother and Father

Contents

Publisher's Note

The history of ideas as a form of scholarly inquiry took shape at The Johns Hopkins University in the first half of the century. The man chiefly responsible was Arthur O. Lovejoy, whose twenty-eight years as professor of philosophy were spent promoting the historiography of the intellect. With two colleagues, George Boas and Gilbert Chinard, he founded the History of Ideas Club, where, in an atmosphere at once congenial and critical, visiting scholars might offer their interpretations of the development of the great ideas that have influenced civilization. Lovejoy was instrumental in founding the *Journal of the History of Ideas* in pursuit of the same end. And in his own writings he persistently and patiently charted the transformations which a seminal idea might undergo—over time, across disciplines, or within the intellectual development of an individual thinker.

When, with Carnegie Corporation support, The Johns Hopkins University inaugurated an imaginatively new program of adult education in 1962, it was a happy inspiration to build it around a set of graduate seminars in the history of ideas, for the History of Ideas Club itself had long before been described as "a sort of seminar where mature men and women learned new and valuable lessons." To be sure, this evening program has followed Lovejoy's spirit of inquiry rather than his own actual practice. Not all the seminars are concerned to pursue in detail the transformations of a single unit-idea. Rather, there is a shared view that no theory—at any time, in any field—is simply self-generated, but that it springs by extension or opposition from earlier theories advanced in the field, or is borrowed from theories in cognate fields, or is derived from the blending of hitherto separate fields into one. To pursue the unfolding of any theory in these terms (so the teachers in the seminars believe) allows a sophisticated and rigorous discussion of contemporary scholarship with an audience lacking previous specified knowledge. These notions are an extension, not an abuse, of Lovejoy's concern; he had never wasted effort on being unduly prescriptive except to call, hopefully, for cooperative scholarship in a venture so clearly beyond the reasonable capabilities of a single scholar.

This series of books, *Seminars in the History of Ideas*, is intended to provide a wider audience with a chance to participate in "a sort of seminar" similar to those in the University's program. Just as the teaching seminars themselves draw on the spirit rather than merely the letter of Lovejoy's original enterprise, so this published series extends beyond those topics already offered in the University's program. But all, nonetheless, reflect that intent with which Lovejoy so long persisted in his own work: "the endeavor to investigate the history, and thereby, it may be hoped, to understand better the nature, of the workings of the human mind."

Preface

Everyone has an organized image of the world surrounding him, and the essence of this abstraction of reality may be considered mathematical or quantitative. To inject some speculation about the logical evolution of mathematical thought, the contents of this book were put together and presented as a course in the program known as The History of Ideas. The audience for this consisted of candidates for the Master's degree and represented a wide variety of specialties and interests; it was this wide range that was of special significance to the author in the preparation of this course.

Only too often the professional mathematician or physicist writes and speaks for and to the professional level in his own field. In this sense, even the students who are filtered into the standard undergraduate programs are professionalized. When faced with the challenge of students having a broad spectrum of interests, the professional scientist is hard pressed to "explain," *in the true sense of the word*, the basic concepts of his field of interest. There is a certain danger inherent in dealing only with professionals; my own experience under these circumstances has been that there is a tendency to avoid elementary explanations and rhetorical description. Using this latter technique, parts of this course could be thought of as historically regressive. Algebra, which has existed for four thousand years or more as a rhetorical science, came to maturity within the last fifteen hundred years as a syncopated and, later, a symbolic system. With the facility and power of the symbolic system well established, the forms of the rhetorical method have disappeared. Certain components of expression have been lost with its suppression—for instance, physics students can seldom discuss Maxwell's equations without recourse to some symbolic representation. The opposite situation can be present with the nonscientist. In this case I have often found that the only significant means of handling elegant equations in theoretical physics is via a rhetorical and geometric method. As an example, the nonprofessional can discuss the linear wave equation for the vibrating string in terms of Newton's second law; concepts of divergence, gradient, and curl have a geometric-rhetorical description which is quite adequate up to a point. By this description I do not intend giving the impression that little can be achieved in the direction of understanding the symbolic methods—on the contrary, this historical approach carries with it a certain amount of obvious logic. I find that after carefully working his way through the various attitudes toward certain universal problems in mathematics, the student invariably finds that certain new ideas—such as the calculus—are rather trivial extensions of the ideas which he has already assembled. Another rather rewarding aspect of this approach is the ability of the student, in step with an historical trend, to appreciate readily the abstract generalization of a field of mathematics. This mastery is usually demonstrated in the theory of groups and then in the creation of non-Euclidean geometries.

One of my initial aims in this course was to delineate clearly the period when frontier mathematics and theoretical physics separated. These were closely allied by the fourteenth century, but one discovers that in the interval between 1820 and 1840 the traditional alliances between physics and mathematics dissolved. Until this time, every great mathematician had some interest in, and made some contribution to, physics. In the same manner, most theoretical physicists were, to some extent, originally mathematicians. After 1840 the formal ties between the two subjects became quite suppressed, and by the turn of the century, any propensity toward science on the part of a mathematician was considered to be a definite handicap. Of course, this is not to deny the existence of exceptions, such as Poincaré and David Hilbert; on the other hand, such men do represent an exception to a rather loose rule.

Another characteristic of the composition of this material is the ordering, or rather nonordering, of topics. Traditionally, histories of either physics or mathematics order the material under topical headings. For instance, one may trace the development of dynamics in time intervals of one hundred years or more. Such a topic might then be followed by a description of the evolution of concepts in thermodynamics. With the material oriented toward topics, one is constantly retracing his steps in history, the result being a certain amount of confusion as to which idea came first. Because of the broad scope of this work, I decided several years ago to orient my discussion in terms of the contributors, and to allow the related topics in mathematics and physics to appear in their natural order. This approach lends some texture to the material, in that the reader is forced to readjust constantly to a multitude of topics which appear in an appropriate time sequence. As a result, a short discussion of the development of one aspect of algebra may be followed by the description of an innovation in electromagnetic theory. The latter may be followed by the story of an advance in the theory of numbers. The technique employed here is concentration upon the important personalities, bringing them into the text roughly in the order of their dates of birth.

It was surprising that this method of laying out the topics yielded more coherence in subject matter than I initially suspected. This fact in itself is quite instructive: it suggests that the great minds of a given period tend to concentrate upon those theories and formalisms which are ripe for exploitation. Without forcing the material, one finds that some topics tend to be expanded in spurts—a particularly striking example of this is the formulation of theories of the magnetism of current-carrying wires.

Although the historical facts of this work are in proper order, they have been presented with little regard for the original notation. This was done for several reasons, one of which being that the historical ordering has been utilized more as a logical device than as a display of scholarly accomplishment. My major aim has been to teach concepts; therefore, whenever possible the modern notations have been employed, with some side references as to the original presentation of the material. I trust that this approach will not offend the purist who may wish to observe the discussions in their original framework of notations. Another reason has been that of efficiency. Because a wide range of special topics is contained in this single volume, I believe that overconcern with the original style and presentation of an idea would invariably lead to an unwieldy manuscript.

The development of chapter titles has been an entertaining game. As the manuscript began to take shape, I found that I viewed the personalities almost as actors in a drama. For this reason the over-all titling began to assume a dramatic character. Most titles are a personal choice, and in one case—"A Man for All Science"—I borrowed the chapter heading from an article written by Thelma Nason about my teacher and colleague, Professor Franco Rasetti. The title of Chapter X is a very bad pun, which I found irresistible in view of Einstein's lofty criticism of the quantum theory.

My indebtedness is great. In the five years during which this course was presented, my students proved vital. Not only was I able to test the material through them, but I learned a great deal from their discussions and from some very superior term papers. Above all, I am most grateful to Mrs. E. Meyerson, who has worked diligently to keep the text in order and who made innumerable beneficial suggestions. My sincerest thanks are extended to Mrs. I. Keesler for her typing and suggestions on the final portions of this manuscript. Last, I wish to acknowledge the insight of Dean Richard Mumma, who created the sequence of courses which now constitute the Master of Liberal Arts program at Johns Hopkins. His pioneering venture into a truly broad educational program at an advanced level should leave a lasting mark upon our system of adult education.

THE UNIVERSE OF THE MIND

I

About the Play

When we consider the world around us, we tend to make some sort of qualitative judgment involving an assignment of a magnitude, and for this we depend upon number, the fundamental basis for all mathematics. One can but surmise that after long and repeated exposure to collections of objects, very early man began to systematize his descriptions of these collections, and evolved a number structure. Since the earliest civilizations number has played a primary role in communication. Basic to any number system are the operations implied between the elements of the system, and records have been discovered which indicate that as early as 3000 B.C.E. the Sumerian civilization possessed a relatively sophisticated number system, within which complex arithmetical problems could be formulated.

Since the most rudimentary and primitive aspects of mathematics are of the highest significance, it is essential to examine those early processes by which the quantitative descriptions of systems and of individual objects were abstracted. We can assume that number was created to facilitate comparisons between groups of objects, like or unlike. This discernment seems to be one with which living things either are endowed or acquire through experience, and the recognition of the numerical essence of a collection was a major advance in the intellectual development of man. At the most primitive level mathematics was perhaps a construct of inequalities; for instance, homoerectus was undoubtedly concerned over the problem of whether one apple was a lesser quantity than, say, five apples, and how the squirrel decides when he has enough food stored for a long winter! At a very early stage, decisions about inequalities must have evolved into a process of labeling, such as "greater than," "equal to," and "less than."

A more sophisticated view is required when one compares more than two collections. Even more difficult is the problem presented when a comparison must be made between collections of unlike objects. It is certainly possible to arrange the comparison as a chain of inequalities, but when the collections are of different objects, sooner or later the question of what is being compared must be raised. One of the triumphs of civilized man was the extraction of the abstract idea of number from a collection. It follows, then, that number is a universal property, and for this reason societies have upon occasion attempted to build their religious structures about it. Perhaps even today, without formalizing the association, we deify number in many attitudes of our daily life.

The extraction of the numerical essence from any collection represents the creation of abstract numbers, and comparisons of civilizations show that the abstract quality of number varied from one to another. The Babylonians appear to have been much more comfortable with the idea of an abstract number than the Greeks. As an example, one finds the Pythagoreans viewing sums as equivalent to perimetric measure, while multiplications were associated with the areas of rectangles. Very early in the history of mankind, then, a correspondence was established between a series of marks on clay or rock and the number associated with a set of real objects.

The introduction of the game of combination, which is called arithmetic, must have occurred very early. Once again our earliest records show that the arithmetic of the Sumerians was quite advanced. Addition as the most fundamental binary operation between numbers should have been self-evident. The early number symbols are, in fact, created by adding in sequence a series of identical marks. Multiplication, being a series of additions, followed shortly. Certainly the pressures brought about in the formation of organized tribes and societies would have forced the development of a workmanlike arithmetic. In the more advanced societies of the Sumerians and Egyptians, multiplication and division were commonplace in the allocation of land, leveling of taxes, assignment of interest, and so forth. Utility, then, must have been the major influence upon the evolution of a system of arithmetic. Basic arithmetic appears to possess universality.

Because early arithmetics were born of human experience, they are readily comprehended and have marked similarities. The conclusions of these basic arithmetics are convincing and unalterable. The experience of ages is built into them, and the repetitive usefulness of the operation of addition provides a uniform aspect to all. Multiplication of integers is an extended addition, and thus the multiplications are quite similar. Dissimilarities appear when the arithmetics incorporate operations involving nonintegral numbers. Before fractions are encountered, arithmetics must deal with problems of division, and it is from division that the field of rational fractions is developed.

It should be emphasized that the act of forming relations between an abstract system and observations in the real world in order to account for and project experiences is basic not only to mathematics but to the sciences as well. At a fundamental level the theoretical physics of today deals in large part with connections between an abstract structure of symbols and observations extracted from physical systems. Utility again plays an important role because the test of an appropriate theory lies in the correspondence between the predictions of the abstract structure and the observations derived from experiments.

The association of a number with a collection of objects is an idealization, in the sense that one can speak usefully about the number of objects without regard to the multitude of other properties which the collection or the individual objects may possess. In the development of physical theory, the concept of idealization assumes an equally important role. One may generalize a system or simplify it until it no longer has any counterpart in the real world. Usually, however, the idealization represents some approximation to a real system; and while a real system may never behave exactly as predicted, a successful idealization will very often anticipate behavior to a high degree of usefulness.

In the evolution of the abstract systems called mathematics, we find many similarities with the evolution of alphabets. Although the constraints of experience and the clarity of logic have provided mathematics with a universality which transcends ordinary language, mathematics can be regarded partly as a communication system. Alphabets develop economically, and arithmetics exhibit the same characteristic. An important feature of alphabet growth is the number of symbols required to represent the phonetics of a language. Given a spoken language with a wide range of sounds, one can ask, "What is the optimum number of letters needed to adequately represent not only the original words but those which will eventually come into being?" The progress of a civilization, to some extent, may be measured by the phonetic representations adopted for the written language. The use of ideograms as graphic symbols to express an idea or an object without expressing the sounds that form its name provide the classic example of cumbersome representation. Granted that most alphabets began as ideograms, those languages which retained such a system of writing—the Egyptian and Chinese, for example—became extremely unwieldy and inadequate for ready extension. While many other factors are involved in the progress of a civilization, the economy and clarity of its language symbols are of fundamental importance.

An essential economic aspect of any number system is its base. Some early systems de-emphasized the base, and effective number systems can employ a broad range of bases, but the economic constraint imposes some limitation. The number base, like the total number of letters in an alphabet, represents a limit on the number group necessary to exhibit all of the numbers required. When the base number is exceeded in counting, the system repeats itself, with or without a carrying symbol to indicate the order of the repetition. Various bases are found in the ancient number systems. The American Indians had a variety of bases, of which 5 was popular; the Mayans had a base of 20, the Egyptians 10, and the Babylonians used a base of 60.

Both mathematics and physics depend upon a useful notation, and often great advances in mathematics occur or are made apparent in the form of a major notational innovation. The Greek renaissance of mathematics was partly a response to the introduction of geometry as a representation, and many significant abstract algebraic problems of Babylon were reformulated, clarified, and extended in terms of Grecian geometry. The Babylonian rhetorical algebra was quite limited; proofs of the methods employed were not demonstrated, and one can assume that the rhetorical method was too cumbersome even to suggest the concept of a rigorous proof. The geometrical notation of the Greeks, on the other hand, opened the way to a profound era in the history of mathematics. Later, we shall observe that when the calculus was developed simultaneously by Newton and Leibniz, it was the superior notation of Leibniz which extended the calculus at a much faster pace in Europe. In fact, by patriotically adhering to the notation of Newton, mathematical advances in England were retarded by almost one hundred years.

The adoption of a geometrical interpretation of number, and the earlier formulation of division, led the ancients to a major puzzle of mathematics: the continuum of real numbers. The first hint of the dilemma arose in the study of irrational numbers. Notation is by nature discrete; further, the one-to-one correspondence between a notation and the field of integers gave arithmetic a semblance of security. Another puzzle inherited with the continuum is the concept of largeness and its limit. This serves as a signpost for the dual systems which arose and led to endless controversy in mathematics and physics. Associated with the Greek period is the creation of the deductive method of proof in mathematics. This remains to some extent the method of mathematical proof today. Deductive reasoning begins with a group premise, and the proof consists of deducing whether the object in question does or does not satisfy the group premise.

Algebra as a rhetorical structure had been present in Babylonia, and early algebraic problems, in contrast with arithmetic problems, involved computations of a higher complexity than those of simple addition and subtraction. Algebra consisted of problems in which an unknown and a set of conditions are stated, and from these the value of the unknown must be extracted. This is not to describe algebra in general; however, it does provide a loose description of the earliest rhetorical algebras. By 250 C.E. algebra was phrased in what has been called a syncopated form, wherein every power of the unknown is assigned a separate symbol. In the period from 300 to 1100 great strides were made, first by the Indians and then the Arabs, and by 1130 the Indian mathematician Bashkara had investigated and produced rules and cautions for multiplication and division by zero.

Until the sixteenth century little attention was directed toward the mathematics of continuously varying magnitudes. The concept of a function which is essential to progress in the physical sciences was present in the form of a few space curves. When the kinematic quantities of velocity and acceleration were examined, European scholars bridged the gap between the discrete mathematics of the past and the analysis of continuous systems. In the sixteenth and seventeenth centuries analytic geometry, calculus (or analysis), and theoretical physics were created.

The relatively late appearance of theoretical physics rests in part on the lack of an adequate representation. It requires a well-understood formalism involving functions, and despite the fact that much can be accomplished by geometry, it alone is ill suited to the task of describing physical events. Descartes' development of analytic geometry was the first major advance toward a workable notation for physics, and a simple and useful advance in algebra, the superscript notation for powers of numbers, was a further significant contribution of the period.

Although the sixteenth century is customarily credited with the rise of theoretical physics, the contributions of antiquity must be acknowledged. Astronomy had held the interest of the early Babylonians. In spite of the failure of early theories based upon geometric figures such as the epicycles, efforts in this direction represent an attempt to establish a physical theory. In the physical thought of Archimedes, a mixed system of intuition and laws of static moments had been applied to solve problems of area and volume. Thus the sixteenth century was rich in attitudes proper to the spirit of inquiry and the conception of models for physical processes.

Nor can the rise of physics be attributed solely to refinements in mathematics. A more careful concern with observation characterizes fifteenth- and sixteenth-century thought, when emphasis is placed on an accurate recording of observations, both in the fine arts and in science. The world's debt to Galileo would be immeasurably vast, even if we considered nothing more than his insistence upon experimental verification. Mankind had spent several thousand fruitless years, unfortunately, in dissociating experience from models of the physical world. The motion of such a simple system as that of a projectile was deemed important for a thousand years, yet in spite of this we find fifteenth-century scientists providing constructs which were never checked by observation of real systems. Here again, one might argue the inadequacies of mathematics as a causative factor because it failed to provide a simple representation against which observations could be checked—a point well illustrated by Kepler's long effort to synthesize the observations of Tycho Brahe into appropriate planetary orbits. Before the appropriateness of the conic sections became apparent to him, Kepler spent virtually a lifetime in trial and error.

Before the sixteenth century, physics was centered largely around static problems—problems concerning systems fixed in space and constant in time. With the heightened clarity of description in kinematics, that is, position, velocity, and acceleration, the men of science came to grips with problems of continuous change. By emphasizing changes in quantities, scientists became more and more aware of functions, of the curves represented by functions, the tangents to points on these curves, and the areas encompassed by the curves. At the time of the complete formulation of the calculus, Newton and Leibniz had the heritage of nearly a hundred years of viable mathematics, and during that time much of the preliminary work upon derivatives and integrals had been performed. Misinterpretation of common experience bore the opprobrium of past error, while the introduction of experimental testing opened vast areas of renewed interest and inquiry. The inverse square law of gravitation is a familiar challenge. It is now apparent that several scientific speculators as well as Newton had suggested that gravitational attraction of the planets to the sun would eventually be shown to depend upon the inverse square of their distance to the sun. We also understand now from symmetry arguments that this hypothesis is the most elementary suggestion one can make; all other force laws for a spherical system lead to an unmanageable set of calculations. The mere hypothesis in itself was of little value because the final test for validity required a demonstration that such a force law would in truth predict elliptical planetary orbits. By applying his own creation, the calculus, Newton was able to derive the form of the planetary orbits from the inverse square law and to supply a number of other supplementary but essential proofs as well. This fortuitous combining of mathematics and physics was characteristic of science from antiquity to the middle of the nineteenth century. Only within the last century have the two disciplines parted into separate and seemingly isolated camps—the future will show what profit such isolation brings to science.

Idealization is a vital element in the construction of the abstract system, mathematics. A more sophisticated idealization is required when describing physical systems by mathematical models. Real systems are so complex that one must be content with a representation sufficiently simplified that, while it can never predict exact behavior, it can predict the behavior of the real system within the limitations of the representation used. Once we are convinced that the approximation of the behavior of the real system is sufficiently close for a successful theory, experience then teaches us that these idealizations are useful and valid.

With the advent of Newtonian mechanics, the world of science was committed to a system of laws and to derivations based on these laws: the concept of the exact sciences was born. From what has been said, however, we are immediately aware that "exact" is not exact relative to correlations between prediction and observation. Furthermore, history demonstrates time and again that exact theories may later be shown to be approximations and special cases of broader theories. Thus the concept of a fundamental law may be questioned. Is it fundamental? Is it a law? Such queries require careful answers. First, the concept "physical theory" should be defined, and the following will be assumed as a definition: "Physical theory is a logical structure based upon assumptions and definitions which permit one to predict the outcome of the maximum number of experiments based upon the minimum number of postulates."* Such a definition does not suggest exactness. It implies that deductive reasoning has proved useful in predicting the outcome of certain physical experiments when applied to certain basic mathematical models. As to the truth or nontruth of a given model, the structures of theoretical physics invariably have only a local validity, since the boundaries of theory are constantly changing and new theories are required. When a theory which has been valid in a limited domain is superceded, it is extended into a newer theory as an approximation. Therefore, we invoke the concept of a domain of validity† for a theory.

Once a theory of physics is postulated, the limits of its validity domain are set, and these cannot be changed without extending the theory. Interestingly enough, the limits of domain may not be discovered until a new theory is created which contains a given theory as an approximation. By these arguments, one suggests that theoretical physics is composed of a hierarchy of theories, each covering the other in ascending order. At any given time, there is one theory which we shall call the covering theory, and the earlier theories which it covers are not to be thought of as invalid, but only as restricted in their domain of validity. A covering theory will therefore have a domain of validity which encompasses and extends beyond all the validity domains of the theories which preceded it within a given hierarchy. Relativity theory is a good case in point. The covering theory at present is General Relativity, and in descending order of validity we find within it the Special Relativity and Galilean Relativity. The Special Theory of Relativity emerged as the covering theory for Galilean Relativity.

The concept of the hierarchy of covering theories is not sufficient, however, to account for what is known as a fundamental law. A law of nature in physics is an invented proposition which is confirmed by relevant experiments. There is no justification for the law other than experiment and the optimum conditions specified by the concept of "covering theory." A theory is confirmed when it has been shown to predict a reasonably accurate outcome of experiments, and inherent in the word "reasonably" is the admission that absolute accuracy is impossible. Thus, we postulate rather than derive theory, and there is no "why" for the postulate. In the early eighteenth century there were two different philosophies concerning the content of an acceptable theory. The Cartesian school held the position that to be acceptable a theory must explain or describe the entire universe. The opposite point of view, and certainly the more reasonable, was presented by the Newtonian school, which concerned itself only with

* F. Rohrlich, *The Dynamics of Charged Particles* (Reading, Mass.: Addison-Wesley, 1965), p. 1.

† *Ibid.*, p. 2.

theories of limited but workable application. Newton did not attempt to explain why the law of universal gravitation was inverse square; once postulated and shown to predict the outcome of experiments with reasonable accuracy, a theory is accepted as a basic law. Newton successfully defended this position, and scientific theory is taken to imply predictability within the limited framework of a covering theory and the basic postulates upon which the theory rests.

It is the mathematical structure of a theory which provides its major rules of manipulation. Indeed, the extension of a theoretical system beyond the boundaries of the domain of validity depends, in addition to insight, largely upon a comprehensive knowledge of mathematics. The pressure of physical problems provided an enormous stimulus in many cases for progress in mathematics, particularly in the eighteenth and nineteenth centuries. Today, however, the introspective and specialized interests of mathematics have divorced it from the area of physical experience.

Theoretical physics combined with mathematical analysis made most of the advances which were possible in their respective domains during the seventeenth, eighteenth, and nineteenth centuries. The science of mechanics reached its peak by the early years of the nineteenth century. By 1900 electromagnetic theory was formulated into the optimum set of equations; thermodynamics was essentially complete, while kinetic theory and statistical mechanics had just been established. As experimental evidence accumulated near the turn of the twentieth century, the old classical covering theories were discovered to be profoundly limited and incapable of anticipating much of what is now known as modern physics.

Physics underwent a dramatic upheaval as the twentieth century unfolded, and exciting changes in domain occurred in almost every area. Relativity theory was superimposed upon the classical mechanics and electromagnetic theories; quantum theory added new concepts and covered the domain of mechanics and the microscopic electromagnetic theory. Needless to say, changes in theory are still taking place, and as a consequence, the future domains of validity and fresh theories are yet to be revealed.

From the beginning of the eighteenth century to the middle of the nineteenth, mathematics was governed by intuition and a rigid formalism. Rigorous proofs of most of the theorems of the calculus were outside the capabilities of the mathematical giants of the time; instead, intuition and physical experience were relied upon to point the way to progress. Elaborate extensions of the calculus took place, while its very foundations were subject to criticism. The theory of limits, the concept of convergence, ideas of continuity—all were virtually nonexistent. The realm of the calculus was vastly extended and came to be called "analysis," while a notation of elegance and sophistication was developed. In part, this speculative attitude was implemented by the purely practical gains achieved by the application of analysis to physics. Rewards for ability in mathematics were readily attained in this atmosphere, as might well be expected when the areas of investigation are new and the domains of possible extension are boundless.

Much of classical analysis was brought into being through its application to analytical dynamics. Infinite series, the calculus of variations, and, in particular, the theory of ordinary and partial differential equations were expanded and formalized to a polished elegance. The expectation of striking gold wherever one looked provided the impetus for such feverish inquiry that much of what could actually be accomplished in these areas occurred within the short space of one hundred and fifty years. One offshoot of this burst of activity was a collection of special functions which is still of profound interest to the modern physicist. Ironically, it holds little interest for the modern mathematician.

Having briefly traced the parallel courses of classical physics and classical analysis, we should examine the differences in the progress of physics and mathematics. As physics continued to reveal new phenomena, which, in turn, gave birth to broader covering theories, physicists in the late nineteenth and into the twentieth century continued to look outward for the boundaries of the old theories and the stipulations of the covering theories. Whatever introspection took place was in terms of relating the broader concepts of new theories to their analogies in those superceded.

Mathematics progressed by an entirely different route. Advances in physics continue to this day through the acquisition of new knowledge, that is to say, knowledge of phenomena which have not hitherto been observed or, if observed, have been neglected. Mathematical innovation, on the other hand, took place through introspection, by reexamining the basic postulates of a given system and by questioning and reevaluating the connections between apparently different fields. Mathematicians at the beginning of the nineteenth century challenged the uniqueness of the postulates upon which the structure itself rested; they began to demand that every step in a mathematical development be logically derivable from the basic postulates, and ultimately structures were divorced from experience to become as abstract as the human manipulators could possibly make them.

The concept of producing a mathematical structure based upon the minimum number of postulates played a role comparable to the analogous concept in physics. In contrast, however, mathematicians discovered that vast new areas could be created by artificially suppressing certain basic postulates in a given set. At this point we perceive a fundamental difference between doing modern mathematics and applying mathematics to the real world of physics. By generalizing and suppressing some basic postulates in mathematics, vast areas of investigation were exposed—so vast that in some cases generations of investigators occupied themselves with the ramifications of a basic set of condensed postulates.

This overly simplified description does not, of course, reveal to any appreciable extent all that took place in mathematics after 1800. One significant concern was with the validity of human understanding. By the end of the century, controversies were under way regarding the concept of infinity and the uniqueness of number as a basis for systems, and it is not clear even today that some of these questions will ever be satisfactorily resolved.

Further challenges arose concerning the self-contradiction of a set of postulates. Hilbert inquired whether it was at all possible to reach contradictions in a finite number of logical deductions by proceeding from the postulates of arithmetic. An apparent answer was provided by Gödel and later modified by Cohen. While these efforts went on, new fields of mathematics were created and existing areas were excessively fragmented. This fragmentation of mathematics into numerous subfields created an attitude of uneasiness among observers concerning the ultimate ability of mathematicians to understand and communicate their own achievements in a useful fashion. The trend, then, in modern mathematics can roughly be characterized as a reexamination of foundations, a creation of rigor, and the subsequent creation of covering theories based on abstract generalization. Only time will provide a reasonable evaluation of the value of these efforts; but to judge from the experience of the mid-nineteenth century, the extended activity of modern mathematics should prove invaluable.

In reviewing the progress of theoretical physics and particularly mathematics, one observes the recurrence of certain fundamental dilemmas and certain mental attitudes. The paradox of Zeno was apparent in the criticism of the calculus at the time of Newton, and the concept of the infinite continued to plague mathematicians at the beginning of the twentieth century. Again it was set aside as only partially understood.

The "Golden Age" of Greek thought ended when the application of the particular prejudice and means of notation of the Greeks was exhausted. In this connection, it is worth observing that a profound change took place in the domain of physics in the twentieth century. From ancient times down to the nineteenth century, physics was concerned with the sensory worlds, macroscopic and astronomical worlds which made phenomena known through seeing, touching, hearing, and so on. The associated models, however abstracted and idealized, provided solutions which were nonetheless observable through the senses. In the twentieth century, the microscopic physical world was discovered and represented, and the connection with the senses became of second order. Objects under investigation were only connected to the macroscopic world through rather elaborate instrumental communication. This is often likened to a situation wherein a black box is interrogated and its corresponding responses recorded by intermediary instruments. Thus, modern experiments consist of controlled and specialized interrogations accompanied by associated responses. Model-building with mathematics has an abstract quality quite unlike the representations of classical physics. However successful these endeavors might be, the majority of mathematical tools employed are extrapolations from the mathematics applied to the classical macroscopic world.

These statements are not intended as a dogmatic assertion that the twentieth century necessarily represents a situation analogous to the second or third century C.E. in Greece. However, such a possibility is as plausible as the possibility that we are on the threshold of even more remarkable successes. Man has achieved the limitation of his particular mathematics in the past and will quite possibly reach the same point at some time in the future. After all, the pride and arrogance of mankind has repeatedly created dark ages of dogma. During these periods, authoritarian rule and a belief in the absolute truth of particular statements about the physical world has held progress in abeyance. It is not inconceivable that science and mathematics could become the medieval church of the future, containing large numbers of ruthless adherents who, in their efforts to maintain power, restrict investigation to those areas which they maintain to be knowable. The practitioners of science and mathematics are gaining power, while their ranks are becoming crowded with zealous adherents. Power over governments and peoples is new to the scientist, and the manner in which this power will be wielded is unknown. Judging from the examples of history, it is clear that both mathematics and physics in the future will undergo advances and changes of the same relative magnitude as in the past. Although it is impossible to anticipate the form which these advances will take, it is essential to recognize that an absolute interpretation of the world by physics or mathematics is not within our power. The only useful questions that can be raised concern the rate of advance or change. For one reason or another, in the past the progress of mathematics and physics has been virtually suspended over long intervals which varied from hundreds to thousands of years. Breakthroughs may come in notation; however, it is very likely that entirely new mathematical methods are necessary if new and concrete advances are to be made.

II

Setting the Stage

Science and mathematics have, of necessity, been molded to an appreciable degree by the military and economic character of a society. This interrelation is not found in all cultures, to be sure, for the Romans, although supreme in military and technical advancement, figured but little in the progress of mathematics. Any comprehensive evaluation of the growth of mathematics, however, must take into account the ebb and flow of past civilizations. The earliest records of mathematical forms belong to the period around 3000 B.C.E. in Sumer. Although Sumerian cuneiform script, believed to date back as far as 5000 B.C.E., has been discovered, mathematical tablets predating 3000 have yet to be unearthed. These early mathematical forms are refined to a level that suggests a long history of mathematical development in Sumer before 3000, but the historical evidence of the last few thousand years teaches us to be wary of such conclusions. A scant hundred years or so has often sufficed to produce profound changes and innovations in mathematics, and we must proceed with caution in our attempts to postulate the state of the mathematical art prior to the first extant records of a given society.

For hundreds of years, western European scholars had assumed that most of mathematics had originated and had come into flower with the Greeks. The few evidences of Egyptian arithmetic seemed poor by comparison. Only recently has it become apparent that the mathematics of the Babylonians anticipated in many respects the outpouring of mathematical concepts of the Greeks. From 3000 to about 600 B.C.E. the two great civilizations upon which ours is founded were those of Egypt and the Middle East. Intellectual attainments and the flowering of knowledge in the Middle East took place among many different interacting societies, beginning with Sumer and continuing to Persia. Throughout man's early folk-wandering period, there was a constant movement of the nomadic pastoral peoples of central Asia into Europe, the Middle East, India, and China. As late as the fourteenth century of the current era, well-organized central Asian peoples continued to move into the Middle East. Our interest begins with the relatively stable civilization of the Sumerians, located at the northern part of the Persian Gulf.

Much of the Middle Eastern culture of the period from 3000 to 1000 B.C.E. is called "Babylonian" only to avoid specifying the multitude of city states, all similar in writing and culture but often economically and militarily at odds, included under that general heading. Thus we shall consider in the Babylonian development Sumer, Akkad, the Assyrians, proper Babylonian dynasties, the Kassite, the later Assyrian, and the neo-Babylonians of 1000 B.C.E. Even the Medes and Chaldeans or Persians will be considered for convenience as a continuation of the Babylonian traditions in mathematics.

The influence of Egypt in the Middle East is also important. During these two and one-half centuries, Egypt was subjected to the pressures of incoming civilizations. That country has, however, a much more unified aspect than does the area of the fertile crescent. Because Egyptian records have been preserved, one can delineate the Old Kingdom, the Middle Kingdom (2000–1800 B.C.E.), the Hyksos invasion, and the New Kingdom, terminating around 1100 B.C.E. Under Amenhotep III Egypt reached the height of her imperialism in 1400, controlling much of the western coastland of the Middle East, for by 1354, the areas of Syria and Palestine had fallen to the Hittites and the Hebrews. The introduction of iron weapons by the Hittites was a factor in ending Egyptian domination, and in the reign of Ramses II, after 1292, his empire slowly disintegrated.

Assyrian domination of the Middle East ebbed and flowed from that people's appearance around 3000 to the climax of their influence around 1100. The Assyrians assimilated much of the Sumerian culture; this is apparent in the evolution of the Babylonian number system from the early Sumerian cuneiform. Constantly expanding, under Sargon II, the Assyrians reached the borders of Egypt at the end of the sixth century. Within a few years the Middle East was invaded by Indo-European hordes of Medes and Chaldeans. In 614 the Medes under Cyaxares took Assur, while the Chaldeans under Nabopolasser conquered Babylon. Two years later they destroyed Nineveh. Babylon was rebuilt by Nebuchadnezzar II (605–562), who overran Judea and defeated the Egyptian armies at Carchemish. In this wise was formed the Persian Empire.

Ionian peoples began to move into the Peloponnesus and the western littoral of Asia Minor possibly as early as 1900 B.C.E. On the basis of the present knowledge of Linear B it is apparent that the Mycenaeans appeared in this area around 1580, conquered Crete in 1400, and settled Cyprus in 1200. The Dorians in 1100 drove the Mycenaeans into the hills of Arcadia, later taking Crete, and then spread into southwest Turkey. Greek influence in the Mediterranean was widely disseminated by 750 B.C.E. The Eubeans had colonies in Sicily and Italy, Achaean settlements were in southern Italy, and the Dorians had cities in Sicily. To the east of the Peloponnesus conflicts arose between the Greeks and the Persians. The fourth century saw a Lydian, Croesus, son of Alyattes, conquer Ionian, Aeolian, and Dorian cities in Asia Minor. The Persian king Kurush (Cyrus) moved against Croesus, who allied himself with Sparta and Egypt. Kurush defeated the Lydians, however, at the Battle of Halys, and then went on to defeat the Babylonians under Belshazzar in 540. Kurush was succeeded by Hystopres, whose son Darius in 512 invaded Greece, opening a two-hundred-year struggle between the Greeks and the Persians. It is against this fluctuating background of upheaval and change that the development of mathematics unfolds.

The earliest known mathematics occurs with the Sumerian and Babylonian texts, the most important of these early records belonging to the period from 1800 to 1600 B.C.E. All the arithmetical systems of this region are sexagesimal. The earliest Sumerians had separate symbols for 1, 10, 60, 600, and 3,600; and these symbols suggest an implicit place value. In the first diagram the gradual abstraction of the early Sumerian symbols may be observed, as they are replaced by the later Sumerian and Babylonian numbers. We observe the confused beginnings of the later

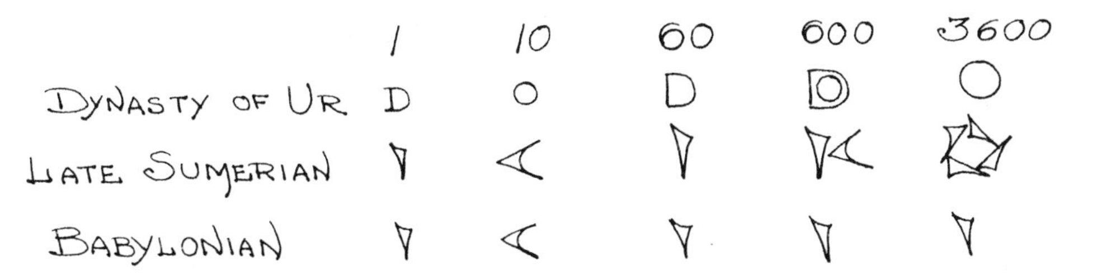

Babylonian number system. In time, usage and practical application led to a refinement and sophistication of the system, and we find that every power of 60, whether positive or negative, is denoted by the same symbol. As a consequence, the leading power of 60 must be defined from the context within which the number appears. Thus the figure-number in the second diagram may be interpreted as 60 + 24 = 84, or as 1 + 24/60 = 84/60.

Although the power of 60 is suppressed, the method possessed the advantage of being a place system—an incalculable advance in Babylonian arithmetic. By presenting an ordered set of numbers representing descending powers of 60, the Babylonians contributed by far the most functional arithmetic known to that time. It is still regarded as the most useful contribution of ancient times. Strangely enough, the Greek and Roman civilizations either overlooked or ignored the superior properties of the Babylonian place system for the most part, and adhered almost exclusively to an additive system. Only in astronomy do the Greeks appear to have recognized the merit of the Babylonian system and employed it in their calculations. An even greater tribute to the men of ancient Babylon is our own use of their base for angular measure. In our system of reckoning degrees, minutes, and seconds, the Babylonian sexagesimal obtains to this day.

No concrete evidence has yet resolved the logical question, "Why a base of 60?" Many early civilizations of Asia, Europe, and America used a digital system with bases of 5, 10, or 20. Numbers which correlate with fingers and toes offer a natural base, but the base of 60 has no such ready explanation. Speculations range all the way from the Chinese sexagenary period to a method of grain measure referred to in ancient Babylonian tablets. The Chinese sexagenary period produced the same magic number of 60 by the twelve earthly branches of five limbs and the ten heavenly stems of six. It is equally probable, on the other hand, that these resulted from, rather than produced, a primary base of 60. One of the most reasonable arguments in favor of 60 is the fact that the circle was divided into 360 parts to correspond to early (albeit slightly inaccurate) estimates of the days in the year. There followed the natural division of 360, which is obtained by dividing the circle into 6 parts by striking off arcs with chords equal to the radius. Then each partition could be divided into subsets of 60 parts each.

Most of the known Babylonian mathematical texts date from 1800 to 1600 B.C.E. and are either problem texts or table texts. On the surface, the multiplication tables appear straightforward, even trivial; however, a closer examination reveals the ingenuity of these ancient arithmeticians and their fundamental appreciation of structure and economy. The tables found and preserved, for instance, give products only from 1 to 20, followed by products for 30, 40, and 50. Economy is thus served, since all 59 products in a given tablet can be calculated by the addition of two of its members. Multiplication tables also appeared for nonintegral numbers,* such as a 1,20; 1,30; 1,40; 3,20; and 3,45, suggesting the existence of 3,600 rather than 59 tables. This puzzle was solved by the discovery that equal numbers appeared in the tables of reciprocals and that the set of existing multiplication tables, together with the reciprocals, provided a complete set. It is interesting that no reciprocals were given for the irregulars 7, 11, and 13.

Another major accomplishment of the Babylonians was the method of approximation. In any number system with a fixed base there are numbers which have an extended representation. To represent these numbers in a reasonable form, the Babylonians often resorted to in-

* O. Neugebauer, *The Exact Sciences in Antiquity* (New York: Harper and Row, 1962), p. 31.

equalities: 1/7 is representative of this technique. It is recorded as less than 8, 34, and 18, and greater than 8, 34, 16, and 59. In fractional notation this appears as:

$$\frac{8}{60} + \frac{34}{(60)^2} + \frac{16}{(60)^3} + \frac{59}{(60)^4} < \frac{1}{7} < \frac{8}{60} + \frac{34}{3600} + \frac{18}{(60)^3}.$$

Their ability to calculate square roots was superior to all civilizations of antiquity, including the Greek. A small tablet in the Yale collection shows the $\sqrt{2}$ as 1;24,51,10. The convention used here is to set off $(60)^\circ$ with a semicolon: 1; means $1 \times (60)^\circ = 1$. Succeeding divisors of $(60)^n$ follow. Therefore, 1;24,51,10 means

$$1 + \frac{24}{60} + \frac{51}{(60)^2} + \frac{10}{(60)^3} = 1 + 0.4 + 0.014167 + .000046.$$

This value is accurate to $22/(60)^4$ and corresponds to 1.414213 instead of 1.414214. One of the methods used for these calculations is similar to the Babylonian problem of the reciprocal sum. If n is the number whose square root is to be taken, and if a is the approximate root, then $a_1 = (1/2)(a + n/a)$ is a better value because one term is larger than the desired number, while the other is smaller. The new approximation, a_1, can now be used with this method to give an even better value, $(1/2)(a_1 + n/a_1) = a_2$.

The Pythagorean theorem was known to the ancient mathematicians long before Pythagoras propounded it: both the Egyptians and the Babylonians have left us ample evidence of its use. In the Plimpton Collection at Columbia University, a tablet lists the triplet of numbers b, d, and d^2/l^2, satisfying the relation $l^2 + b^2 = d^2$. l, b, and d can be called the Pythagorean numbers. The Babylonians not only were interested in the three numbers but also investigated the ratios of these numbers; the numbers satisfying $d/l = (1/2)(p/q + q/p)$ are tabulated. These relate to the Pythagorean numbers by $l = 2pq$, $b = p^2 - q^2$, and $d = p^2 + q^2$ with p greater than q. Tables have been found for squares and cubes as well, but apparently the importance of prime numbers was not recognized.

Based upon our present knowledge (which is always subject to revision by future disclosures), it would appear that in general geometric consideration of a problem played a secondary role in Babylonian calculations. The fact that clay tablets were the means of keeping records may have served to conceal whatever geometric insights they may have had. Certainly the clay tablet lends itself most readily to an abstracted shorthand notation, whereas, while well-drawn geometric figures may well have had their aesthetic appeal, they could not be consistently produced in clay. Algebraic problems were well suited to the writing media.

To solve a problem in two unknowns, the Babylonian mathematician would reduce it to a problem in one unknown; this is called the method of the complementary unknown. Consider the two equations $xy = p$ and $x + y = a$. In solving a problem containing $x + y = a$, x was assumed to be $a/2$ plus an increment w, and y to be $(a/2)$ minus w. Thus the first condition, $xy = p$, with this substitution would be reduced to a quadratic in w.

The reciprocal sum $(x + 1/x) = b$ was solved for special values of b by the method of completing the square. The expansions $(a + b)^2$ and $(a - b)^2$ were probably performed with a geometric interpretation.

Strictly speaking, it cannot be concluded that geometry was completely ignored by the Babylonians. The 1936 discoveries in Susa indicate that they were intensely interested in polygons inscribed in or circumscribed about a circle. From the hexagon they found a value for the $\sqrt{3}$ of 1;45 or $1 + 45/60$. The area of a regular polygon of n sides is given as a characteristic number times the square of the length of a side. The circumference of a hexagon was given as 0;55,36 times the circumference of a circle, which implies 3;7,30 or $3\frac{1}{8}$ as a value of π. Generally we find that the Babylonians used the value 3 for π; in this case, however, we observe that a more accurate value was known to them. For zero they had no symbol but implied its existence in calculations by leaving a gap. The evaluation of π and the representation for zero are two of the most significant aspects of all ancient mathematical endeavors. Our respect for the mathematics of Babylon has been heightened with each new discovery, and it is not too much to say that Babylonian mathematics may rise to an even higher plane in our esteem.

Early man's most lively scientific interest was astronomy. As the drama of scientific development unfolds, we observe that the inquiring mind of man addresses itself first to the astronomical world, then to the macroscopic world, and finally to the microscopic world. In our own century, man has completed the cycle and returned anew to the astronomical world.

Space and time—man's ability to perceive a displacement between two points and his ability to distinguish the time ordering of two events—have always been his two chief scientific intuitions. They characterize man's earliest scientific concern. It is natural, then, that early man turned his attention to the heavens, for the recording of time has always been vital to human society. And the passage of time over long intervals is most readily and visibly discernible in the apparent motion of the sun, moon, and stars.

A theoretical model for astronomical motion cannot be attributed to the Babylonians from our present knowledge of the time, but clay tablets dating from 1700 to 100 B.C.E. suggest that Babylonian scientists were keen observers of the heavens and careful compilers of data. The use of data was mainly empirical. The records indicate, for example, that the solar velocity was represented as a series of step functions which averaged out to an appropriate value. Lunar velocities were analyzed as though the variations were continuous linear functions. These linearized data roughly resemble a sinusoidal variation. Notwithstanding these rough approximations, the Babylonian astronomical tables of 200 B.C.E. take into account several perturbations of the moon's orbit and of other celestial bodies. A distinction is made in science between a model of the astronomical system and an abstract empirical method; to the best of our present knowledge, the Babylonian system, successful as it was, was not based upon anything other than a set of rules for manipulating the numerical data.

The twenty-four-hour day seems to be of Egyptian origin; the 365¼-day year developed in both Babylonian and Egyptian civilizations. These two early cultures were, after all, not entirely independent. It seems to be a rule of life that when men do not fight they trade, and both wars and commerce fostered a great deal of intellectual exchange among ancient cultures, as they have done down to modern times. The coincidence of the motion of astronomical bodies and their variations from uniformity were explored to establish reliable clocks and calendars. Egyptian development of the time scale and the calendar was of incalculable value for all later civilizations. The twenty-four-hour day was accomplished by separating the night into twelve intervals and the daylight into ten, with one hour each for sunset and sunrise. The twelve-hour night was based on twelve sections of the night sky, the divisions being established by star locations. As each section rose and set below the horizon, one hour was defined.

The year was reckoned at 365¼ days, normally twelve months and five holidays. By synchronizing their calendar to the helical rising of the bright star Sothis (Sirius), the Sothic cycle was established. This cycle has a phase change of one-quarter day per year; thus in 4 × 365 years, the cycle is complete and the calendar begins a new period. Records of the cycle, or of correlations with the cycle, have proven invaluable in establishing the dates of ancient events.

The Egyptian number system was additive, with an implied base of ten. Hieroglyphic symbols for the basic number units are:

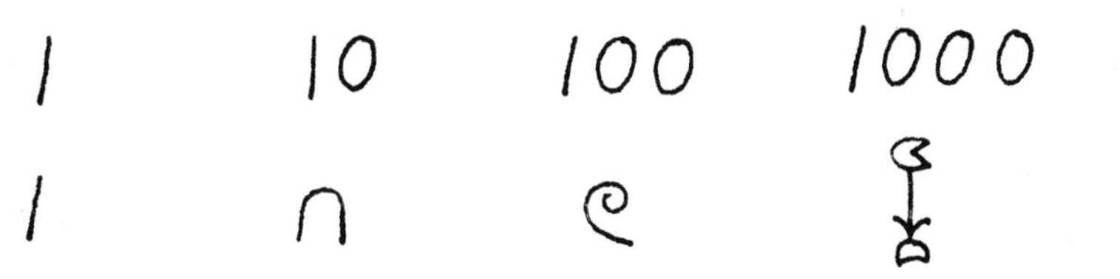

Addition was quite simple: one merely accumulated symbols and contracted when any subset exceeded its limit:

III and I∩∩ gives IIII∩∩ ; 3+21 = 24

II ∩∩∩∩ and I ∩∩∩∩∩∩ gives III@ ; 42+61 = 103

Multiplication was accomplished by doubling; the multiplier is broken down into a power of 2 plus 0 or 1. Multiplication is then done in steps which involve either multiplications by 2 or by 1. Consider as an example 21 × 9:

	21 taken once
	21×2 = 42
	42×2 = 84
	84×2 = 168
	168+21 = 189

This system is quite primitive and precludes calculations of any degree of sophistication. On the other hand, it served quite well for practical applications; the fact that it was easily understood and readily mastered accounts in some measure for its influence on Greek and Roman cultures.

The total influence of the Egyptian civilization upon later peoples was, of course, considerable. The invention of writing in ideograms provided at once a clear and facile method of communication. Pictorial writing seems to have been a natural accomplishment of cultures which possessed convenient media such as silk or papyrus. The kind of writing materials available to a people may well have a marked influence on the direction their written expression takes. Abstraction in writing and in mathematics may arise in response to some exigency, and an early race not faced with the necessity to abstract may, in the final analysis, produce only mathematics which is practical and pictorially appealing.

Unquestionably the Egyptians were quite advanced in practical applications of mathematics. In the Rhind papyrus we find time and again that mathematics was applied to practical problems. The Pesu calculations involve algebraic problems in one unknown, in which for instance, the amount of grain required to produce n loaves of bread is computed. Geometry also has the same practical cast: in Rhind 36, the reader is asked to compute the inclination of a pyramid, given the length of a side of the base and the height.

Egyptian mathematics is characterized by the use of unit fractions. In the Rhind papyrus we discern a well-worked-out method for handling a large number of common fractions in terms of an expansion in unit fraction. An exception was the fraction 2/3, which was referred to as "two parts." The symbol used, for example, was , which is 1/11. Writing $\bar{n}$ for $1/n$, we can list several combinations: $\bar{3} + \bar{6} = \bar{2}$, and $\bar{2} + \bar{6} + \bar{3} = 1$. The fraction 3/8, for instance, would be broken down as: $\bar{4} + \bar{8}$. In the division* of 2 by 31, one begins by determining a method for

* B. L. Van Der Waerden, *Science Awakening* (New York: Oxford University Press, 1961), p. 25.

producing 2 when 31 is divided by the numerator of the leading fraction. Consider 2/31; then 30/20 gives $1 + \overline{2}$; but this is less than 2, and the unit fractions must be found which, when added to $\overline{2} + \overline{20}$ give 1. Therefore, we take 1/4 of 1/31, or 1/124 plus 1/5 of 1/31 or 1/155; this, in turn, gives 2/31 as $\overline{20} + \overline{124} + \overline{155}$. Manipulations with 1/3 are a favorite. As an example, dividing by 3:

$$\overline{3} + \overline{3} = \overline{2} + \overline{6},$$

$$\overline{9} + \overline{9} = \overline{6} + \overline{18},$$

dividing by 5,

$$\overline{45} + \overline{45} = \overline{30} + \overline{90}.$$

Thus,

$$\frac{2}{45} = \frac{1}{30} + \frac{1}{90}.$$

Another aspect of their practical use of mathematics was the Egyptian determination of π. The common value of π employed by them was $4(8/9)^2$, which gives 3.1605. It is believed that this value was reached by comparing the weights of a vessel of square cross section (d on a side) filled with water to a height h, and a cylindrical vessel (of radius d) filled o a height h. In practice, water-containing vessels are unnecessary because solids of like material having square and circular cross sections with equal heights would have weights in the ratio of π; i.e., $\pi d^2 h$ vs. $d^2 h$.

Although mathematics and applied calculations were relatively advanced by 600 B.C.E., we do not look upon these efforts as a primitive heoretical science. The collection of astronomical data, the empirical ules for using the data, and the experimental aspects of data collection are certainly science, but the lack of a model from which the projections can be extracted prevents these efforts from being considered as theoretical. The truly great achievements of the ancient men of the Middle East have been obscured by the Greek method. Certainly the Greek tradition s firmly rooted in the heritage of Babylon, but while there is ample evidence of lively speculation on model concepts of the nature of things, one does not find successful theoretical predictions until the time of Archimedes and, later, Ptolemy.

III

The Universe of the Mind

Under the weight of an ancient and overly simplified tradition, we are wont to think of Greek mathematics as having sprung full-grown from the minds of Thales and Pythagoras. Recent translations of Babylonian tablets suggest, however, that the glorious edifice of Greek mathematics was built within the scaffolding of early Babylonian algebra and number theory. Indeed, to judge from all historical testimony, it could have hardly been otherwise. In the course of our inquiry we shall find that great discoveries and inventions are, more often than not, a fortuitous climax to centuries of excellent groundwork. This evaluation in no wise detracts from the merit of the inventor or the significance of his contribution—on the contrary, it provides us with a more comprehensive appreciation of the manner by which all human progress is made.

Without a doubt number theory and many of the mathematical problems propounded by the Greeks were Babylonian derivatives. These two cultures were contiguous, and it is therefore not surprising to discover early mathematical concepts which were characteristic of both. Cuneiform tablets show that the Babylonians could calculate the sum of a general series of the type $1^3 + 2^3 + 3^3 + \cdots + n^3$, and we shall presently discuss the ease with which this is done with the Pythagorean figurate numbers. As more and more effort is directed toward the discovery and deciphering of cuneiform tablets, it becomes evident that many of the earliest Greek contributions may have been based on Babylonian antecedents. The influence is apparent not only in Greek mathematics, where we observe it in the data of Ptolemy, but in geography and astronomy as well. The Babylonian method of determining latitude and longitude survived into the Greek period. This method fixed geographic positions in terms of the ratios of the longest and shortest days of the year. Alexandria was known to be in zone 7/5, or a zone in which the longest day was fourteen hours and the shortest was ten. By comparison, Babylon was in zone 3/2. The astronomical tables of Babylon exercised a profound influence on Greek astronomers, who adhered throughout to the sexagesimal system in their calculations. Conversely, one of the arithmetic tragedies of the Greeks was their dependence on the Egyptian-derived additive numbers in every-day computations, rather than the more powerful place system of the Babylonians. On the other hand, the Greek invention of a geometric notation opened the way for the creation of deductive reasoning and the idea of proof in mathematical thinking. These, indeed, were Greek contributions of unqualified merit.

Early Greek historians did not look elsewhere for the foundations of their mathematics; they asserted that the science of land measurement led them to geometry, and Aristotle himself attributed the birth of Greek mathematics to the presence of a leisure class. Both theories doubtless have some basis in fact, but the presence of many Babylonian rhetorical algebraic problems in the standard Greek geometry suggests a much older tradition as the wellspring for the Greek period in mathematical development. The lack of an accurate record of original sources is not necessarily a failing of the early Greeks alone; there is, in fact, no evidence for supposing that credits were ever an issue in prior periods. They did take on importance for the Greeks themselves, however, as we may judge from the reply attributed to Thales when asked what he would take for a particular discovery: "It will be sufficient reward if, when telling of it to others, you do not claim it as your own discovery but will say that it was mine." Greek historians began very early to keep a careful account of their own authors associated with specific mathematical contributions. With the rise of Arab mathematics, around 700 C.E., the preservation of authors' names takes on a signal importance in the history of mathematics. Arab scholars attached particular importance to a careful record of the original authors, especially those who were not Arabs. Civilization is therefore indebted to the Arab culture not only for preserving much of Greek mathematics itself but for the names of its authors as well, thus providing us with a early scientific historicism.

The history of Greek mathematics begins with Thales and Pythagoras, the Babylonian masters. I have chosen to call them this in order to emphasize their heritage in the older traditions of Asia Minor. Born in Miletus, a Greek city-state on the western coast of Asia Minor, Thales (624–546 B.C.E.) was indeed a cosmopolitan. His mother is believed to have been Phoenician, and it is generally thought that he studied in both

Egypt and Babylon. Thales was a man of many talents—mathematician, astronomer, engineer, statesman, hard-headed businessman, and philosopher. His highest view of human conduct, he is reported to have remarked, consisted in "refraining from doing that which we blame in others."

As an astronomer, Thales is credited with the prediction of the solar eclipse on May, 28 585 B.C.E. Such a prediction, however, presupposes a knowledge of centuries of astronomical data. Since there was no mathematical model of the lunar system, the period of 223 lunations for eclipses could only have been conjectured from empirical data. For the present, we must assume that the only known data at that time were Babylonian.

Our knowledge of Thales' contributions to mathematics comes mainly from Proclus. In his *Commentary on Euclid, Book I*, many general theorems in elementary geometry are attributed to Thales, a few of which are: Euclid I.5, the base angles of an isosceles triangle are equal; Euclid I.15, if two straight lines intersect, the opposite angles are equal; Euclid I.26, if two triangles have two angles equal respectively to two angles and one side to one side, the triangles are equal; Euclid III.31, the angle inscribed in a semicircle is a right angle. The last theorem is also attributed to Thales by Pamphile, and it poses a problem. It is proved by means of I.32, which states that the sum of the angles of any triangle is two right angles. Proclus states, on the other hand, that I.32 was first proved by the Pythagoreans, who came later. Assuming that all statements are correct, we must then conclude that Thales proved III.31 by a device other than I.32. T. L. Heath demonstrates a possible solution, in which the circle is completed with the mirrored image of the inscribed angle in the second semicircle.* The resulting parallelogram can be shown to be a rectangle, proving that the original angle is a right angle.

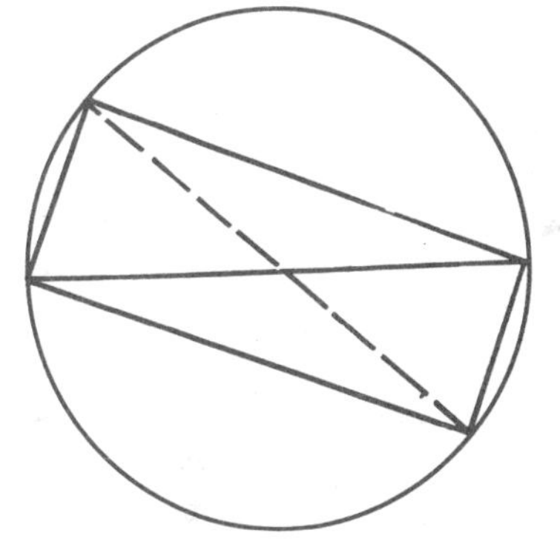

From this it becomes apparent that Greek geometry is much more than pure geometry. It is a form of algebraic geometry, and, moreover, we observe that the Greeks rephrased and proved many ancient Babylonian algebraic problems by ingenious geometric constructions. Such application is more than geometry—it is geometry used as a powerful notation, and with it a systematic deductive logic was invented. Strict adherence to expression solely by geometry inevitably led to the stagnation of Greek mathematics. By 300 C.E. all that could be done within this framework had been thoroughly investigated, and because of the magnitude of success achieved by geometry, the western world for a thousand years was incapable of crossing the frontiers of new mathematical thought. Such an obvious lesson should alert modern mathematicians and physicists to the fact that a modern mathematical gestalt could well be our undoing.

During the interval between Thales and Pythagoras, only one mathematician bears mentioning, and he is Anaximander. According to the scanty reports available, Anaximander introduced the gnomon, a right-angled figure resembling a carpenter's right angle, and he produced a

* *A Manual of Greek Mathematics* (Oxford: Oxford University Press, 1931).

geography—a map of the world as he conceived it to be. Anaximander modeled the earth as a flat, cylindrical disc of a height equal to one-third the diameter of the base. It was suspended unsupported in the center of the universe. The sun, moon, and stars were encapsulated in opaque rings of compressed air, the rings being concentric with the earth and filled with fire. Circular vents in these opaque rings allowed the firelight to shine through, thereby accounting for our light from these bodies. Anaximander also tried to provide estimates of the distances between the earth and these heavenly vents, initiating the idea of planetary distances.

Much of what is known to us about those responsible for the rise of Greek mathematics comes from Eudemus' *History of Mathematics*. Although this is an excellent contemporary source, it has suffered some alterations and additions. Certainly the section at the end concerning Euclid must have been added, for Euclid was born after the book was written. We may assume that some myths crept into the legend of Pythagoras, and our knowledge of him is derived largely from Eudemus; there is even some question about his very existence.

Born about 573 B.C.E. on Samos, according to legend Pythagoras became the ideal of mathematics, a philosopher and a prophet, apotheosized in his own lifetime by his own society. His rise to fame and leadership followed an extensive exposure to the Egyptian and Babylonian cultures. The orator Isocrates relates that Pythagoras journeyed to Egypt, where Cambyses, the Persian conqueror, captured him and took him to Babylon. There he remained for seven years, learning the theory of numbers and of music and many of the mystical rites. He is said to have received training from the Chaldeans and at one time came under the influence of Zarathustra, or Zaratas, the Chaldean. It is therefore not surprising to find that a strong element of the later Pythagorean societies was an Oriental mysticism combining numerology and astrology. Their famous and somewhat mystical rule of "golden proportionality," $A/H = R/B$, where $R - A = B - R$, and $(H - A)/A = (B - H)/B$, is reported to be Babylonian in origin.

Pythagoras returned to Samos after his Oriental travels, only to flee from the tyrant of Samos, Polycrates. With his followers he emigrated to southern Italy and there founded the Society of Pythagoreans. Membership in this group was highly selective, and even then was possible only after a period of the most rigorous and extensive testing. The Pythagoreans adhered to a monastic life, vegetarianism, and common ownership, and to their mystery rites. Above all, this society believed that the elevation of the soul to union with the deity took place by means of mathematics. The universality of numbers led them to espouse the doctrine that God had ordered the universe by means of number. Harmony subtended their religious doctrines because it consists of numerical ratios.

In their ideal theory of numbers, the Pythagoreans rejected the number 17, for it separates 16 and 18, the only numbers whose area equals their perimeters: $4 + 4 + 4 + 4 = 4 \times 4$, and $3 + 6 + 3 + 6 = 3 \times 6$. They created the concept of perfect numbers, those numbers which equal the sum of their proper divisors. Whatever mystical fascination these early concepts may have held for their religious devotees, the modern scholar is struck by the fact that certain elements of the fundamental theorem of arithmetic were intuitively implied in Pythagorean number theory.

Nicomachus gives four examples of perfect numbers: 6, 28, 496, and 8,128. He further implies that the general rule that $2^n p$ is a perfect number if p is a prime and is formed from the geometric series,

$$\sum_{m=0}^{n} (2)^m = 1 + 2 + 2^2 + 2^3 + \cdots + 2^n = p.$$

By inserting the sum of the geometric series,

$$\sum_{m=0}^{n} 2^m = 2^{n+1} - 1,$$

one discovers that p must be a prime equal to $2^{n+1} - 1$.

Another favorite of the Pythagoreans was the amicable number. These were number pairs such that each is the sum of the proper divisors of the other: 220 and 284 form an amicable pair. In Nicomachus one finds many links with the Babylonians. He is particularly interested in the figurate numbers, both plain and solid, and because they play a role in the mathematics of the Renaissance, they will be outlined here. In the writings of Galileo, one finds references to the sum of the odd integers; and in the development of the early integral calculus the Babylonian and Greek series are also employed for several of the simplest of integrals.

A totally different kind of superior number was also stressed by Pythagoras—such a number is 10, the sum of 1, 2, 3, and 4. This set of four numbers is called *tetractys*, and adjacent pairs of this set form the ratios corresponding to the fundamental musical intervals: 2, the octave; 3/2, the fifth; and 4/3, the fourth. The tetractys form a perfect triangle and thereby became a fundamental symbol in the oath of the Pythagoreans.

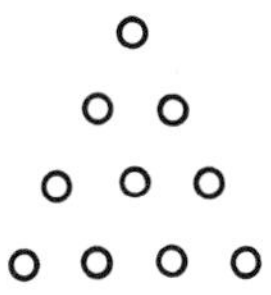

On his deathbed Pythagoras urged his followers to apply themselves to the monochord, a musical instrument formed by a string stretched over a straight keyboard which is divided into twelve parts. By shortening the string from 12 to 6, 12 to 8, and 12 to 9, one obtains frequencies in the ratios 2, 3/2, and 4/3.

In the illustration we use the modern concept that the frequency of vibration, f, times the wavelength, λ, of the tone equals the constant velocity of propagation, c, of the wave on the string. In each of the configurations, the distance spanned by one loop of the stopped string equals one-half the wavelength of the note. Let the initial length be L. Then the wavelength λ_0 of the fundamental is $2L$. Because $f_0\lambda_0 = c$, then $f_0 = c/2L$.

The octave is created by the length, $L/2$, corresponding to $\lambda_1/2$; therefore, $f_1\lambda_1 = c$, giving $f_1 = c/L = 2f_0$. In the same manner the fifth has $(2/3)L = \lambda_2/2$, giving $f_2 = (3/4)c/L = (3/2)f_0$. Finally, the fourth has a frequency which is $(4/3)f_0$.

The figurate numbers which play an important role in early number theory are formed from geometric arrangements of points. This representation has an immediate usefulness arising from the grouping of the points making up any given figure. The most obvious characteristic of each number is that it pictorially represents the sum of a simple arithmetic progression. Familiar to us are the plain figurate numbers, the triangular numbers, the square numbers, square plus an additional column, and the pentagonal numbers.

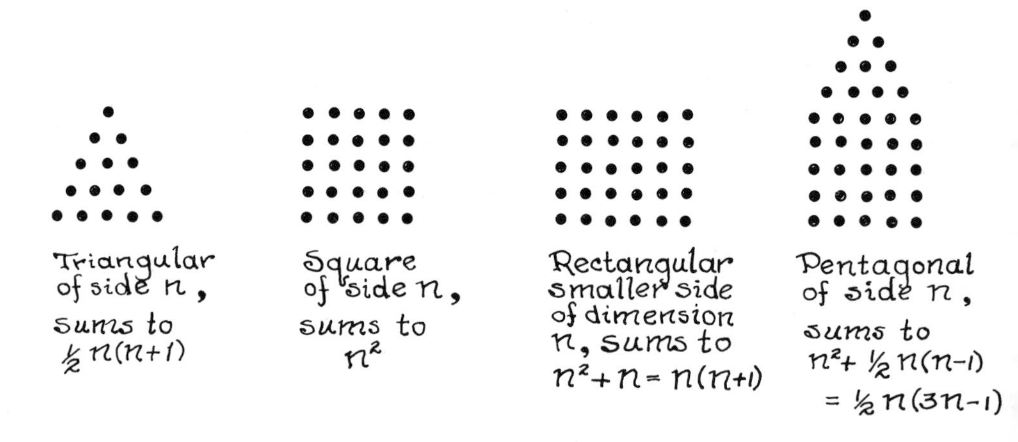

To achieve the arithmetic series associated with each of these arrays, a knowledge of areas can be utilized. For instance, the square number is the simplest to analyze. By subdividing the square array into a series of gnomons, the number of points in succeeding gnomons increases as the odd integers. The first gnomon is "one"; the second contains "three points"; the third contains "five." The last gnomon in a square array of points contains $2n - 1$, where n is the number of points on a side of the square. Consequently, the total number of points is the sum of the odd integers ending with $2n - 1$. Of course the total number is also n^2.

The square number of side $n = 1 + 3 + 5 + \cdots + (2n - 1) = n^2 = \sum_{x=1}^{n} (2x - 1)$. In this last symbolic form we have introduced the instruction symbol $\sum$, which means sum whatever appears to its right between the values shown below and above the symbol. Thus the instruction in the equation is to sum $(2x - 1)$ (a variable), letting x take on all integer values between 1 and n. It is understood that the variable will take on each integer value once and only once, giving $1 + 3 + 5 + \cdots + (2n - 1)$.

From the square number one can construct the rectangular number, having sides n and $n + 1$, by adding a column of n points. The area is then $n^2 + n$, or $n(n + 1)$. To obtain the series expansion, 1 is added to each gnomon of the original square. In other words:

$$(1 + 1) + (3 + 1) + (5 + 1) + \cdots + (2n - 1 + 1) = 2 + 4 + 6 + \cdots + 2n.$$

Summarizing these results: the rectangular number of sides n, and $(n + 1)$ is

$$2 + 4 + 6 + \cdots + 2n = 2 \sum_{x=1}^{n} x = 2(1 + 2 + 3 + \cdots + n) = n(n + 1).$$

Once the rectangular number has been evaluated, the triangular number can be derived from it. By distorting the rectangle, one observes that it is composed of two triangular numbers of side n, giving its total as one-half the rectangular number.

The triangular number of side n is

$$\sum_{x=1}^{n} x = 1 + 2 + 3 + 4 + 5 + \cdots + n = \frac{1}{2} n(n + 1).$$

The triangular number of side n is the sum of the integers from 1 to n.

The pentagonal number is a composite being formed of a square of side n and a triangular number of side $(n - 1)$. By adding the two series, the pentagonal number of side n is

$$1 + (3 + 1) + (5 + 2) + \cdots + (2n - 1 + n - 1)$$
$$= 1 + 4 + 7 + 10 + \cdots + (3n - 2)$$
$$= n^2 + (1/2)(n - 1)n = (1/2)n(3n - 1).$$

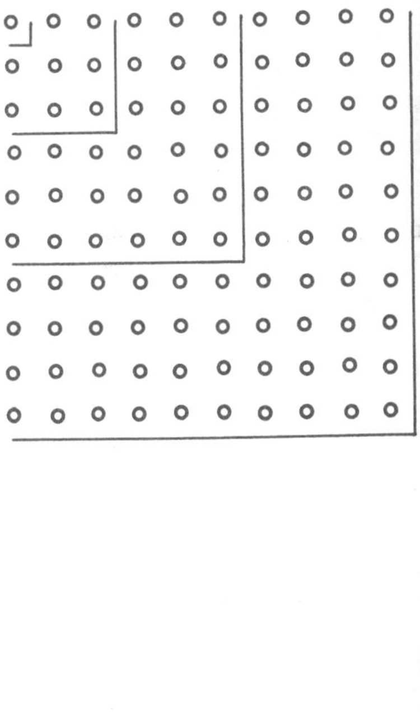

From these basic examples a method is developed which depends upon the geometric truth of the figures. More sophisticated series can be summed using the elementary results of the plain figures. Consider the sum of the cubes of the integers from 1 to n: $1^3 + 2^3 + 3^3 + 4^3 + \cdots + n^3$. The cubes of the integers are multiple gnomons of certain square numbers. These square numbers must have a side of length $(1/4)(l^2 - 1)$, where l^2 is an odd integer: 1^3 is the first gnomon of width 1; 2^3 is the second gnomon of width 2, which has the value $3 + 5$; 3^3 is the third gnomon of width 3, of value $7 + 9 + 11$. The last gnomon is the $(1/2)(l - 1)$ term with a width $(1/2)(l - 1)$. Then we observe that

$$1^3 = 1$$
$$2^3 = 3 + 5$$
$$3^3 = 7 + 9 + 11$$
$$4^3 = 13 + 15 + 17 + 19$$

and

$$n^3 = \{n(n + 1) - (2n - 1)\} + \{n(n + 1) - (2n - 3)\} + \cdots + \{n(n + 1) - 1\}$$

$$= \sum_{x=1}^{n} \{n(n + 1) - (2x - 1)\}.$$

Summing the right- and lefthand sides gives the result as an odd series, terminating in the term $\{n(n + 1) - 1\}$. Since the odd series is a square number of side $(1/2)n(n + 1)$, with the last term going as $\{n(n + 1) - 1\}$, we find that

$$\sum_{x=1}^{n} x^3 = 1^3 + 2^3 + 3^3 + \cdots + n^3$$
$$= 1 + 3 + 5 + 7 + \cdots + [n(n + 1) - 1]$$
$$= \left[\frac{1}{2} n(n + 1)\right]^2.$$

An older series which is known to be Babylonian is the sum of the squares of the integers:

$$\sum_{x=1}^{n} x^2 = 1^2 + 2^2 + 3^2 + 4^2 + \cdots + n^2 = \frac{1}{6} n(n + 1)(2n + 1).$$

The result shown on the left can be derived by modern symbolic algebra; however, this can also be done with figurate numbers. Each square is the sum of a set of odd integers, and the sets increase by one term at a time. This series is then a solid square pyramid number.

The solution becomes more apparent by stripping away one face at a time. The first face is a triangular number of order n, which we designate as Δ_n. We are left with the figure shown on the right. The next face is a Δ_n minus a triangular number of order 1. The third face is a Δ_n minus a triangular number of order 2. In all there are n faces and $(n - 1)$ subtractions, giving

$$1^2 + 2^2 + 3^2 + 4^2 + \cdots + n^2 = n\,\Delta_n - (\Delta_1 + \Delta_2 + \Delta_3 + \cdots + \Delta_{n-1}).$$

By writing out a few cases, we observe that

$$\Delta_1 + \Delta_2 + \Delta_3 + \cdots + \Delta_{n-1} = 1 + (1 + 2) + \cdots + (1 + \cdots + (n - 1))$$
$$= (1/3)(n - 1)\Delta_n;$$

thus,

$$1^2 + 2^2 + 3^2 + \cdots + n^2 = (1/6)n(n + 1)(2n + 1).$$

Many more fascinating examples can be arranged by using pyramids of triangular cross-sections and cubes. The examples to which we have devoted some attention here are especially useful, and these results will appear again in the period preceding Newton.

It is said that Pythagoras discovered the means—the arithmetic mean, the geometric mean, and the harmonic or subcontrary mean, as it was originally called. The arithmetic mean of three terms implies that the first exceeds the second by the same amount as the second exceeds the third. The geometric mean occurs when the ratio of the first to the second is equal to the ratio of the second to the third. The harmonic mean implies that by whatever part of itself the first exceeds the second, the second

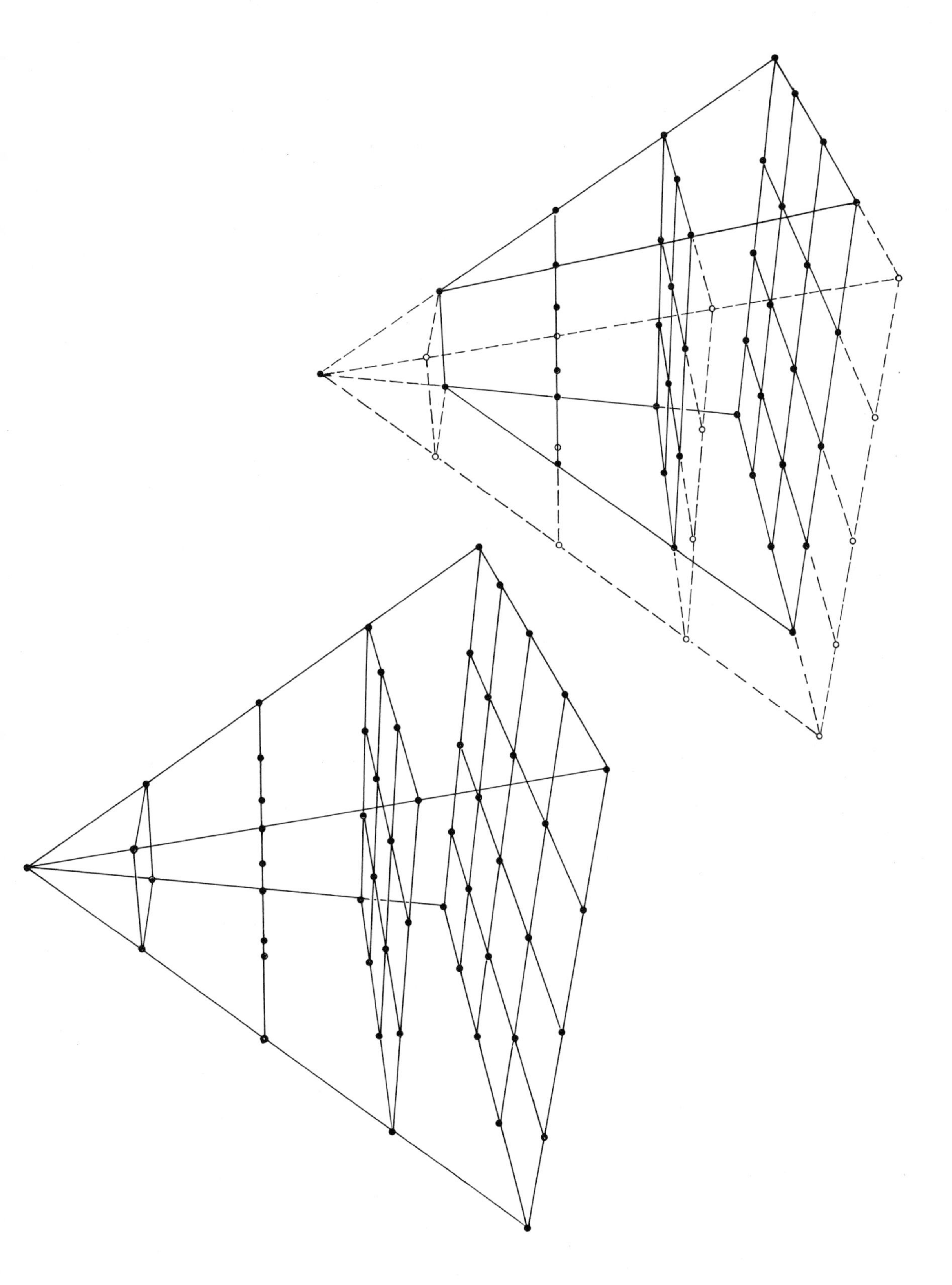

exceeds the third by the same part of the third. To illustrate the last definition, consider B as the mean between A and C; then $A = B + A/N$, and $B = C + C/N$. Solving for N, we obtain $(A - B)/A = (B - C)/C$. All in all, ten means are described by Nicomachus.

Having founded their order on the purity of number, the Pythagoreans were dismayed to discover the existence of the irrational number. Such numbers may well have played an important role in their mystery rites. Soon after the death of Pythagoras, Hippasus not only contributed innovations to the sacred doctrines but, more important, communicated his views and some of the Pythagorean doctrines to outsiders. For this he and his followers were expelled from the society, and they in turn founded their own society, which they called the "Mathematikoi."

The extensive study of irrational numbers by the Greeks represents an important contribution to mathematics. Their view of this number was geometric rather than arithmetic because of the limitations of the notation. Possibly the discovery of irrational numbers was brought about by attempts to represent the diagonals of squares and rectangles in terms of the ratios of integers. It must be remembered, however, that the Babylonians had computed the $\sqrt{2}$ in their place notation. Their calculations suggest that the $\sqrt{2}$ could not be represented in closed form down to four or five places. According to Plato, Theodorus of Cyrene was the first to prove that $\sqrt{3}$, $\sqrt{5}$, $\sqrt{7}$, etc., are irrational. The $\sqrt{2}$ was proved to be irrational before Theodorus. The traditional proof of the irrationality of the $\sqrt{2}$ is given by Aristotle (Euclid Book X). This proof employs the method of *reductio ad absurdum*, as follows. Consider a square of side A and diagonal D. Assume that D is commensurable with A, and let a/d be their ratio expressed in least integers. Because $d > a$, then $d > 1$. Here the sign $m > n$ means m is greater than n. The squares of the ratios are equal, or $d^2/a^2 = D^2/A^2$. Because $D^2 = 2A^2$ (Pythagorean theorem), $d^2 = 2a^2$; hence d is even. Since d/a is a ratio in the least possible numbers, "a" must be odd. One may then set $d = 2m$. Consequently, $d^2 = 4m^2 = 2a^2$, and $a^2 = 2m^2$, indicating that "a" must be even. Since d/a cannot be in the ratio of least integers and also both be even, d/a does not lie in the field of rational numbers.

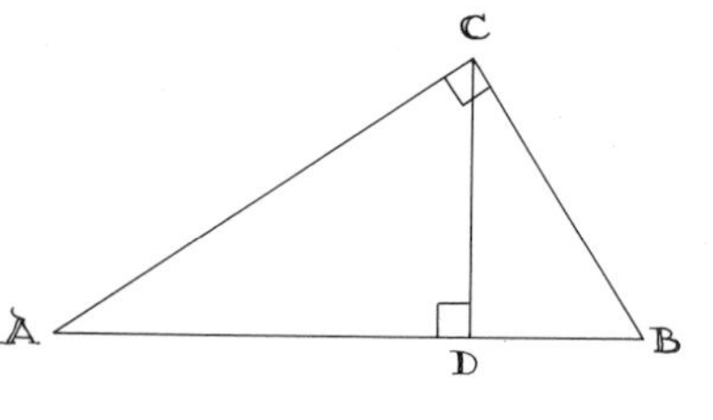

In this demonstration the Pythagorean theorem has been used as part of the development. It is now quite clear that the theorem itself was known a thousand years or more before the time of Pythagoras, although a proof before his time has not been encountered. There are many ways of proving this theorem. A proof by proportion can be constructed with the figure shown.

Consider ABC, a right triangle. Drop a perpendicular from C to AB, intersecting at D. From the three similar triangles (i.e., triangles whose angles are equal), $AC/AD = AB/AC$, and $CB/DB = AB/CB$. Then $(AC)^2 + (BC)^2 = (AB)[(AD) + (DB)] = (AB)^2$.

Another well-known derivation employs the geometric transformation of the two squares on the sides to give the square on the hypotenuse. This can be performed in a series of steps, as shown in the next figure.

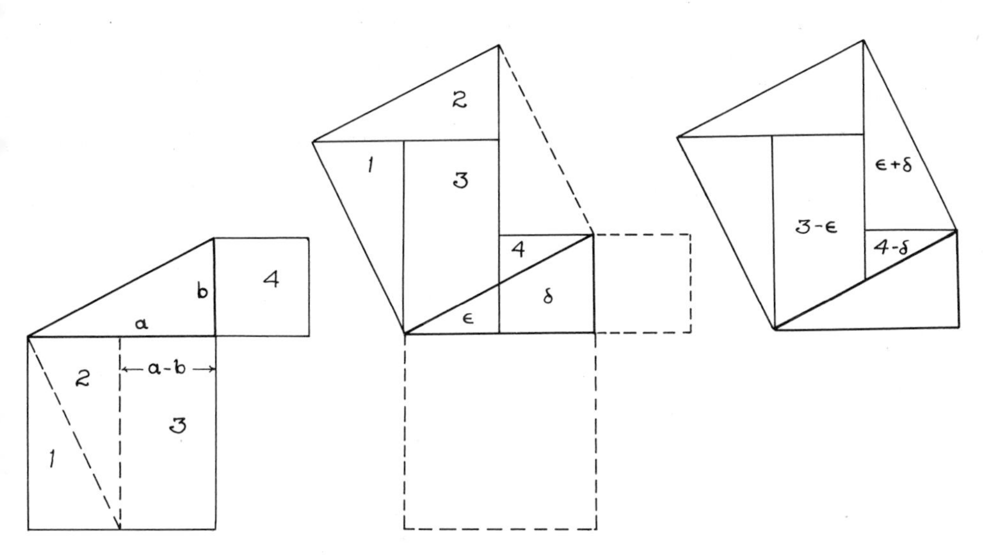

Associated with this theorem are the Pythagorean numbers. These numbers have an importance in mathematical development which precedes the Grecian culture by a thousand years, since, as we have pointed out earlier, the Babylonians compiled extensive tables of numbers which satisfy the modern algebraic relation $l^2 + b^2 = c^2$.

Processes by which the roots of algebraic polynomials can be found have been of fundamental interest from the beginnings of mathematics. The Babylonians were able to solve first- and second-degree polynomials (the latter in very special cases). Pythagorean algebraic geometry focuses upon this problem in several instances. The first-degree polynomials appear in Euclid I.44,45, where there is a treatment of the problem of applying to a specified line as a base a parallelogram having a given angle and an area equal to a specified triangle or rectilinear figure. Second-degree equations appear when parallelograms are required which are either greater or smaller in area than a given similar parallelogram. These problems give rise to solutions which in modern terminology can be represented by $(a \pm x)x = b^2$.

Typical of Greek geometry is a problem in which the parallelogram to be found is a rectangle and the excess or defect is a square. As an example, regard Euclid II.5,6, illustrated here.

Let $AB = a$ and $BC = x$. Then $(a - x)x =$ rectangle AC, which in turn is equal to the gnomon MNO. If the area of the gnomon is equal to b^2, then $(a - x)x = b^2$, or $x^2 - ax = -b^2$. To solve this geometrically, AB is bisected at D. Draw DE at right angles to AB and of length equal to b. With E as a center and $(1/2)a$ as a radius, strike the arc intersecting AB at C. If $(1/2)a > b$, the circle will cut DB at C. From this diagram we observe that $(DC)^2 = (EC)^2 - (ED)^2 = (1/4)a^2 - b^2$. Therefore, by finding C we obtain $DC = (1/2)a - x$. Thus geometrically the square has been completed to give

$$x^2 - ax + (1/4)a^2 = (1/4)a^2 - b^2,$$

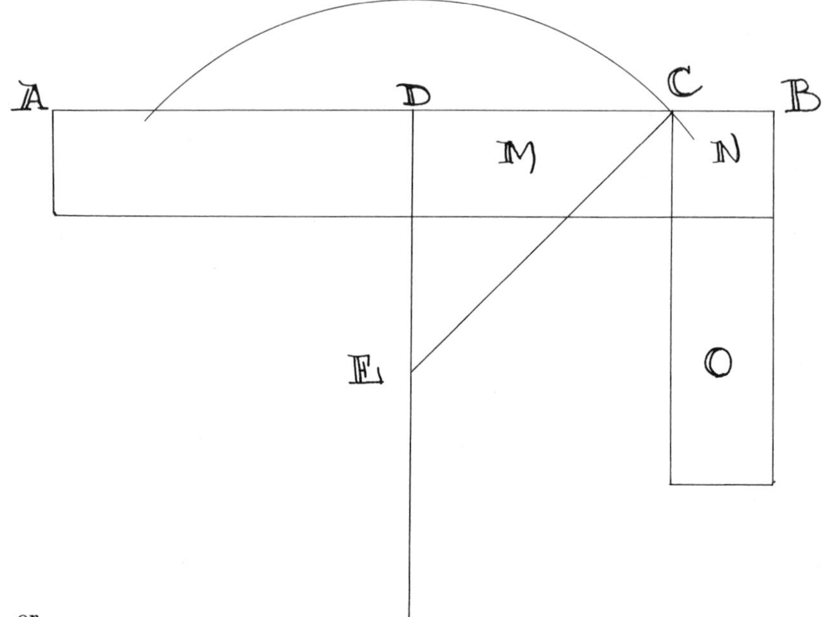

or

$$[(1/2)a - x] = \sqrt{(1/4)a^2 - b^2}.$$

The limitations of this solution are that $(1/2)a$ must be greater than b, and that only the positive root is taken. We shall find that acceptance of the negative roots and recognition of the complex roots does not occur in this famous problem for many centuries. In 1629 Albert Girard suggested recognition of the negative roots, while complex roots were not completely acknowledged until Descartes distinguished between real and imaginary roots. Newton gave full recognition to imaginary roots and further observed that in the case of a quadratic equation with real coefficients, complex roots always occur in conjugate pairs.

A detailed investigation of the contributions of the early Pythagoreans to geometry will not be attempted within the scope of this study. To mention a few, we show the famous construction which demonstrates the sum of the angles in a triangle as equal to two right angles.

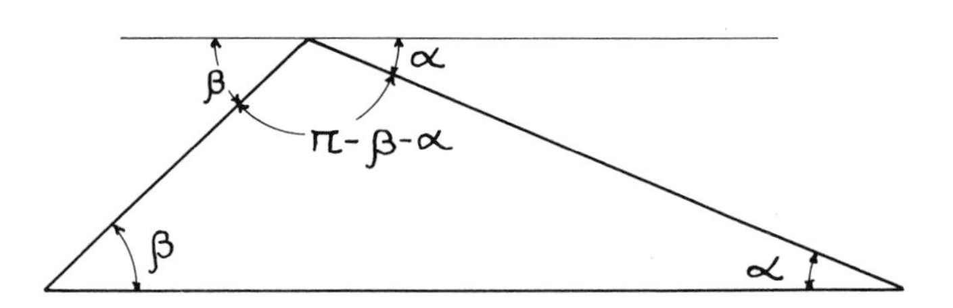

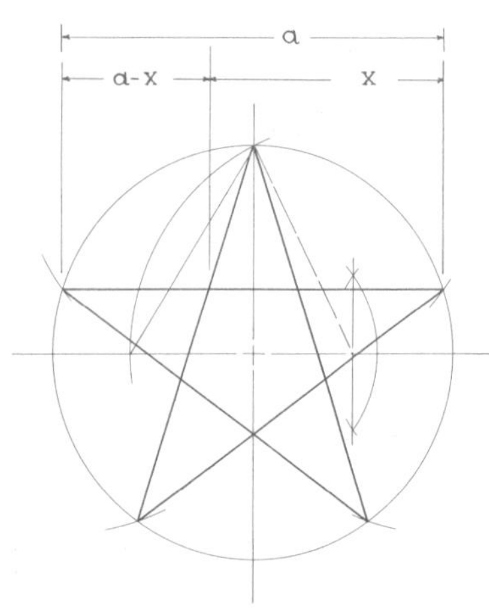

A star pentagon was created by observing that each of the five lines divides every other one in the mean and extreme ratio. From the notation on the diagram, this implies that $(a - x)/x = x/a$. Euclid I, in Proposition 42, and Euclid II deal mainly with the transformation of areas into equivalent areas of a different shape by means of application. The use of the gnomon in Euclid II is derived exclusively from the Pythagoreans. An interesting application of the Pythagorean theorem appears in the problem of finding a square equal to the product of the sum and difference of two numbers, r and s. The appropriate construction indicates the clarity of their procedure; the problem is solved by constructing a square equal in area to a given rectangle (Euclid II.14): $x^2 + s^2 = r^2$, giving $x^2 = (r - s)(r + s)$. Here use is made of an earlier proof that $r^2 - s^2 = (r - s)(r + s)$.

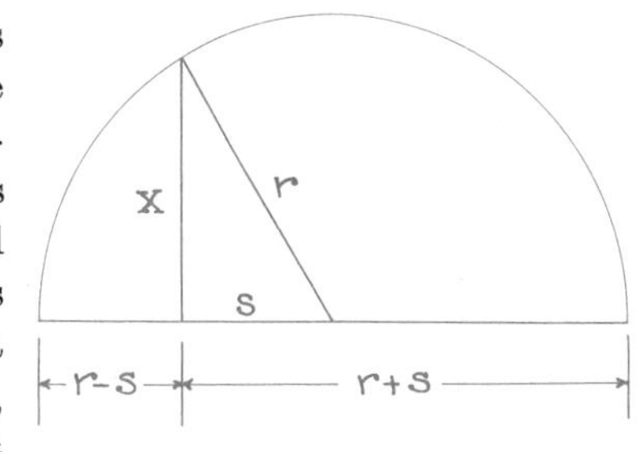

The mean proportional between two numbers, a and b, is Pythagorean. This proportion plays an important role in the question of duplicating the cube using only straight edge and compass, a problem which was to challenge mathematicians for some two thousand years. As late as the eighteenth century it was demonstrated that only algebraic numbers which are derived from polynomials of even degree may be solved by straight edge and compass. Duplication of the cube involves algebraic numbers of third degree; consequently, the original proposal is impossible. Between a and b, the mean proportion x/y is $a/x = x/y = y/b$. Solving this in various ways for x or y leads to a cubic equation, i.e., to a polynomial of odd degree.

The period between the early Pythagoreans and the Academy (about 385 or 386 B.C.E.) reveals a gradual development of mathematics toward a climax, the great academicians of the Golden Age of the fifth century—the Age of Pericles, when the Athenian empire reached its zenith in art, philosophy, and drama. This period is introduced by the unsuccessful revolt of the Ionian cities against the Persians, and ultimately led to the Persian Wars, from which Athens and Sparta emerged as masters of the Peloponnesus. During this period, the Pythagorean society was split by dissension within its ranks. Hippasus, creator of the Mathematikoi, was driven out of Italy around 530 B.C.E. A century later another major group left Italy, and late in the fourth century a group which apparently carried on the traditions of the Mathematikoi was established in Italy. From this group came Archytas of Tarentum, according to legend a friend of Plato. Many of the books of Euclid are thought to have been produced in this period—the contents of the ninth book concerning the arithmetic of even and odd numbers, the material in Book X on incommensurables, Book VII on the theory of proportions—as well as the theory of numbers used by the Pythagorean school.

The method of approximation which the Greeks employed to calculate irrationals such as the $\sqrt{2}$ was, on the surface, geometric. Side numbers a_n and diagonal numbers d_n were created which obeyed the relations: $a_{n+1} = a_n + d_n$, and $d_{n+1} = 2a_n + d_n$. These numbers suggest that the ratio of d_n/a_n is an approximation to the $\sqrt{2}$. This is a consequence of $d_n^2 = 2a_n^2 \pm 1$. According to Van Der Waerden,* the initial equation

* *Science Awakening*, p. 127.

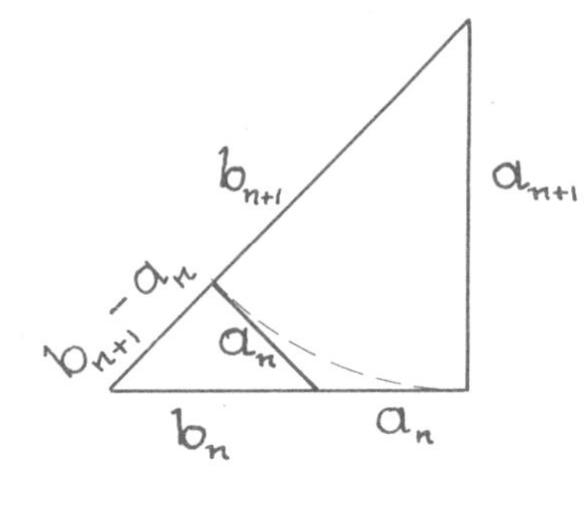

follows from a geometric construction which relates the side and diagonal of a square in a series approximation. Consider a_{n+1} as the side and b_{n+1} as the diagonal of a square. The figure indicates the relationship. From this result we then obtain: $2a^2_{n+1} - b^2_{n+1} = \pm 1$. Plato refers to 7 as the rational diagonal of the side 5. In other words, with a side number 2 and a diagonal number 3, we obtain a new side number, $2 + 3 = 5$, and a new diagonal number, $2 \times 2 + 3 = 7$.

The Pythagorean school definitely set the stage for Greek mathematics. It was responsible for much of the basic work in plane geometry and discovered three of the regular solids: the cube, the pyramid, and the dodecahedron. Its contributions to the theory of number, commensurable and incommensurable, form the foundation of later number theory. Through geometric algebra it reexamined many of the ancient Babylonian problems and provided proofs of the methods for extracting roots and for solving the general quadratic equation.

Outside the school of Pythagoras, we have little more than a brief list of names to account for the mathematical advances before the Academy of Plato. Anaxagoras of Clazomenae (500–428) was a teacher of Pericles. He taught that the sun and stars were rocks glowing in outer space and that the moon received its light from the sun. He provided a correct description of solar and lunar eclipses. It is said that while he was in prison Anaxagoras attempted the quadrature of the circle; that is to say, he searched for a means of expressing the area of the circle in terms of the square of the diameter. This problem was to become one of the primary interests of mathematicians until the age of Newton and Leibniz. The search for an accurate value of the transcendental number π pervades the investigations of Eudoxus, Archimedes, Aryabhata, and John Wallis, to mention but a few. Oenopides of Chios, younger than Anaxagoras, was primarily an astronomer. He estimated the length of time it took the planets to resume their initial positions as fifty-nine years, giving a single year of $365\frac{22}{59}$ days. This must be weighed against the fact that the Babylonians had previously listed the period of Saturn as fifty-nine years.

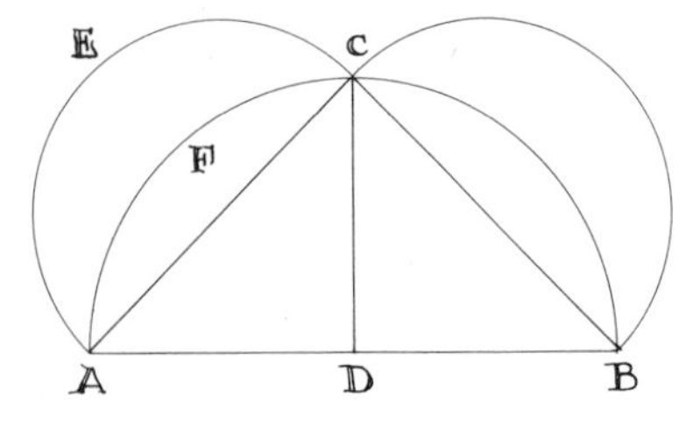

Among the most famous geometers of the fifth century B.C.E. is Hippocrates of Chios. He attempted the quadrature of the circle and became famous for the development of the lunals of a circle in the process. A lunal is the excluded part of a semicircle which has been constructed with one side of a regular polygon, which itself is inscribed in a larger circle, as a diameter. The simplest case is encountered when an isosceles right triangle is inscribed in a semicircle.

AC is the side of a square of area $(AC)^2$. AB is the diagonal of the square and is the diameter of the circle within which it is inscribed. AEC is a semicircle constructed with AC as its diameter: $(AB)^2 = 2(AC)^2$. Because the areas of the circles are proportional to the squares of their diameters, $\frown ACB = 2 \frown AEC$, where $\frown$ means semicircle. This quadrant $AFCD = \frown AEC$. Subtracting the part common to $AFCD$ and $\frown AEC$, one obtains lune $AECF$ = triangle ADC.

Finally, the sum of the two lunes equals the triangle ACB. This example is particularly simple. Hippocrates then proceeded to execute a series of quadratures of increasing complexity—a lune in which the outer circumference is greater than the semicircle, one in which it is less than the semicircle, and a particular lune plus a circle. From this series of demonstrations, he introduced a rigorous use of inequalities. Such concepts represent an early mathematical sophistication which was lost after the Greeks, not to be recovered for over a thousand years. Fourth-century Greece carried this sophistication forward, as the work of Eudoxus and Archimedes testify, and Hippocrates discovered that the Delian problem of duplicating the cube could be reduced to the general problem of finding two mean proportionals in continued proportions between two straight lines. Much more on this problem appears later, both in Greek mathematics and in the more modern period, when considering the algebraic numbers.

Democritus (430 B.C.E.) was both mathematician and physicist. His contributions to mathematics are mentioned by Archimedes and Plutarch, and a list of his works is provided by Laertius. Democritus wrote on astronomy and geographic surveys. In contrast to Anaximander's round earth, he envisaged an elongated earth. Evidently, he held an atomistic view; with proper interpretation, references to his work imply that he conceived of taking the volumes of cones and pyramids by slicing them into thin squares or discs.

At the end of the fifth century the irrational number was subjected to renewed scrutiny by Theodorus of Cyrene. Plato provides our most reliable information about him in the dialogue dedicated to the memory of Theaetetus, a mathematician who fell in battle in 369. In the dialogue, Theaetetus contemplates some special constructions of Theodorus and generalizes them; particularly fascinating is the diagram of his incommensurables. The diagonals of successive right triangles are employed to provide a side of a succeeding triangle of higher order. These are invariably constructed with one unit side and a second side which is the hypotenuse of the previous triangle. Thus, by starting with an isosceles right triangle, all of the square roots can be constructed in order.

In Greek thought the $\sqrt{3}$, $\sqrt{5}$, etc., were not numbers in the sense that they were viewed as line segments which were incommensurable. Van Der Waerden* suggests that Theodorus would have dealt with the approximation of the $\sqrt{3}, \sqrt{7}, \ldots, \sqrt{17}$ separately because the process runs differently in each case. He further suggests that Theodorus' method of proving the incommensurability consisted of a subtraction procedure which leads to a pair of lunes whose ratio is the same as that of the pair from which one started. In such a case, apparently, the subtraction process will never end. The subtraction process was shown in the case of side and diagonal numbers. From this point of view, Theodorus would stop, as he certainly did, at the square root of $\sqrt{17}$, for $\sqrt{18}$ is not interesting, while the $\sqrt{19}$ is abnormally complex.

* *Ibid.*, p. 143.

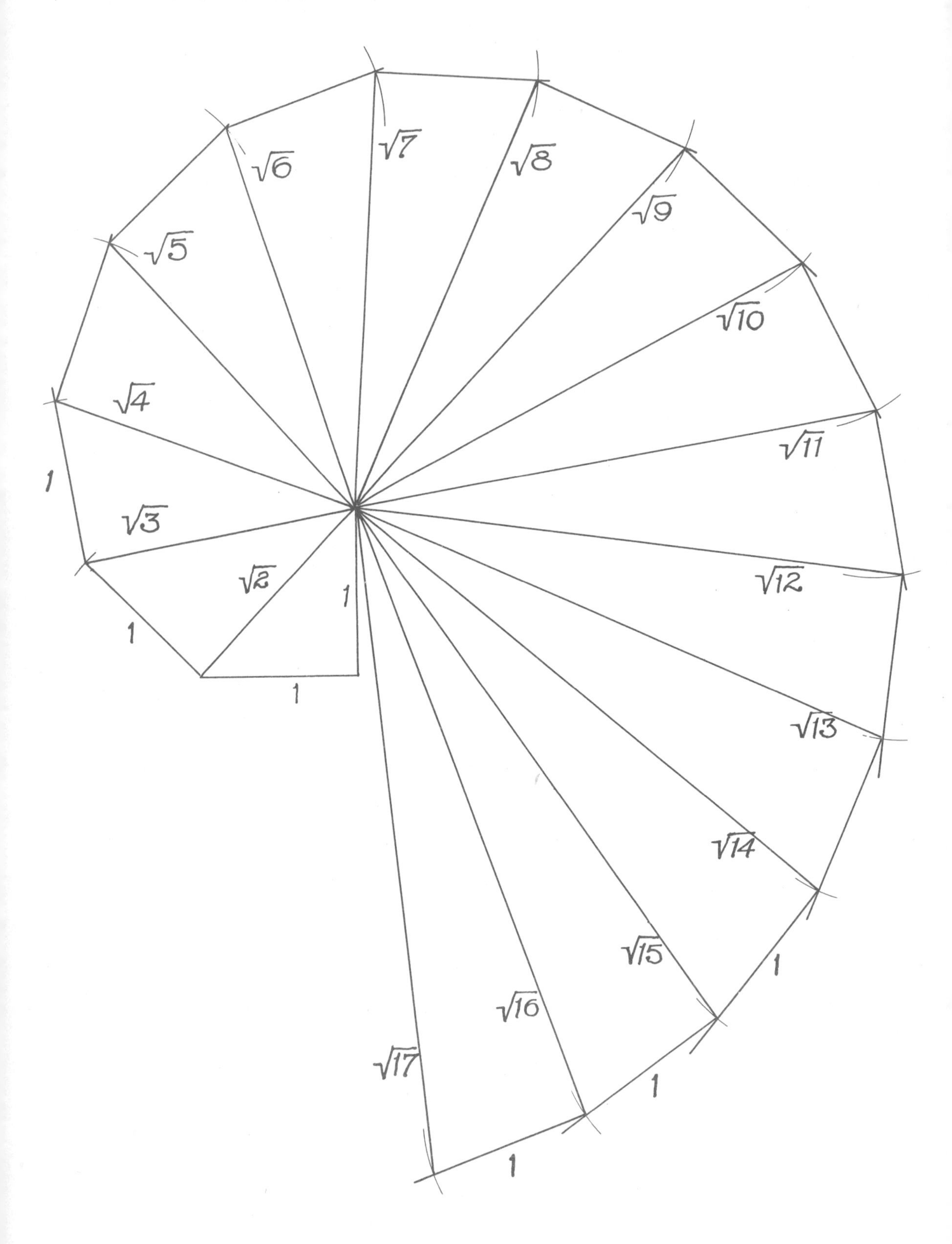

1
1
1
1
$\sqrt{2}$
$\sqrt{3}$
$\sqrt{4}$
$\sqrt{5}$
$\sqrt{6}$
$\sqrt{7}$
$\sqrt{8}$
$\sqrt{9}$
$\sqrt{10}$
$\sqrt{11}$
$\sqrt{12}$
$\sqrt{13}$
$\sqrt{14}$
$\sqrt{15}$
$\sqrt{16}$
$\sqrt{17}$
1
1
1

In the same dialogue, Plato tells us a great deal about the mathematical thought of the fallen hero Theaetetus, whose greatest contributions were not only in the theory of numbers but perhaps in his analysis of incommensurable line segments which produce commensurable squares. Van Der Waerden analyzes this dialogue in the light of Book X of *The Elements* and arrives at the reasonable conclusion that Theaetetus must have been the author of the first part of Book X.* This connection is further strengthened by observing that Book X may well have been compiled with a view to its application to Book XIII, wherein the five regular solids are constructed: the cube, the pyramid, the dodecahedron, the octahedron, and the icosahedron, the last two of which are the work of Theaetetus.

The construction of the icosahedron is truly a beautiful example of Greek geometry. On a semicircle of diameter d, erect a perpendicular at $(1/5)d$, creating a chord r of length $d/\sqrt{5}$. Next, a regular pentagon is inscribed in a circle of radius r. A second circle of radius r, with its plane

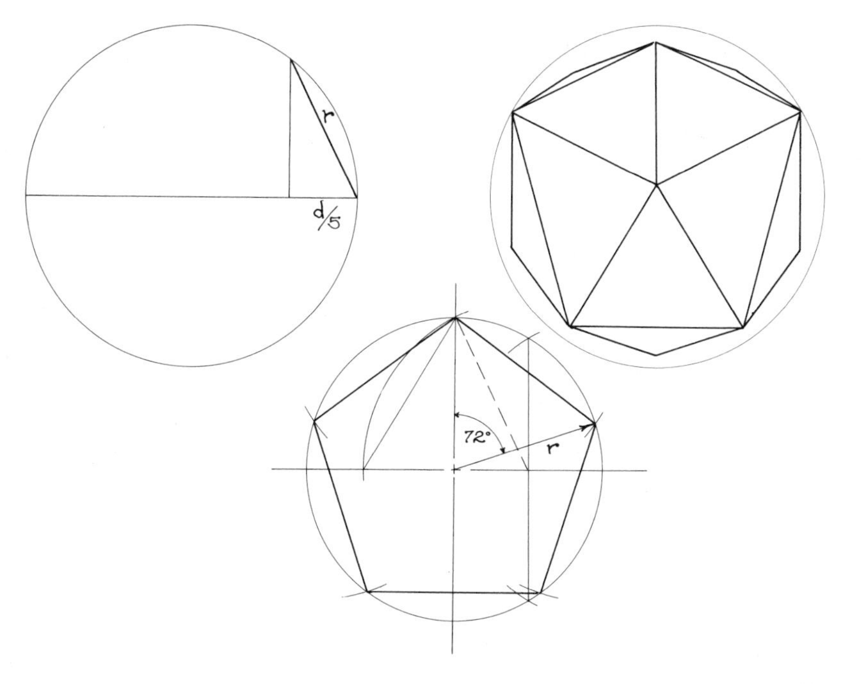

parallel to the first, is laid off at a distance r above the first. A pentagon inscribed to the midpoints of the arcs of the first circle is placed in the second. Finally, two regular five-sided pyramids of height equal to $\sqrt{l^2 - r^2}$ are placed at the top and bottom. Then,

$$r + 2\sqrt{l^2 - r^2} = \sqrt{5}\, r = d\,.$$

The era beginning with the death of Socrates in 399 is usually referred to as the period of the Academy, or the Century of Plato. Reason without

* *Ibid.*

the prerequisite of experience was held in the highest esteem, and mathematics was therefore assigned a position of eminence. As Plato himself had expressed it, "The study of mathematics develops and sets into operation a mental creation more valuable than a thousand eyes, for through it alone can truth be apprehended." The method of proof known as *reductio ad absurdum* was employed as a general method.

The Delian problem was taken up by a friend of Plato, the Pythagorean Archytas of Tarentum. Although for a scientist his ideas were poorly expressed and his reasoning often faulty, he made a major contribution to the Delian problem and in his proof of the existence of incommensurable line segments. The duplication of the cube cannot be achieved solely with straight edge and compass; it requires the intersection of a cylinder and a cone. Several solutions to this problem will be mentioned, in each of which a mechanical artifice was needed to execute the solution.

Theon of Smyrna refers to the Delian problem as a story of Eratosthenes. According to the tale, God instructed the Delians to create an altar twice as great as (twice the volume of) the existing one, and from this puzzle the Delian problem was created. The importance of the problem is a dual one, and until very recent times it intrigued the minds of some of the greatest mathematicians. As it turned out, the characteristics of modern algebraic numbers must be understood before a clear proof can be offered that the construction is not possible with straight edge and compass.

Perhaps the greatest mathematician of the era of the Academy was Eudoxus of Cnidus, born about 400. Arriving in Athens as a poor youth, he carried on his studies with Plato, later studied astronomy in Egypt, and formed a school at Cyzicus on the Sea of Marmara. He returned to Athens around 365 and was acclaimed as a great scholar. As a philosopher, he taught that pleasure and joy were the highest good, states that all beings strive for; on this and many other points, Eudoxus and Plato were in disagreement.

Eudoxus modeled a planetary system composed of a set of concentric spheres with the earth at the center, and accounted for motion by the relative rotations of the various spheres. Although the historical accounts of his contributions to mathematics are not as extensive as one might wish, they suffice to suggest that the theory of proportions in Euclid V are his. The mathematics of inequalities, introduced here, is highly sophisticated. Of the importance of this contribution no more need be said than that its definitions of equal ratios are the same as those in the theory of irrationals of Dedekind, more than two thousand years later, and that in structure it is identical to Weierstrass' definition of equal numbers.

The method of exhaustion was the creation of Eudoxus. We find in the beginning of Euclid XII a proof that the ratio of the areas of two circles is equal to the ratio of the squares of their respective diameters. The mathematics of inequalities is employed, and we further observe that the space-filling figures are introduced here, and with them the fundamental concept that one can make the difference between the original figure and the space-filling figure *as small as one pleases*. In one phrase, then, Eudoxus in practice avoids all of the semantic imperfections of the concepts of zero and infinity. His proof in the original is phrased somewhat differently than will be shown here; for clarity we have changed the notation while retaining the basic steps.

Consider two circles, 1 and 2, with areas A_1 and A_2, and diameters d_1 and d_2. Suppose that the area of a circle is not proportional to the diameter squared. Then the ratio d_1^2/d_2^2 would be equal to A_1 divided by U, where U is an area not equal to A_2. In this case, U may be greater or less than A_2. Take U smaller than A_2, as an example; now inscribe in A_2 a square of area S_2.

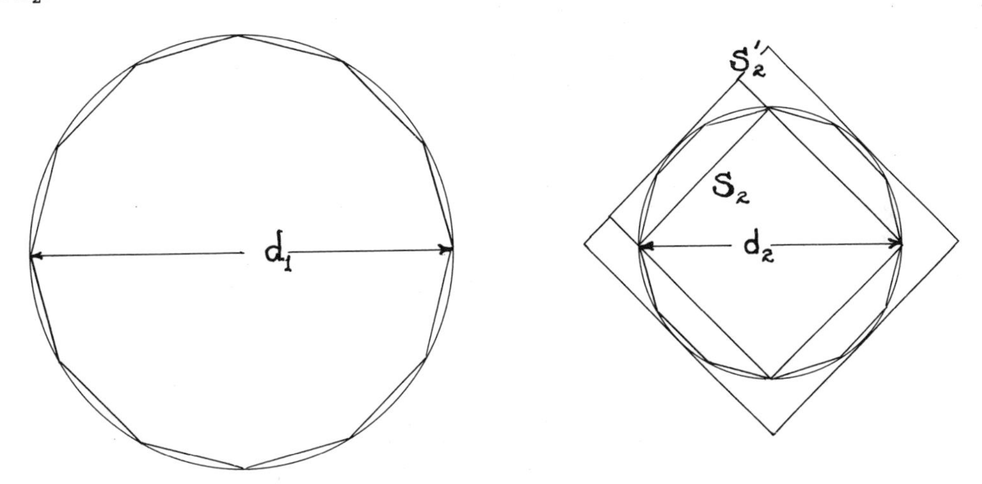

S_2 is greater than $(1/2)A_2$ because the circumscribed square, S_2', is greater than A_2, while $S_2 = (1/2)S_2'$. Bisect each arc bounded by the chord which is a side of S_2. Each triangle added to the square to form the inscribed octagon O_2 is greater than 1/2 the circular segment formed between the chord and the arc. Again this is so because the triangle is one-half the rectangle R_2 circumscribed about the circular segment, while the circular segment is less than the circumscribed rectangle R_2.

The process of bisecting the arc and inscribing a triangle to form a $4n$-sided figure can be performed n times to give an inscribed polygon of $4n$ sides. Repeated often enough, this process can make the difference between the inscribed polygon P_2 and the circle A_2 smaller than the difference $A_2 - U$. Remember that we choose $U < A_2$; thus the inscribed polygon can be constructed to have an area $P_2 > U$.

Now construct a similar polygon in circle 1 of area P_1. The ratio of the polygons P_1/P_2 is equal to d_1^2/d_2^2: $P_1/P_2 = d_1^2/d_2^2$. We assumed that $A_1/U = d_1^2/d_2^2$; therefore, $A_1/U = P_1/P_2$, giving $A_1/P_1 = U/P_2$.

Since A_1 is greater than P_1, we have a contradiction because we have shown that if $U < A_2$, we are able to construct $P_2 > U$. As a result, A_1/A_2 must be equal to d_1^2/d_2^2.

From the works of Archimedes, one discovers that Eudoxus contributed lemmas 1, 2, 3, and 4 of Book XII. These lemmas stipulate the relation between the areas of circles and their radii, the volume of spheres and their radii, and the volumes of cones and pyramids in terms of their altitudes and bases. Many samples of Eudoxus' work are found as well in the similarity and volumes of pyramids of various numbers of sides. In Book V of Euclid there is a scholium by an unknown author who attributes the general theory of proportions to Eudoxus. This is reasonable, since a rigorous use of approximations and inequalities are employed in Book V to prove equalities by the method of *reductio ad absurdum*.

In the work of Eudoxus we gain the concept of continuously varying magnitudes which can approach certain values in the limit. This approach toward infinitesimals is a prelude to the integral calculus, and one cannot help but wonder at this point why Greek mathematics stopped short of this invention. No single factor provides a truly satisfying explanation; several suggest themselves and no doubt there were many more. For one thing, a convenient notation was lacking: no reasonable symbolic representation of a continuous function was available. For another, the idea of the infinitesimal was severely attacked on philosophical grounds, the polar dictum having been pronounced long before by Zeno in his four famous paradoxes.

Zeno's four paradoxes were not independent; each was centered about one major premise, at first sight appealing. However, he built into his argument a contradiction in terms: he allowed processes to take place whose spatial dimensions could be made as small as one wished, while, on the other hand, implicitly assuming that the corresponding time intervals must not be subdivided in the same manner. Consequently, the arbitrarily large number of divisions of the spatial interval necessarily required an arbitrarily large time interval in which to take place. Zeno denied by implication that the time required to traverse an arbitrarily small spatial interval might itself be arbitrarily small in such a manner that the ratio of the space interval to the time interval has some finite value. In the Dichotomy, Zeno argued that there can be no motion because that which moves must cover half the distance before it arrives at the end; furthermore, before it traverses the second half it must cover half the half, and so on until an infinite number of intervals are traversed. Hence, the motion cannot begin. Such premises were firmly adhered to by most Greeks, and no mathematician was inclined to violate them.

A pupil of Eudoxus, Menaechmus, discovered the conic sections, although, as has been mentioned, the names parabola, hyperbola, and ellipse were not introduced until later by Apollonius. Menaechmus employed these curves, which he called "the section of a rectangular cone," "the section of an obtuse-angled cone," and "the section of an acute-angled cone," respectively, to solve the Delian problem. His brother Dinostratus used the quadratrix, originally discovered by Hippias, to obtain the quadrature of the circle or to find a value for π. The construction is performed with the first quadrant of a circle. By sectioning the square containing the quadrant along the vertical axis with lines parallel to the horizontal, and by dividing the right angle, $\pi/2$, in the same relative proportions, with radii extended to designate each fraction of $\pi/2$, the intersection of the horizontal line representing the fraction f of the side of the square and the radial line representing the fraction f of the right angle $\pi/2$, then describe the locus of all points falling upon the curve known as the quadratrix. In demonstrating this curve, Pappus proved that arc $CB/AC = AC/AO$.

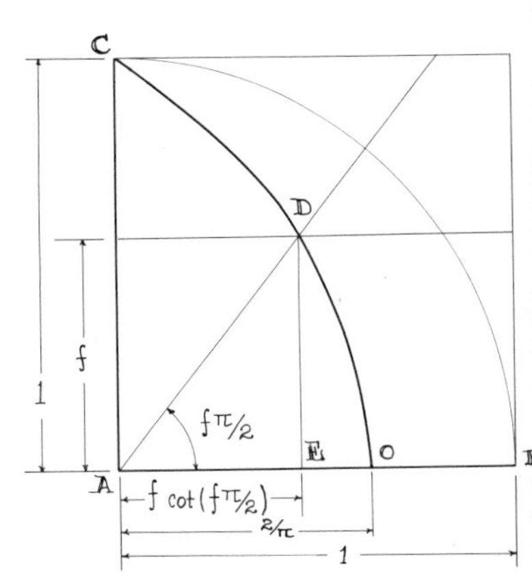

If the radius of the circle and the side of the square AB is 1, we see that the intersection of the quadratrix with the horizontal axis AB strikes off a distance $AO = 2/\pi$, or, for an arbitrary square, $AO/AB = 2/\pi$, giving $\pi = 2(AB/AO)$. The demonstration of this equality in modern algebraic form* is:

$$AE = f(AB) \text{ cotangent } f(\pi/2), \text{ where } f \text{ is a number less than } 1$$

$$\begin{aligned} AO &= \text{the limit of } AE \text{ as } f \text{ approaches zero} \\ &= \lim_{f\to 0} f\cdot(AB)\cdot\frac{\{1 - 1/2[f(\pi/2)]^2 + \cdots\}}{\{[f(\pi/2)] - 1/6[f(\pi/2)]^3 + \cdots\}} \\ &= 2(AB)/\pi. \end{aligned}$$

According to pure Greek reasoning of the time, the limiting process was inadmissible; a careful drawing, however, gives an extrapolation of the quadratrix to the point O which will provide relatively good values of π. As mentioned, the interest in accurate values of π extends over several millenia and ultimately leads in part to the binomial expansion for non-integral powers and to investigation of the general properties of infinite series.

In 338 B.C.E., Philip of Macedon defeated the Athenians. Six years later his son, Alexander the Great, conquered Egypt and went on to subjugate a good part of eastern Asia, imposing upon the conquered peoples the cult of Hellenism. With the death of Alexander and the splitting up of his empire, Egypt emerged as a powerful kingdom under Ptolemy Soter, a former student of Aristotle, and his successors, Ptolemy Philadelphus and Ptolemy Euergetes. In this period from 322 to 305, the arts and sciences were revered as supreme activities, and Alexandria, with its famed library, became the focal point for the scholars of the time. It is here that Euclid lived and taught during the reign of Ptolemy Soter.

Euclid represents the apogee of the Athenian Academy and its traditions. Actually, what has come down to us about Euclid, mostly from Proclus, is very little, yet his impact upon mathematics has been immeasurable. It may even be said that Euclid was more of an editor than a mathematician; his *Elements* were the conclusion and summary of a sequence of earlier works, and ultimately superseded its predecessors.

* The reader must accept the fact that the cotangent of an angle expressed in radians can be expanded as the ratio of two infinite series. In the numerator the leading term is 1, while in the denominator the leading term is the angle. Thus *for small angles* the cotangent behaves as 1 divided by the angle.

Book I begins with definitions, postulates, and axioms or self-evident proofs. Words are defined, such as a "porism" or corollary, which is something revealed in the course of a demonstration; a "lemma" is defined as merely an assumed statement but very often is used as an auxiliary proposition requiring proof. Book I sets forth five postulates, the first three of which are postulates of construction: (1) a straight line can be drawn to join two points; (2) a straight line can be extended in either direction; (3) a circle can be described with a given center and a given radius; (4) all right angles are equal; and (5) the postulate of parallels: if a straight line falls upon two other straight lines so as to make the sum of the two interior angles on one side of the cutting line less than two right angles, then the two straight lines will meet on that side of the cutting line on which are the two angles which are less than two right angles.

The last postulate is open to question; in fact, by inserting the contrary postulate instead, mathematicians of the nineteenth century were able to create non-Euclidean geometry. The books of *The Elements* can be briefly described in terms of their major topics:

Book I: the geometry of straight lines and plane rectilinear figures
Book II: algebraic identities
Book III: the geometry of the circle
Book IV: the properties of figures inscribed in and circumscribed about circles
Book V: the theory of proportion applied to commensurable and incommensurable magnitudes
Book VI: applications of the theory of proportion to plane geometry and the application of areas
Book VII: number theory, definitions
Book VIII: numbers in continued proportions
Book IX: number theory; products, primes, perfect, etc.
Book X: irrationals
Book XI: the geometry of three dimensions
Book XII: solids and the method of exhaustion
Book XIII: inscription of the five regular solids in a sphere

There are two additional books, which are supplements written by other authors and seem to be of less value: Book XIV, by Hypsicles, second century B.C.E., is on the division of the circle and the harmony of the spheres, and Book XV, probably by Damascius of Damascus, concerns solids.

Much of our information about Aristarchus is derived from his book, *On the Sizes and Distances of the Sun and Moon*, and from the writings of Archimedes, who tells us that Aristarchus measured the apparent angle subtended by the sun as 1/720 of the zodiac circle. He viewed the moon as receiving its light from the sun. His estimate of the distance of the sun from the earth is >18, <20 times the distance of the moon from the earth, and the ratio of the sun's diameter to the earth is given as $>19/3$ but $<43/6$. From this last result, we realize not only that Aristarchus was a master of inequalities but that he obviously held the earth to be a sphere. Here, perhaps, was one of the first scientists of imagination and mathematical power.

Archimedes, who undoubtedly studied in Alexandria under the successors of Euclid, belongs to this same period. He may have been the greatest mathematician and scientist of antiquity. Inventor, astronomer, mathematician—his treatises are models of classical mathematical exposition. He virtually created the theory of elementary mechanics and hydrostatics. His greatest contributions in geometry consisted in the quadrature of curved plane figures, that is, in finding the area under certain curves and the quadrature (or area) of curved surfaces. These investigations were, in fact, the forerunners of the integral calculus. Using the method of exhaustion and the mathematics of inequalities, Archimedes calculated approximate values of π and the square roots of large and small nonsquare numbers. His manuscript *The Sand Reckoner* demonstrates an ability to deal with numbers of $8(10)^{16}$ ciphers.

With the aid of mechanics, particularly his Law of Levers, Archimedes provided many ingenious solutions to geometric problems. In hydrostatics he created methods for finding the centers of gravity and buoyancy of paraboloids, conics, and hemispheres, to name but a few. The extant treatises of Archimedes are:

On Plane Equilibriums I
On Plane Equilibriums II
The Quadrature of the Parabola
The Method
On the Sphere and the Cylinder
On Spirals
On Conoids and Spheroids
On Floating Bodies
The Measurement of the Circle
Psammites (*The Sand Reckoner*)
Stomachion (a fragment)

References in his own works as well as in those of Pappus indicate that many other works of Archimedes have been lost—works on thirteen semiregular solids, on the naming of numbers, and a theory of mirrors among them. But those which survived the vicissitudes of time and chance served as the basis for mathematicians and scientists of post-Renaissance Europe.

In *The Method*, a typical problem was that of finding the area of a parabolic line segment by the technique of moments. The method of exhaustion is employed with great effect in the manuscript on the sphere and the cylinder. To find the area of a surface of a sphere, Archimedes took the great circle of the sphere and inscribed and circumscribed a regular polygon of an even number of sides, bisected by the plane of the great circle. By allowing the polygons to revolve about an axis of symmetry, AA', a given diameter of the circle, he obtained two surfaces, the areas of which bracket the area of the sphere. The lines formed by BB' to LL' are parallel. By similar triangles and adding antecedents and consequents:

$$\frac{A'B}{AB} = \frac{(BB' + CC' + \cdots + LL')}{AA'} .$$

After revolving the polygon about AA' the surfaces of the cones and the frustra of cones is, for example, $\pi(BC)(BM + CM)$. Thus the surface of the inscribed solid is

$$\pi(AB)[(1/2)BB' + 1/2(BB' + CC') + \cdots + (1/2)LL'] = \pi(AB)(BB' + CC' + \cdots + LL').$$

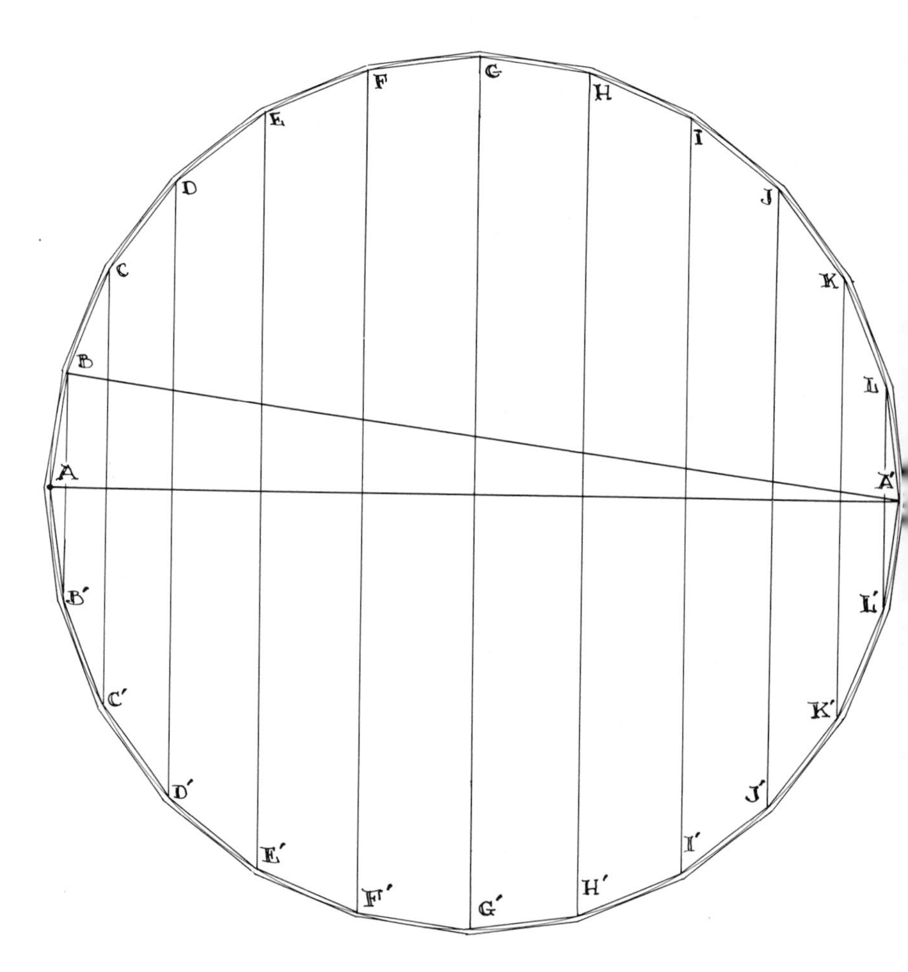

Finally, by using the first equation, the surface area is:

$$\pi(A'B)\cdot(AA') < \pi(AA')^2.$$

Performing the same operations for the circumscribed surface of revolution demonstrates that the area of the circumscribed surface of revolution is

$$\pi(a'b)\cdot(aa') > \pi(AA').$$

Because the surface of the sphere is less than $\pi(a'b)(aa')$ and greater than $\pi(AB)(AA')$, it is neither greater than $\pi(AA')^2$ nor less than $\pi(AA')^2$, and therefore must be equal to $\pi(AA')^2$. By this logic, Archimedes in essence created a procedure equivalent to integration.

In *The Measurement of the Circle*, he computed an approximate value of π using polygons of 96 sides inscribed in and circumscribed about a circle. Once again the elements of the method of exhaustion were employed, although Archimedes did not go to the limit of an arbitrarily large number of sides. His final value can be given as $3\frac{10}{71} < \pi < 3\frac{1}{7}$.

On Spirals contains two concepts which illustrate simple ideas that later became quite important. The spiral illustrates a function plotted in polar coordinates, but more significant is the fact that Archimedes introduces the idea of two uniform motions: the rotary motion of the radius vector at a constant angular velocity, and the radial motion of a point with constant linear velocity.

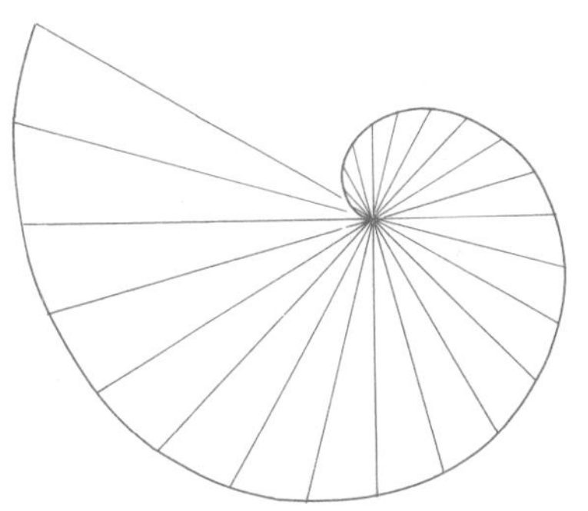

Archimedes computed the area of any segment of the spiral by the method of exhaustion. In Propositions 18, 19, and 20, he derives the tangents to the spiral at any point, and here again he anticipates the calculus by regarding what would be the derivative of the curve.

A famous problem demonstrated in *The Method* is the calculation of the volume of a sphere in terms of the volumes of a circumscribed cylinder and a cone having a circular base equal to that of the cylinder. The volume of the cone of base radius $2R$ and height $2R$, together with the volume of a sphere of radius R are balanced through their centers of gravity on a moment arm of length R against a cylinder of radius R and height $2R$ on a moment arm of length $2R$. In modern notation, then:

$$[V_{\text{sphere}} + V_{\text{cone}}] = [V_{\text{sphere}} + (4/3)V_{\text{cylinder}}] = 2V_{\text{cylinder}},$$

and

$$V_{\text{sphere}} = (2/3)V_{\text{cylinder}} = (4/3)\pi R^3.$$

A problem similar to this one is alluded to in the seventeenth century, in Galileo's *Two New Sciences*. Galileo utilizes the fact that a bowl formed by a cylinder of height R and radius R minus a hemisphere of radius R has a volume equal to the right circular cone inscribed in the cylinder. In addition, Galileo uses Archimedes' principle that any plane passed parallel to the base of the cylinder cuts the bowl and cone in such a fashion that the truncated volumes are equal. This construction is employed in the seventeenth-century work to indicate that the volume of a circle equals the volume of a point.

A young contemporary of Archimedes, Eratosthenes of Cyrene, set about quite successfully to measure the circumference of the earth. Observing that during the summer solstice at Cyrene the sun cast no shadow from an upright and that at Alexandria, assumed to be on the same meridian, during the same period the inclination of the sun's rays with the vertical was 1/50 of four right angles, he calculated the circumference of the earth at 252,000 stadia, or 24,662 miles. The true average circumference is 24,888 miles, so that Eratosthenes erred by very little. He estimated the distances of the tropic and polar circles, as well as the size of the sun and the moon, and by lunar eclipses calculated the distances between the moon, the sun, and the earth. By this time the duplication of the cube was a problem which all great mathematicians took up at some point in their careers, and Eratosthenes was no exception. He gave a mechanical solution, which is engraved in stone in the temple of Ptolemy. Evidently, a model existed in bronze consisting of three plates which could be adjusted back and forth between a fixed and rotating straight edge.

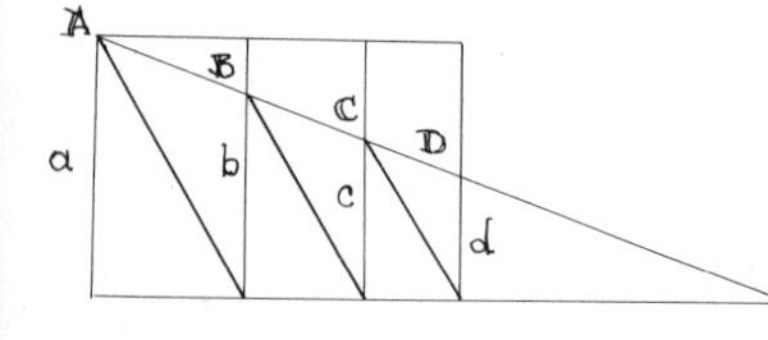

The aim of the arrangement shown in the figure above is to obtain the two mean proportions in continued proportion to two given lines, a and d. By adjusting the plates so that the points $ABCD$ lie on one line, and carrying out the ratios of the sides of similar triangles, one obtains $a/b = b/c = c/d$.

In the theory of numbers, Eratosthenes created the process which is known as "the sieve of Eratosthenes." All of the integers less than N are written in order. By striking out all numbers which are multiples of $2, 3, 4, \ldots, N - 1$, until all the composites have been eliminated, all of the primes up to N are accounted for.

Two ingenious constructions, one for trisecting the angle and the other for the duplication of the cube, were given by Nicomedes, who is also known for the creation of a curve known as the cochloid, a curve used in both of the constructions mentioned above.

The last great geometer of antiquity was Apollonius of Perga, whose major work was done in the period around 210 B.C.E. He studied in Alexandria under the successors and former pupils of Euclid. His astronomical models were to have a long-range effect upon theoretical models of the system of the planets and the sun. The climax of Kepler's study of planetary orbits came when he discovered that the epicyclic and eccentric curves of Apollonius were not appropriate as descriptions. Remarkably, Kepler found that the ellipse was the proper mathematical curve, and since Apollonius was a major figure in the study of conic sections, Kepler merely discarded one of Apollonius' curves for another.

Epicycles and eccentric curves are formed quite simply by allowing the center of one rotating circle to revolve about the circumference of another. The relative rates of rotation of the two circles determine the apparent linear motion of the planets. In one proposition, quoted by Pappus, the relationship between angular velocities is found which must be satisfied if the planet appears stationary. At a more elementary level,

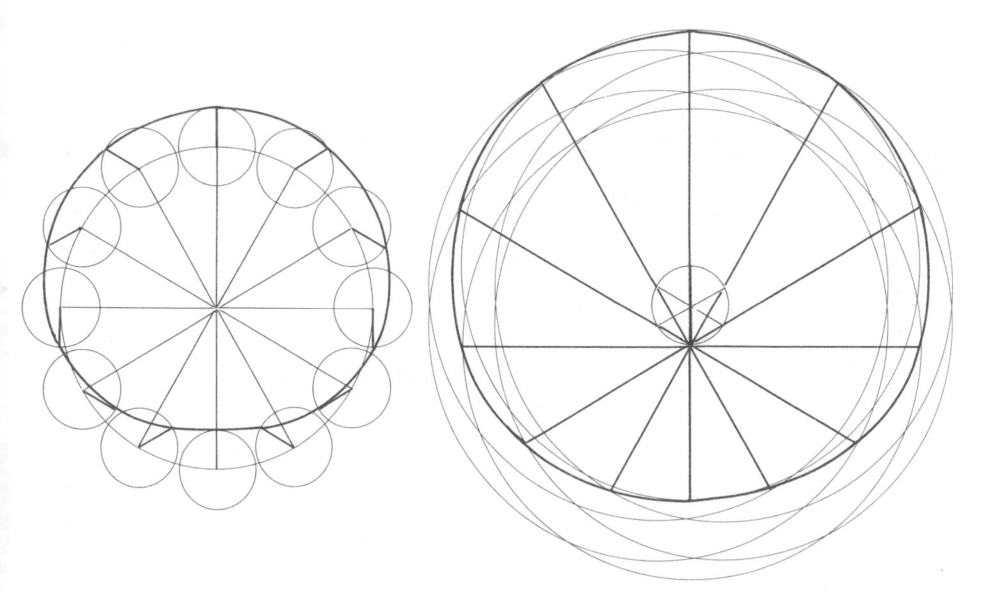

a few diagrams are shown which suggest the similarity between the epicyclic curve and the ellipse.

One importance of these diagrams lies in their suggestion that by 210 B.C.E. the Greeks had a clear description of rotary motion at a constant angular velocity. In other words, the concept of sweeping out the same length of arcs on a circle in equal intervals of time leads to a concept of the linear tangential velocity—the distance traveled on an arc divided by the corresponding time interval. The idea of a stationary point in the epicyclic motion implies that the linear vector composite motion of the point was understood. The relation between the magnitude of the tangential linear velocity v and angular velocity ω is $v = R\omega$, where R is the radius of the circle and ω is given in fractions of π per second, or radians per second. A complete traversal of a circle encompasses an angle of 2π radians. Thus, as an example, 60°, which is 1/6 of 360°, corresponds to $1/6 \times 2\pi = \pi/3$ radians.

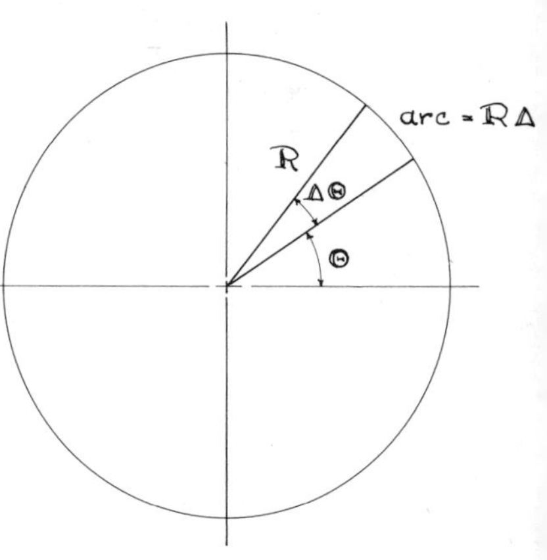

In the accompanying diagram we observe that a change in angle $\Delta\theta$ (expressed in radians) sweeps out an arc $R\Delta\theta$. If this change is sufficiently small, the arc R$\Delta\theta$ approximates the chord to a high degree of accuracy. The velocity along $R\Delta\theta$ is the distance divided by the time interval, Δt, during which the distance is covered. Thus, the magnitude of the linear tangential velocity at any instant is given by:

$$v = R\frac{\Delta\theta}{\Delta t} = R\omega,$$

where

$$\frac{\Delta\theta}{\Delta t} = \frac{\text{change in angle}}{\text{change in time}} = \omega.$$

This illustration is not meant to imply that Apollonius employed this direct symbolic notation; on the other hand, his writings show that the concept was well understood.

Apollonius' work on the conic sections was one of his greatest contributions to geometry. As has been pointed out, the conic sections were known long before. Archimedes and the earlier men represented the conic sections in terms of equations and referred to rectangular coordinate axes and occasionally to oblique axes. For instance, Archimedes gave the equations in the "two abscissas form." Let $AB = a$ be the major axis of the conic. The perpendicular $PQ = y$ from a point on the conic AB is called the "ordinate." The distances $AQ = x$ and $BQ = x_1$ are known as the abscissas. In the case of the ellipse, $x_1 = (a - x)$; for the hyperbola, $x_1 = (a + x)$. The "symptom" of the curve, a condition which is to be satisfied at every point P of the curve, is then $y^2/xx_1 = \alpha$, a constant. The constant α determines the class of the curve; $\alpha = 1$ gives a circle.

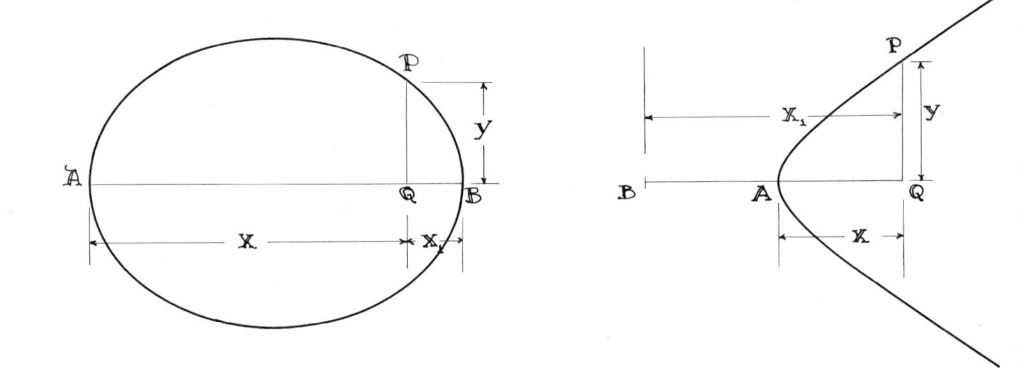

Symptoms were derived in a classical manner.* Apollonius used similar methods which were simpler and more general. The older geometers restricted themselves to planes which were perpendicular or parallel to the symmetry axis, or to the generators of the cone. Apollonius dropped these restrictions when he developed more general methods, and considered any plane of intersection. The names "parabola," "ellipse," and "hyperbola" were his creations. Conjugate hyperbolas were discovered, and a method for constructing a tangent to a conic from a given point was shown.

All in all, these are impressive contributions. The development of the coordinate methods, together with the treatment of tangents, approach modern analytic geometry. Without minimizing the contribution of Descartes, one need only observe how close his invention of analytic geometry comes to the classical works of Archimedes and Apollonius. Similarly, the methods of Fermat and Barrows for computing tangents bear more than a passing likeness to the works of Apollonius. A lesson which we ultimately learn is that there are seldom single, unique contrib-

* Archimedes, *On Conoids and Spheroids*, Propositions 7 and 8.

utors in mathematics and science; virtually all innovation invariably has some antecedent in the manifold creations of earlier men.

Greek mathematics came to a virtual standstill after Apollonius. Greek civilization itself weakened and declined; wars and poverty eliminated the leisure classes which, in Aristotle's opinion, were a prerequisite to the furtherance of knowledge in any society. Equally important, however, was the cumbersome nature of the geometric algebra. In the final analysis, almost all that could be done with the geometric notation had been exhausted. As an example, algebraic equations of degree higher than 4 are not available in the Greek notation, whereas the history of mathematics demonstrates time and again the importance of a new notation which allows a more powerful expression of an idea. Similarly, the communication of ideas from one generation to the next requires a handy and easily transmitted mathematical language.

Upheavals in Greek society naturally affected the normal processes of education, disrupting the continuity of knowledge passed on by succeeding generations of teachers. From 150 B.C.E. to 150 C.E., little of note was accomplished. During this interval, beginning with Hipparchus (150 B.C.E.) and culminating with Ptolemy (150 C.E.), a trigonometry of arcs and chords was given impetus by the demands of astronomy. This trigonometry prevailed for a thousand years, as it was taken over first by Indian astronomers in the fifth century and then by Arabic mathematicians in the tenth.

Not having the definitions for sines and cosines at his disposal, Ptolemy worked out the rules of trigonometry in the relations between chords and arcs on a circle. In most respects his work represented a continuation of the ancient Babylonian tradition in astronomy. He employed the ancient astronomical tables of the Babylonians for data and also used the ancient sexagesimal number system. This is not surprising because the Greek additive number system would have made most of his calculations impossibly complex. His major work, the *Almagest*, was written to determine astronomical relations, and for these, precise rules were indispensable.

Actually, our *modern angular measure* in terms of radians is nothing more than the *measure of a length of subtended arc on a unit circle*. Ptolemy's methods for relating arcs and chords are actually methods by which angles are related to the subtended chord. As an example, the angle π is the length of arc along the circumference of a semicircle of unit radius. In Ptolemy's degree notation this angle is represented as 180°. A 60° angle subtends an arc of $\pi/3$ on a unit circle. This digression is only meant to

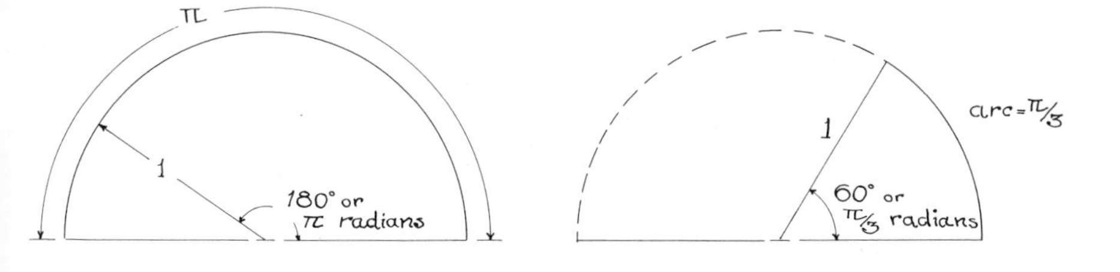

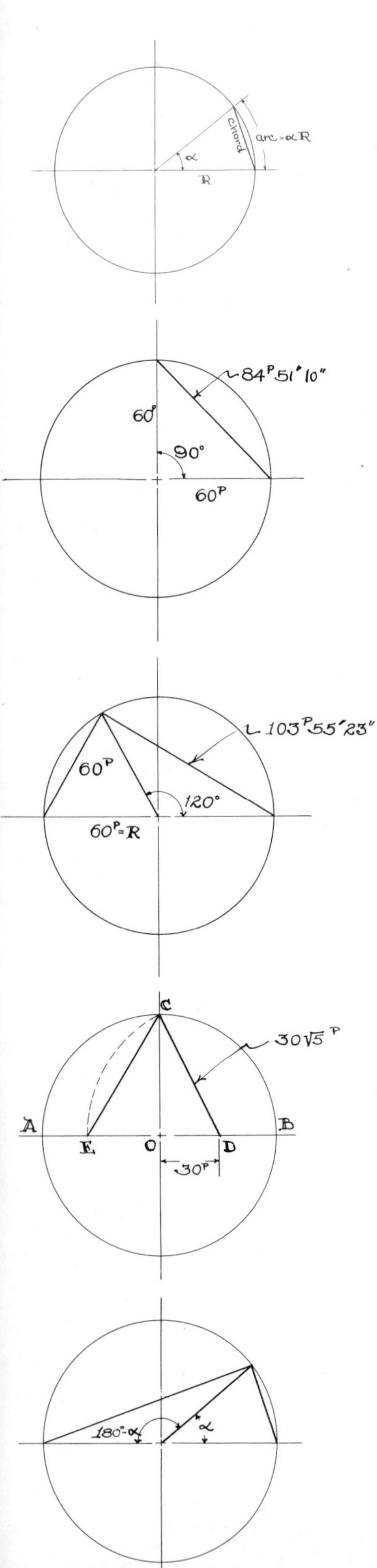

substantiate the logic of Ptolemy's method. It should be kept in mind that Ptolemy measured angles in degrees, minutes and seconds, the sexagesimal system of angular measure.

The basic arc-chord relation of Ptolemy is shown in the diagram. In modern trigonometric notation on a unit circle, the chord of α would be twice the sine of α.

Ptolemy divided the circumference of a circle into 360 parts, with subdivisions of any part being taken in minutes (1/60) of a degree, and seconds (1/60) of a minute. The diameter was used as a standard for measurements of length and was divided into 120 parts. Thus chords are given in diameter parts. The elementary relations can be built up from obvious diagrams. Consider the chord of 90°:

$$\text{the chord of } 90^\circ = \sqrt{(60^P)^2 + (60^P)^2} = 60\sqrt{2}^P$$
$$= 84^P51'10''.$$

Here occurs a numerical result which is not often encountered in classical Greek works: an accurate value for the $\sqrt{2}$ given in the sexagesimal system. The 120° chord can be obtained from the right triangle inscribed in a semicircle (knowing that one side is equal to the radius [60^P]):

$$\text{the chord of } 120^\circ = \sqrt{(120^P)^2 - (60^P)^2} = 60\sqrt{3}^P$$
$$= 103^P55'23''.$$

Chords of 45° and other angles can be found in a similar fashion. The pentagon and decagon subtend 72° and 36°, respectively. Using the construction from Euclid, which divides a line in extreme and mean ratio, Ptolemy obtained the $\sqrt{5}$ in a sexagesimal representation.

In the figure here BE is divided at O into extreme and mean ratio. CE is the side of a pentagon, while OE is the side of a decagon.

$$OE = CD - OD = 30^P\sqrt{5} - 30^P,$$

and

$$(EC)^2 = (60^P)^2 + (OE)^2.$$

Therefore,

$$\text{the chord of } 36^\circ = 30^P(\sqrt{5} - 1) = 37^P4'55'',$$

and

$$\text{the chord of } 72^\circ = 30^P\sqrt{10 - 2\sqrt{5}} = 70^P32'3''.$$

To extend his tables Ptolemy employed relations equivalent to the modern equations $\sin^2 \alpha + \cos^2 \alpha = 1$, and $\sin(\alpha - \beta) = \sin\alpha\cos\beta - \cos\alpha\sin\beta$. Most of the trigonometric relations appear in an equivalent form in the *Almagest*. Because the triangle ABC is inscribed in a semicircle,

$$[\text{chord } 2\alpha]^2 + [\text{chord }(180^\circ - 2\alpha)]^2 = (120^P)^2.$$

This is equivalent to $\sin^2 \alpha + \cos^2 \alpha = 1$, on a unit circle.

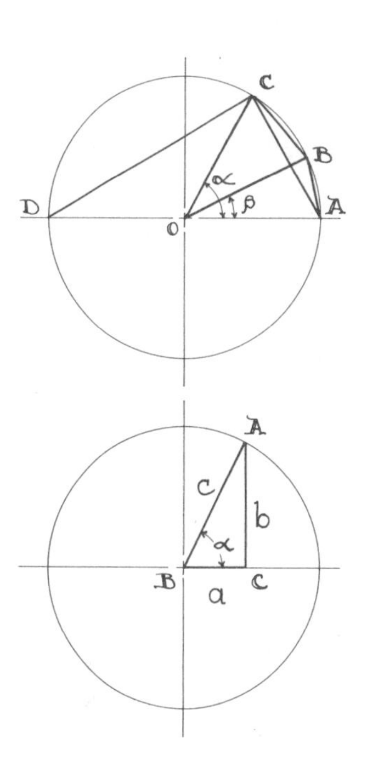

The use of this theorem can be illustrated by finding the chord of 108°, knowing that of 72°. The chord of the difference is obtained again by triangulation:

$$\frac{AC}{BD} = \frac{BC}{AD} + \frac{AB}{CD}, \quad \text{or} \quad \frac{BC}{AD} = \frac{AC}{BD} - \frac{AB}{CD}.$$

Rewriting this:

$$[\text{chord } (\alpha - \beta)][\text{chord } 180°]$$
$$= [\text{chord } \alpha][\text{chord } (180° - \beta)] - [\text{chord } \beta][\text{chord } (180° - \alpha)].$$

Because the sine and cosine, in a somewhat different form, become quite important in the history of the development of complex numbers in the eighteenth century, it is appropriate to mention here the triangular definitions. Consider the right triangle ABC with a unit hypotenuse:

$$\cos \alpha = \frac{a}{c} \xrightarrow[c=1]{} a,$$

$$\sin \alpha = \frac{b}{c} \xrightarrow[c=1]{} b,$$

$$\text{tangent } \alpha = \frac{b}{a},$$

$$\text{cotangent } \alpha = (\text{tangent } \alpha)^{-1} = \frac{a}{b},$$

$$\text{secant } \alpha = (\text{cosine } \alpha)^{-1} = \frac{c}{a} \xrightarrow[c=1]{} \frac{1}{a},$$

$$\text{cosecant } \alpha = (\text{sine } \alpha)^{-1} = \frac{c}{b} \xrightarrow[c=1]{} \frac{1}{b}.$$

With this elementary diagram the Pythagorean theorem is a statement that $\sin^2 \alpha + \cos^2 \alpha = 1$.

Spherical trigonometry was also quite advanced in this period. Although the emphasis in modern times is placed upon plane trigonometry, spherical trigonometry was actually developed at the same time.

There is still a great deal of doubt about the exact dates of many of the Greek masters. Historians place Heron (or Hero) of Alexandria at various dates from 150 C.E. to some time after Ptolemy; the present consensus places him just after Ptolemy. His work falls into two classes, geometric and mechanical; of most importance are his calculations of areas and volumes. Known by his name, Heron's formula gives the area A of a triangle of sides a, b, and c in terms of the semiperimeter $2s = a + b + c$: $A = \sqrt{s(s - a)(s - b)(s - c)}$. Heron employed a form of vector addition known today as the parallelogram method. A vector quantity has two properties, direction and magnitude, and in modern graphic representation it is often shown as a directed line segment whose length is proportional

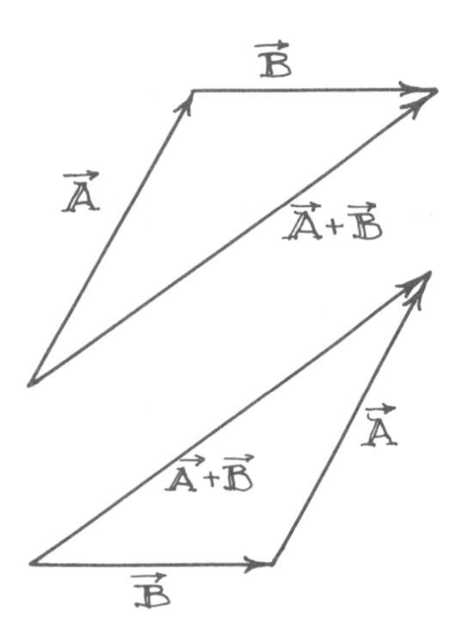

to the magnitude of a vector. Vector addition of two directed line segments can be achieved pictorially by drawing the two connected so that the tail of one vector touches the head of the other. The resultant vector or the vector sum of the two is then achieved by connecting the free head and free tail of the combined pair. In the facing illustration, a vector $\vec{\mathbf{A}}$ and a vector $\vec{\mathbf{B}}$ are added in two equivalent configurations to give the vector sum $\vec{\mathbf{A}} + \vec{\mathbf{B}} = \vec{\mathbf{B}} + \vec{\mathbf{A}}$.

An equivalent method was employed by the Greeks wherein the two vectors formed the sides of a parallelogram, while the resultant (or vector sum) was the appropriate diagonal of the figure. Using the same two vectors $\vec{\mathbf{A}}$ and $\vec{\mathbf{B}}$, the equivalence of the parallelogram method to the direct additions is pictorially obvious.

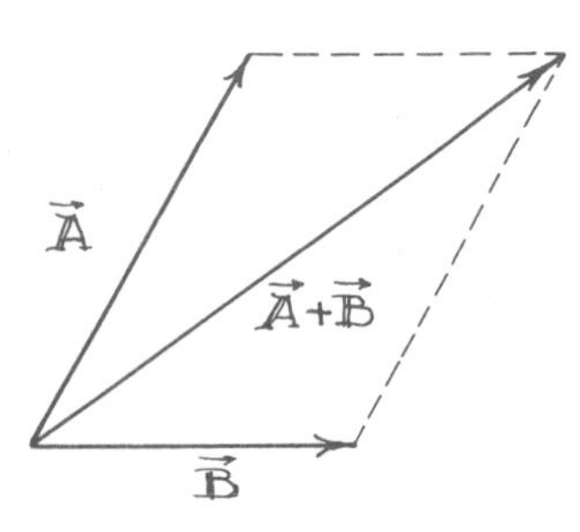

It is interesting to note that because the geometric order in which the sum is formed is not unique, the addition commutes. In the study of planetary motions the Greeks used the parallelogram method to combine two independent vector motions in order to obtain the resultant velocities. A typical epicyclic motion is shown in the accompanying diagram. The independent velocities at any instant are the tangential velocity $\vec{\mathbf{v}}_\Omega$ of the center of the small circle about the main center, plus the tangential velocity $\vec{\mathbf{v}}_\omega$ of the point P about the center of the small circle. If the rotation of O' about O proceeds with constant angular velocity Ω, the tangential velocity $\vec{\mathbf{v}}_\Omega$ has a magnitude equal to ΩR. By the same reasoning, if P rotates about O' with a constant angular velocity ω, the tangential velocity $\vec{\mathbf{v}}_\omega$ has a magnitude given by ωr, where r is the radius of the smaller circle. An observer at O does not see the individual components $\vec{\mathbf{v}}_\Omega$ and $\vec{\mathbf{v}}_\omega$ but records the vector sum $\vec{\mathbf{v}}_\Omega + \vec{\mathbf{v}}_\omega$. This concept, which was

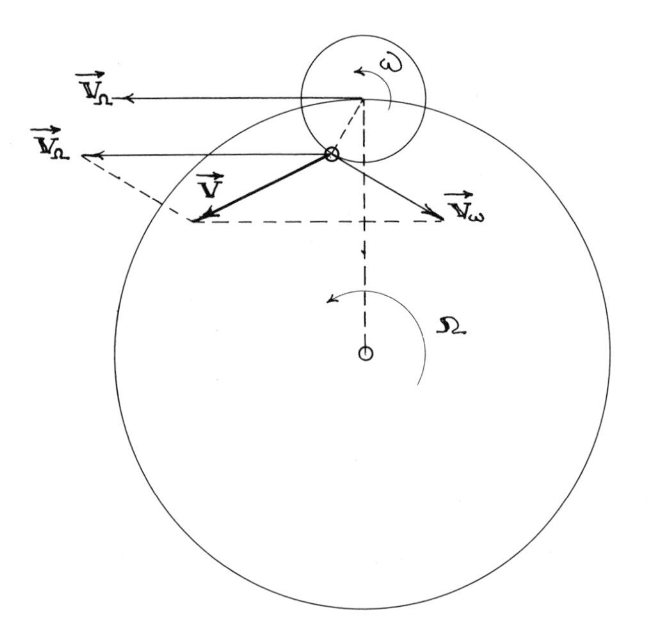

clear to the Greeks, proves to be fundamental to physics. Even though the methods for handling vectors have become quite sophisticated, the basic concept of addition remains much the same. Here for the first time the intuitive conviction of linear superposition is applied with success. In other words, the Greeks assumed without proof that the instantaneous motion of a point could be resolved into two independent vector components which were combined in a linear fashion to form the final vector.

Optics and, in particular, the reflection and refraction of light had fascinated the minds of men since antiquity. The fact that the angle of reflection from a plane surface is equal to the angle of incidence must have been recognized very early. These angles are measured relative to a line perpendicular to the reflecting surface, as seen in the diagram below. Heron of Alexandria made a major contribution to the theory of reflection by observing that when light is emitted from a point A and is reflected from a plane surface to a point B, the path corresponding to equal angles of incidence and reflection is the shortest path. He assumed that the ideal path, i.e., the shortest, represented the physical situation, and in this assumption he was quite correct.

Proof of this statement is quite straightforward. If A is reflected through the plane to A', the path from A to B is equal to $A'OB$ and is a straight line. Consider another possible path APB. The mirror path $A'PB$ is equal to APB and forms two sides of the triangle $A'PB$. Since the straight line $A'OB$ is also a side of the triangle $A'PB$, the path APB is less than the sum of the two sides $A'P + PB$. Therefore, AOB is the shortest distance between A and B under reflection at the plane surface.

This approach to the question of reflection has much greater significance than the result shown above. The implication is that the laws of nature obey some ideal principle—in this case, that the time for a ray to proceed via a reflection from A to B is a minimum, although in Heron's

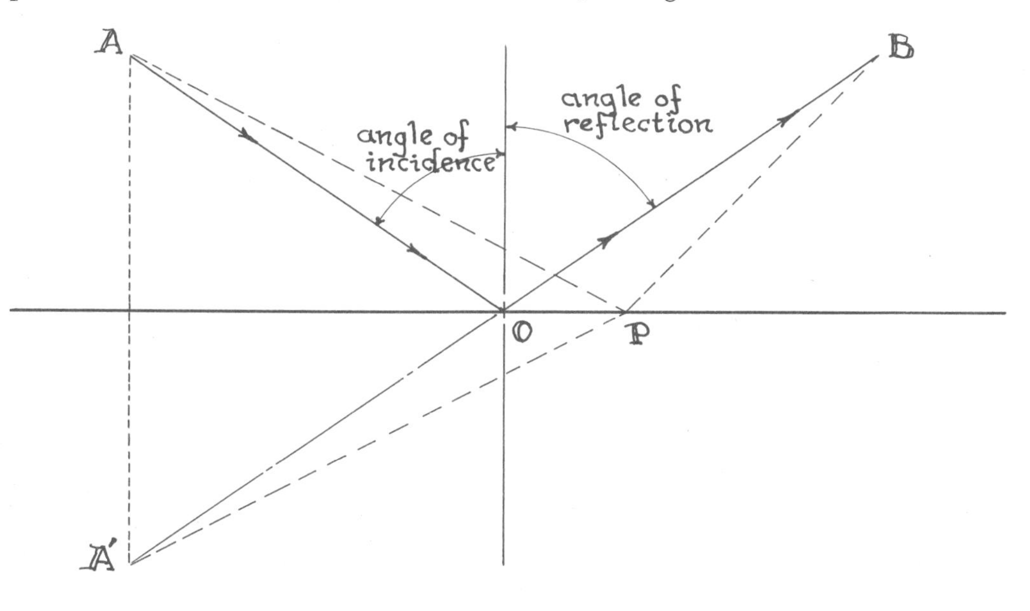

age it was not recognized that the velocity of light is finite. When one incorporates the finite velocity of light and the fact that the velocity of propagation along both segments is the same, the result of Heron's construction implies that the transit time of a light ray along the reflected path is a minimum. Recognition of this fact in the seventeenth century led Fermat to the Least Time principle, in deriving the law of refraction of light rays.

One of the last Greek mathematicians, and one who had a dramatic influence on later mathematics, was Diophantus of Alexandria, who lived around 250 C.E. He wrote thirteen books, but only six have survived. Here for the first time we find a sign for the unknown quantity. The unknown was written as ϛ, although it also appears in some editions of his work as, among other things, S, and some difference of opinion surrounds this symbol. The square of the unknown was Δ^{Y}, the cube K^{Y}, and the unit $\dot{M}$. To write x^5 he employed the symbol ΔK^{Y}. Reciprocals of the unknown were formed by adding χ to the superscript. $\Delta^{Y\chi}$ represented $1/x^2$ in modern symbolic notation.

The subtraction sign was ⊂⊃, or a wanting; addition was understood, a forthcoming. "Equal to" was written as ι^{σ}. Only solutions in rational numbers were admitted; this excluded surds—imaginary quantities and negative quantities. When numbers were considered by themselves (not as roots), Diophantus gave the rule that a minus times a minus is a plus, and a minus times a plus is a minus. His fame rests upon his treatment of indeterminant equations. Determinant equations were treated as they had been in the past; all were of first and second degree except one equation of third degree. Simultaneous equations involving quadratics were handled in the style used by others, which converted to one unknown, i.e., the method of the complementary unknown. His work on indeterminant equations is quite thorough and sophisticated, so much so, in fact, that it is difficult to appreciate it without reading the complete work. As an example of equations of second degree, one might encounter $Ax^2 + Bx + C = y^2$. Diophantus treats single equations of this type; then he classifies types in which A and/or B and/or C is zero, and so forth. Cubics are treated as special cases. Interestingly enough, Omar Khayyám, in his listing of cubic equations, used the same technique in introducing coefficients. Diophantus provides an extensive method for approximating to limits as well as an interesting set of number theorems. It would be difficult to convey the ingenuity and variety of the devices he employed in his mathematical constructs; some measure of his worth may be gleaned from the profound effect he had on such later mathematicians as Viète, Bachet, Fermat, Euler, and Lagrange.

After Diophantus there are a few able mathematicians, but only one, Pappus, made any major contribution, and that was his compilation of the works of others, along with his own, in eight volumes. His own work, as mentioned earlier, was concerned mainly with solid geometry, higher plane curves, and isoperimetric problems.

Thus one of the truly great ages of mathematics slowly died. Some pretense of learning was maintained in the eastern Roman empire, but Europe remained a relatively primitive society, while contacts, travel, and the exchange of ideas were taking place between the Arab peoples of the Middle East and the Asiatics of India and China. In the pendulum play of history, the cultivation of mathematics oscillated from India to the Arabs, and through these peoples, particularly the Arabs, the wealth of Greek mathematics was treasured and preserved. Only after almost a thousand years did Europeans turn to the study and advancement of mathematics.

IV

A Passage to the Orient

At least three thousand years before the Christian era, mathematics is known to have commanded serious interest in the Orient. The Chinese had a highly developed culture which dated back to antiquity, and such a culture, by its very nature, required a knowledge of astronomy and geometry to serve the needs of agriculture and land measurement. The Chinese are reputed to have known the Pythagorean properties of the right triangle long before Pythagoras expounded them. Scholars have argued that the number 60 was derived from the duality of Yang and Yin, and it has been pointed out that there is evidence that the number 60 was regarded as having remarkable properties as early as the twenty-seventh century B.C.E.

The long Chou dynasty, which lasted from 1122 to 255 B.C.E., was characterized by an emphasis on learning and attendant cultural growth. It was during this interval that Confucius lived (500–478). Toward the end of the third century, however, the nation suffered setbacks because of internal unrest. Emperor Chin Shih Huang, no doubt alarmed by the influence of the schools and their participation in the social upheavals, ordered the random destruction of books. Although the order was directed mainly against humanistic works, it was carried out with a vengeance, and most mathematical and astronomical works in the realm were also destroyed.

The year 202 B.C.E., which initiated the Han dynasty, saw a revival of the earlier concern with learning. Those works which had been somehow salvaged from the recent pyre were restored to their places, and Greek learning began to filter into China. It is not at all improbable that the Chinese had long had some knowledge of the Babylonian contributions to mathematics, judging from the fact that similar and, in some cases, identical problems appear in the literature of both cultures. During the Han dynasty a book appeared entitled *Arithmetic in Nine Sections*, by Chian Ts'ang, which purported to be based upon material dating back to 1000 B.C.E.* The work covered the measurement of plane figures, square and cube roots, measurement of solids, fractions, linear equations having more than one unknown, and the Pythagorean theorem—an odd and interesting assortment. In it, π was assumed to have the value of 3, while in the solution of linear equations negative quantities were recognized. A resemblance to ancient Egyptian and Greek works can be seen in the measurement of the volume of a truncated square pyramid, as well as in a knowledge of quadratic equations.

Indo-Iranian tribes invaded the Indus Valley around 1200 B.C.E. and, as with other cultures, must have developed some practical mathematics. The first achievements of note in this area, however, do not appear in India until after Alexander's eastern conquest of the Indus Valley in 327 B.C.E. From this time onward, interchanges between East and West were greatly broadened. From later efforts in mathematics, however, it may be observed that the Babylonian and Persian influences in India were equally as important, if not indeed more so, than that of the Greeks. A very strong aptitude for arithmetic and algebra was exhibited by the Hindus, but apparently they lacked interest in the geometric and pictorial aspects of the mathematics of their time. The beauty of the deductive method appears lost in Indian mathematics. Definitions were suppressed and logical order was quite neglected; rules were created without regard to fundamental principles.

Perhaps precisely because they were not bound by many of the restraints of Greek formalism, Hindu mathematicians made remarkable innovations. Their contributions to arithmetic alone are immeasurable, as our own present number system, which is a derivative of the Hindu system, attests. The place value principle was reinstated, and this heritage too has had a profound effect upon mathematical invention and its forward movement.

In algebra the Hindu masters solved quadratics by methods which are essentially modern: they did not discriminate between rational and irrational numbers, and because of this, the irrational roots of quadratic equations were accepted without question; negative numbers were adopted on an equal level with positive numbers, and operations with zero were incorporated into the number field. Greek influence is apparent in

* J. F. Scott, *A History of Mathematics* (London: Taylor and Francis Ltd., 1960), p. 81.

the Hindu concern with indeterminate equations—in this field alone they went quite far, anticipating some of the findings of modern algebra.

The earliest known Hindu works are the *Siddhantas,** the astronomical works, and these manuscripts represent much more than the compilations of the Babylonians. In them one finds a certain amount of theory in which there is a strong Greek influence.

Aryabhata, one of the great early Hindu mathematicians, flourished at the beginning of the sixth century C.E. In addition to understanding simple linear and quadratic equations, his major work, *Aryabhatiyam*, treated indeterminate equations by continued fractions, a method modern in itself. The work contains sixty-six verbal rules which are indeed quite complicated and difficult for modern man, conditioned as we are by the use of symbolic notation. All in all, the effect was chaotic, since the refined Greek logic was not present. In these rhetorical problems one was instructed to follow a set of steps, and the assumption was that the final result was meaningful. The emphasis was on computation, and therefore the correctness of the method was gauged by the usefulness of the answer.

Much attention was devoted to trigonometry. Aryabhata deserves credit for the invention of the sine and versed sine—certainly a great advance beyond the half chords of Ptolemy. Significant too was his meticulous application of the method of exhaustion to the circumference of the circle. He divided the circle into 21,600 parts (or minutes of arc), with a diameter divided into 20,000 parts. The circumference of the resulting polygon was given by adding 4 to 100, multiplying by 8, and adding 62,000. The result, 62,832 divided by 20,000 gave a value for π equal to 3.1416, certainly the most precise value obtained up to this date. Unfortunately, the Hindus neglected Aryabhata's value for π.

Another eminent name in Hindu mathematics is that of Brahmagupta, who was born in 598. The work of Aryabhata provided the foundation for many of his contributions. His work on arithmetic series, the sum of a limited sequence of integral squares and integral cubes, is hardly original, these problems having been long ago solved by the Babylonians and then the Greeks. From the surviving texts one might assume that Brahmagupta arrived at his sums of the series empirically; however, it is highly probable that these rules had arrived in India from Greece. The Diophantine equations were also given an elaborate treatment, although without the discipline and logic of the ancient Greeks.

Brahmagupta's interest in geometry was superficial—π was taken as the square root of 10. The problems with which he concerned himself indicate a concentration upon practical mathematics and upon a surprisingly elaborate set of rules, without any sources given for the rules employed. A typical problem is stated as follows: "500 drammas were a loan at a rate of interest not known. The interest on that money for four months was loaned to another, at the same rate, and it accumulated in ten months to 78. Tell me the rate of interest on the principal."†

* *Ibid.*, p. 67.
† *Ibid.*, p. 69.

In geometry, Heron's formula for the area of a triangle is given, along with several other minor Greek rules. It is in his analysis of equations that Brahmagupta demonstrated invention, beginning with a set of rules for handling negative numbers and zero. He stated that negative numbers multiplied by negative numbers give positives and that negatives subtracted from zero give a positive number. Because of the empirical nature of the rules, he assumed that zero divided by zero was zero. Thus, with Aryabhata and Brahmagupta we discover the first serious efforts to incorporate operations with zero into the system of arithmetic. Because the Hindus avoided the formalistic approach, they were evidently not afraid to introduce these subtly dangerous operations into their system. When dealing only with the real numbers, excluding zero, every product has a unique result. Only when multiplying by zero is the result not unique, in that any finite number multiplied by zero provides zero. Some six centuries later, Bhaskara indicated that the zero resulting from a number $n \times 0$ is a curious quantity, in that one must always keep track of its origin. This dilemma becomes apparent if one considers $n \times 0 = 0$, which in turn is divided by zero. In such a case the original number n must be restored. Division by zero is now regarded as leading to an undefined result. In this sense Bhaskara did, in fact, suggest that definition may be reestablished in some cases. The concept of the arbitrarily large and the infinitesimally small persists to plague mathematicians of all generations. Thus it is quite intriguing to consider that at the very outset of the incorporation of zero into the number system Brahmagupta was aware of some of the ambiguities of this operation.

The Hindu origin of the Arabic numbers and the place system may be traced back to the third century B.C.E. in India, where various primitive forms of our numbers have been found. Kharosuthi numbers are similar to Roman numerals; however, the Brahmi symbols show a marked likeness to some of the Arabic numbers, for instance, to the six, seven, eight, and nine. None of these early numbers had a positional notation. From the Brahmi system evolved the Gawalior numbers, and these later gave way to the West Arabic or gobar and the East Arabic numerals. The West Arabic, except for modifications of four and five, essentially represent our modern Arabic numerals. By the fifteenth century these numerals were employed in Europe in the modern form, with our present symbols for four and five. In spite of some differences of opinion, it is generally held that the nine number symbols and the zero were employed in India, with a place system, before 600 C.E.

All major mathematical advances seem to have been associated with a relatively stable and flourishing culture. As with the Babylonians, the Egyptians, and the Greeks, so was this true with the Arabs. Until the seventh century C.E. one cannot view the Arab world as anything more than a multitude of contentious, divided peoples. The appearance of Mohammed and the subsequent unification of large parts of the Middle East and North Africa produced a culture and an intellectual brilliance which lasted for almost eight hundred years. Strangely enough, as the

Muslim world slackened its pace, Renaissance Europe grasped the initiative, and the new science and the new mathematics were born.

Mohammed's religious revolution provided the impetus for a period of renascence for the Middle East, from which Europe was to reap enormous rewards—the Renaissance. Although the prophet died in 632, the impulse of the new movement under one God swept over the richest part of the world, from the Indus Valley to Spain. With it came a new ethic, tolerance, a virtue more often forgotten than remembered. Muslim law gave religious freedom to the Jews and Christians; tolerance, wise government, and concern for their subjects were the great contributions of the Muslim movement. Such attitudes gave back rich dividends. Because the ancient cultures of the subjugated peoples were not destroyed, Islam was able to absorb much of the greatness of her subjects. The arts and sciences flourished. Damascus, rich in the cultures of Greece, Rome, and the Semites, became the capital of the caliphate in 635, and by 766 the Caliph Al-Mansur had built Bagdad near the ruins of Babylon. To this center of learning came Jewish, Syrian, and Persian scholars, perhaps attracted less by the rich pay than by the academic freedom—a daring concept for those times.

Soon after the creation of Bagdad, the *Siddhanta*, one of the early Hindu astronomical treatises, was presented to Al-Mansur. He had it translated into Arabic immediately. Not only were the Hindu manuscripts assembled and translated, but most of the famous Greek works were studied and preserved in Arabic. Al-Mansur and Al-Mamun, caliphs in 754–775 and 813–833, collected the works of Euclid, Archimedes, Ptolemy, and many others, and this zeal for scholarly study and collection typifies the Arabic period. Major translations were made by Al-Hajjajb and Yusuf b. Matar during the reign of Harun-al-Rashid (786–809), and by Ishaq b. Hunain b. Ishaq (d. 910). Those of the last seem to be the best translations.

Much of our knowledge of Greek masterpieces is, in fact, due to the existence of the Arabic manuscripts. One notices here a remarkable change in the scientific style of acknowledgment, for although the Greeks preserved an internal history of the development of their own mathematics, they did not record or acknowledge any early debt to outside sources. Evidently the Hindus in like manner reworked some areas without recognizing earlier sources. In keeping with their tolerance in other areas, the Arabs established a tradition of scientific acknowledgment, and it is ironical that this great tradition has quite often led historians to denigrate their contributions. Certainly a culture appears more inventive if it has created rather than merely inherited its mathematics.

The Caliph Al-Mamun was the son of Harun-al-Rashid, well known as one of the major characters in the Arabian work, *One Thousand and One Nights*. Al-Mamun encouraged learning and research, and to this end established an academy, an astronomical observatory, and a library. From these halls emerged the first of the great algebraists, Muhamed ben Musa, or Al-Khwarizmi. This man's thought influenced mathematics in the Middle Ages to a greater extent than that of any other. He knew both the West Arabic, or gobar, and the East Arabic numerals. Although, as mentioned earlier, the modern Arabic numeral was derived from the gobar, the East Arabic numerals are still used in some parts of the Middle East. Al-Khwarizmi produced the first work on algebra and also wrote on arithmetic, utilizing as models the works of Brahmagupta. Perhaps his most significant contribution to algebra lies in his recognition of the fact that a quadratic equation could have two roots. For solving the quadratic, he used a method similar to the earlier techniques of Diophantus and the Hindus.

One of the earliest geodesic surveys to determine the length of a degree of the meridian of the earth was conducted under Al-Mamun's caliphate. The Greeks knew the earth was round, and the Arabs were convinced of it: they measured its radius accurately some six hundred years before the great age of the Spanish and Portuguese explorers. The influence of Eratosthenes of Cyrene is patently obvious in this enterprise.

Following Al-Khwarizmi came Tabit ibn Korra (830–901), a translator, and, around 900, Al-Battani, who constructed many of the early tables of Arabic trigonometry. Using Ptolemy's *Almagest* and the sine and versed sine concept of the Hindus, he compiled a table of sines and then like tables of cotangents. He worked out the lengths of the shadows cast by a vertical rod of length 12 in one-degree intervals of the sun. Mathematicians of this age were thoroughly schooled in the work of their predecessors: we find in the works of Al-Battani a knowledge of the various trigonometric combinations, plane and spherical, including the angle between two points on the unit sphere when the locations of the points are given by the set of three primary angles.

Toward the end of the tenth century Abul-Wefa invented a method of repeated division wherewith he computed a table of sines in intervals of half a degree. So accurate was his method that his value of the sine of 30′ was good to the ninth decimal place.

One mathematician of consequence appeared in India around 1150, Bhaskara. He wrote the *Siddhanta Siromani,* which was a complete rendition of Hindu mathematics. In the first chapter, entitled "*Lilivati*," one finds much of the works of earlier Hindu writers, including a discussion of the arithmetic operations with zero. In the third chapter, however, Bhaskara deals with the unusual problem of dividing by zero: "A quantity divided by zero becomes a fraction of which the denominator is zero. This fraction, of which the denominator is cipher, is termed *infinite* quantity. In this quantity, consisting of that which has cipher for its divisor, there is no alteration though it be inserted or extracted; as no change takes place in the infinite."* Bhaskara had indicated earlier that in multiplication by zero the number multiplied must be retained even though the product was zero, and, further, that this must be done if, in a second operation, this zero is divided by zero. In such a case the original number must be

* *Ibid.*, p. 73.

returned. All of this demonstrates an agile mind which readily recognized the dangers of operating with zero.

Bhaskara came quite close to a symbolic notation in his algebra: a negative quantity appeared as a symbol with a superior dot, while a positive quantity had no dot*—operations were written in words, the initial syllables of colors were used as symbols for unknown quantities, and initials of the words "square" and "solid" denoted the powers. Bhaskara also recognized that the square root had both a positive and a negative value. He stated that the squares of numbers are positive, implying, of course, real numbers.

Besides the treatment of quadratics, planes, and circles, Bhaskara provided a rather complete summary of indeterminate equations, although there is no evidence for what is original with him and what he borrowed. In Diophantus' works fractional solutions were acceptable; however, Bhaskara confined his interest to integer solutions. His problems are rhetorical, dealing with numbers of horses, geese, cranes, etc., and very colorful: "The horses belonging to four persons are 5, 3, 6, and 8. The camels pertaining to the same are 2, 7, 4, and 1; the mules belonging to them are 8, 2, 1, and 3; and the oxen, 7, 1, 2, and 1. All four persons being equally rich, tell me the price of each horse and the rest." This problem represents equations of first degree in three unknowns: $ax + by + cz = d$. Bhaskara arrives at the answer: 85 horses, 76 camels, 31 mules, and 4 oxen. "Tell me if thou knowest two such numbers such that the sum of them multiplied severally by 4 and by 3 may, when 2 is added, be equal to the product of the same two numbers." The equation $4x + 3y + 2 = xy$ represents the problem, for which he gives several solutions, such as 17 and 5 or 10 and 6, etc. "What square number multiplied by 8 and having 1 added shall be a square?" This is known as Pell's equation and has always intrigued mathematicians: $ax^2 + 1 = y^2$. For $a = 8$, it is satisfied by $x = 6$ and $y = 17$.

Bhaskara lived during the climax of Arab intellectualism. Arab cultural influence extended well into the twelfth and thirteenth centuries. Contacts between Europe and the East came about in many ways: the invading crusaders from Europe were exposed to the advanced culture of the twelfth-century Arab world; there was widespread communication with Moorish Spain; and commerce with the Byzantine world brought Eastern culture into Europe. By the time that Bagdad was threatened in the east by Genghis Khan, Córdoba was becoming the intellectual center of Islam in the west. Ahmed al-Leiti, a native of Córdoba, wrote on astronomy and arithmetic. Many of the learned scholars of eleventh- and twelfth-century Spain were Jews. These men, whose intellectual inquiries flourished in the spirit of free inquiry and tolerance encouraged by the Moors, contributed much to the advancement of mathematics in Europe.†

The first of the prominent Jewish scholars was Abraham bar Chiia, who is chiefly known for an encyclopedia which included arithmetic and

* *Ibid.*, p. 76.
† D. E. Smith, *History of Mathematics* (New York: Dover, 1958), 1:206.

geometry. The next great name in the Hebrew scholarship of the period was Abraham ben Ezra, who wrote on the theory of numbers, the calendar, magic squares, astronomy, and the astrology. He was followed by Moses Maimonides (1135–1204) of Córdoba, physician to the sultan, rabbi, philosopher, and astronomer, whose name today holds a place of eminence in the history of Spanish as well as Hebrew scholarship. The great Maimonides was followed by Johannes Hispalensis, who wrote on arithmetic and astronomy and painstakingly translated Arabic mathematical works into Latin.

The presence of highly skilled Jewish mathematicians in Spain proved highly fortuitous for Europe. As the Moors were driven out of Spain in the fourteenth century, so the Jews were forced to flee the relentless zeal of the Inquisition in the fifteenth. Many found refuge in Turkey, but larger numbers fled to central and northern Europe, carrying with them the heritage of the ancient world, a well-established mathematical tradition.

The fate of Islamic culture was sealed in the East as well, first by invasions of the European crusaders and then by the Mongol invasions of Genghis Khan between 1206 and 1227, followed by those of Kublai Khan (1216–1294) and of Timur, or Tamerlane, as he was called in the West (1336–1405). During this destructive period, Al-Kaŕkhi was writing on Hindu and Greek arithmetic and algebra. In algebra he summarized the known results concerning equations of first and second degree, and among his quadratic equations are some special quartics, such as $x^4 + 5x^2 = 126$, and various indeterminate equations, apparently taken from Al-Khwarizmi and Diophantus.

Omar Khayyám (1100), although best known to the western world for his *Rubáiyát*, was an astronomer of no little fame and achievement. He also wrote on Euclid, on algebra, and is reported to have calculated the general binomial coefficients for the expansion $(a + b)^n$ with an integral positive exponent. Since this was a fundamental problem in the development of the calculus, we may profit by examining the details of this puzzle. Exponents were not known; therefore the expansion was probably referred to as $(a + b)$ multiplied by itself n times, giving:

$$(a + b)^n = \sum_{m=0}^{m=n} C_{n,m} a^{n-m} b^m = C_{n,0} a^n + C_{n-1,1} a^{n-1} b + \cdots + C_{0,n} b^n.$$

The individual terms arising from the expansion $(a + b)^n$ are formed of all the products of the form $a^{\alpha_1} b^{\beta_1} a^{\alpha_2} b^{\beta_2} \ldots$, subject to the constraint that the sum of the powers of $a(\alpha_1 + \alpha_2 + \cdots + \alpha_j)$ is $n - m$, while the sum of the powers of b $(\beta_1 + \beta_2 + \cdots + \beta_r)$ is m. Finally, the sum of all the powers of both a and b is n: $\alpha_1 + \alpha_2 + \cdots + \alpha_j + \beta_1 + \beta_2 + \cdots + \beta_r = n$.

The heart of the problem lies in discovering how many times the product $a^{n-m} b^m$ appears. If we examine the number of ways that $n - m$ and m can be arranged in groups of terms, this number is the binomial coefficient. To illustrate the method, consider the special case $n = 4$, with

$m = 2$. Then $n - m = 2$. The possible arrangements are *aabb, abab, abba, baba, bbaa, baab*.

This counting is equivalent to mapping the total a's and b's as a grid. The number of combinations then represents the total number of separate paths to get from one corner, 0, to an opposite corner, 0′, as shown.

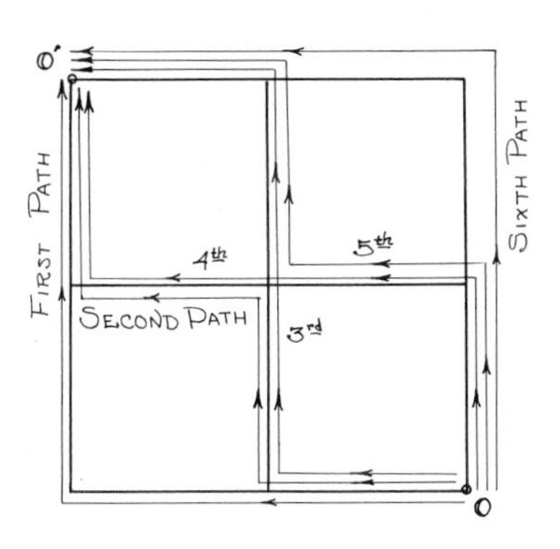

Now consider a more complicated term, say $a^{10}b^7$ in the expansion of $(a + b)^{17}$. We draw a rectangular network of 10 × 7 compartments, 10 on the horizontal and 7 on the vertical. The path $a^2b^3ab^4a^7$ is shown. The number of permutations of n elements, among which there appear $n - m$ identical elements of one kind and m identical elements of another, is the total number of arrangements of n things, $n!$, divided by the number of indistinguishable arrangements $(m - n)!\,m!$. This, of course, is the binomial coefficient $C_{n,m}$, where

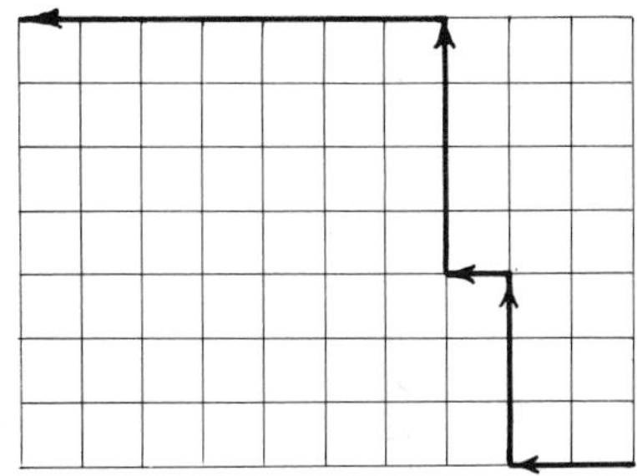

$$C_{n,m} = \frac{n!}{(n-m)!\,m!} = \frac{n(n-1)(n-2)\cdots 3\cdot 2\cdot 1}{(n-m)(n-m-1)\cdots 2\cdot 1\cdot m(m-1)\cdots 2\cdot 1}.$$

Here the sign ! means factorial and $n! = 1\cdot 2\cdot 3\cdot 4\cdots(n-1)\cdot n$.

During the exposition above we have employed modern factorial notation and have related this coefficient to the number of separately distinct paths between one corner of a grid and an opposite corner. This is not meant to imply that Omar Khayyám solved the problem in this fashion; however, if he did solve it as claimed, his solution must have been in many ways equivalent.

While concerned with classical geometry, Khayyám showed a remarkably inventive mind. In one of his texts he replaced the parallel postulate of Euclid with a different set of propositions, arriving at concepts such as the "hypothesis of the obtuse, acute and right angles,"* which in modern times would be classified as belonging to non-Euclidean geometry. He reviewed Euclid's theory of proportions and created as a substitute a numerical theory bordering upon the concept of the irrational in real number theory.

His work on the cubic equation was to have a decided influence upon Scipio del Ferro and Cardano in the sixteenth century. Cubics had been singled out for special attention since antiquity—such puzzles as the duplication of the cube are common. By the seventh century in China problems appear which suggest that the cubic was an accepted part of the rhetorical algebra. There is a written record of work by Wang Hs'iao-tung in which the reader is given the following problem: "There is a right triangle, the product of whose sides is $706\frac{1}{50}$ and whose hypotenuse is greater than the first side by 309/60. It is required to know the lengths of the three sides." The answer is provided, along with the rules by which it was obtained.†

Omar Khayyám classified all the cubics with a positive root. Not only

* D. J. Struik, "Omar Khayyam, Mathematician," *The Mathematics Teacher* 51 (1958):280.

† Yoshio Mikami, *The Development of Mathematics in China and Japan* (New York: Chelsea Publishing Co., 1913).

did his classifications appear in Cardano's *Ars magna*, but the order of Khayyám's list was maintained. His classification of cubics was made logically in terms of the number of non-zero coefficients,* listed as compound, trinomial, or quadrinomial. He began with forms of the type $a = x^3$ to $ax^2 = x^3$, and completed his list with arrangements of the type $x^3 + bx^2 + cx = d$ to $x^3 + dx = bx^2 + c$. By permuting the positions of terms in this last group, he demonstrated that the coefficients were to be considered as positive integers. In the examples given, of course, symbols were not employed. Instead, equations were implied by graphic constructions reminiscent of the Greek constructions.

Bagdad was sacked in 1256, during an invasion of the Mongols, probably led by Kublai Khan, and Arab culture began to decline. Later a new university was established by the Seljuk Turks, and here one of the last of the great Arab scholars, Nasir-Eddin, taught and studied. The *Shakl-al-gatta* of Nasir-Eddin separated plane trigonometry from astronomy and presented it as a science in its own right. His influence was widely felt later in Renaissance Europe.

The invention of decimal fractions is attributed to many different civilizations—Chinese, Turkish, and that of Renaissance Europe. The earliest claim† places the invention in China and then India around 595 C.E. The Turkish scholar Jemshid al-Kashi (d. 1436) exhibited a versatile use of decimal fractions in his numerical work, comparable to the sophistication achieved in sixteenth-century Europe. He solved cubic equations by iteration and by trigonometric methods. For the solution of general algebraic equations, he used what was later known as Horner's method, while his computation of the ratio of the circumference of a circle to the radius was carried to sixteen decimal places.

The decline of Persian and Arab scholarship was followed by the Renaissance in Europe. Amply supplied with Greek and Arabic works flowing from Byzantium, from the strongholds of the crusaders, and from Moorish Spain, the European scholars took the intellectual lead and have held it to this day.

* G. Cantor, *Vorlesungen uber die Geschichte der Mathematic* (Leipzig: 1892); David Kasir, *The Algebra of Omar Khayyam* (New York: Columbia University Press, 1931).

† D. J. Struik, *A Concise History of Mathematics* (New York: Dover, 1967), p. 71.

V

Observation as a Fine Art

From the collapse of the Roman Empire at around 410 C.E. until the first indications of an awakening around 1000, Europeans lived in a primitive and static world. The Church gradually assumed the responsibility that had been the Romans'—that of civilizing the known world of the West. During these dark ages few lights continued to shine. Boethius, a Roman citizen born in 475, contributed little more than work based upon Nicomachus and Euclid, and in the centuries that followed scant interest was evidenced in the literature and science of the Greek and Roman periods. Except for a few rudimentary works by the Venerable Bede (673–735) and Alcuin of York (735–804), learning remained static. By 1000, however, some interest was aroused in arithmetic through contacts with Arab Spain. From 1095 to 1270 the Crusades brought large groups of Europeans into direct contact with the higher culture of Asia Minor. So impressed were many early crusaders with the culture they found that they settled in their conquered lands and took on the role of oriental princes. It is reported that later crusaders arriving in the Middle East were greatly distressed at the sight of so many conversions among their countrymen.

Adding to these contacts between Europe and Asia Minor, the Spanish Inquisition in 1232 drove many Jewish scholars into Europe, and with them came a rich tradition in oriental mathematics. While at no time was the study of mathematics entirely abandoned in Europe, until the beginning of the fifteenth century it remained in its infancy. From 1200 to 1224 the universities of Paris, Cambridge, Oxford, Padua, and Naples were established. The thirteenth century in Europe was in some respects the analogue of the Moslem revival of 700. Although little original mathematics was accomplished, there was a flurry of translations from Arabic works and from the Greek.

Leonardo Fibonacci, or Leonardo of Pisa, was the first significant translator of the thirteenth century. His early travels to Egypt and Greece resulted in an evaluation of arithmetic, in which he introduced the Hindu-Arabic integers, as well as works dealing with geometry and algebra. In his own lifetime his efforts went unnoticed, and of the early English scholars, the most prominent was Roger Bacon (1214–1294), who demonstrated a knowledge of Euclid, Ptolemy, and Archimedes.

The Hundred Years' War between England and France (1337–1453) and the ten years of the Black Death decimated the population of Europe and sapped the vitality of its people. But from this period emerged Dante (1265–1321), Petrarch (1304–1374), and Boccaccio (1313–1375), and by the fifteenth century all of Europe was astir. Printing was invented in 1450, navigation and the subsequent discovery of new worlds demanded more and better mathematics and astronomy, and commercial expansion led to new and more efficient means of computation.

By and large, science had been confined to the study of astronomy since the time of Archimedes. Exceptions had occurred; as an example, Ibn al-Haytham (or Alhazen, 965–1039) wrote a work on optics which came to exert a great influence on the West. The importance of the scientific method, together with the mathematical representations of scientific theory, were of secondary interest up to the fourteenth century. Basic to all scientific theory is the experimental method. Philosophically the Greeks had negated the significance of correlating organized observations with theories concerning the world about them. One of the earliest records of a science based upon experiments appears at Merton College, Oxford. Robert Grosseteste (1168–1253) and his disciple Roger Bacon foresaw the need for an hypothetico-deductive method of experimental testing. In their scheme a tentative hypothesis, after being postulated, was subjected to appropriate experimental tests. On the basis of the experimental evidence, it could then be either accepted or rejected. Their chief interest was in optics, and they made extensive use of Al-Haytham's work on this subject. Their experimental method is so basic that it is difficult to comprehend the reasons for its delayed appearance.

Following Grosseteste's lead, Merton College flowered in the invention of a scientific method. Fundamental to all physics are the definitions of the kinematic quantities—position, velocity, and acceleration. Kinematics was traditionally associated with Greek astronomy. In order to introduce these concepts, particularly those of terrestrial motion, one must consider a quantification having a dual aspect: first, a language of quantity which is applicable, and, second, theories and techniques of measurement which can produce numerical results.* The early language of quantity was based upon the Euclidean proportions. Modern quantification in terms of standards is a form of proportion, but it is of a more universal nature. In older theories velocities and accelerations are expressed as ratios rather than as metric statements.

Notions of instantaneous changes, mean velocities, and constant acceleration are associated with four natural philosophers present at Merton College between 1328 and 1350: Thomas Bradwardine, William Heytesbury, Richard Swineshead, and Thomas Dumbleton.† All of the Merton developments are of an arithmetic and logical nature, whereas the proof of Galileo, which occurs some three hundred years later, is strictly geometric. It is instructive to observe the mean speed theorem as constructed by Heytesbury in 1335: "For whence it commences from zero degree or from some [infinite] degree every latitude [i.e., increment of velocity or velocity difference] as long as it is terminated at some finite degree [i.e., of velocity], and as long as it has acquired or lost uniformly, will correspond to its mean degree. Thus the moving body acquiring or losing this latitude [increment] uniformly during some assigned period of time, will traverse a distance exactly equal to what it would traverse in an equal period of time if it were moved uniformly at its mean degree of velocity."‡

By 1350 the Merton College studies of kinematics had spread to France, where a system of graphing of movements utilized a kind of coordinate geometry. The most complete treatment of this system was developed by Nicole Oresme. Whole two-dimensional figures were applied to represent the quantity of a quality, resulting in a rectangle as a uniformly intense quality and a right triangle as a uniformly nonuniform quality (i.e., uniform acceleration). A further advance in mechanics was made by Jean Buridan, who considered the motion of a massive projectile. In his description the impetus (i.e., momentum) was proportional to the amount of matter in the body and to its velocity. Gravity was believed to add continuously to the impetus of a falling object. Here are all of the concepts necessary to form an organized dynamics. Although Galileo was unaware of the specifics of these works, he must have been affected by the subtle changes in the minds of the natural philosophers preceding him—these changes would indeed be correlated with the works of the Merton and Paris groups.

Nicole Oresme is often dismissed as little more than an authority on classical Greek mathematics. He was, in fact, one of the most important innovators before the age of Galileo. He developed the "latitude of forms," or the graphic description of kinematics associated with the Merton group.

* M. Clagett, "Novel Trends in Science," in *Art, Science, and History in the Renaissance*, ed. Charles S. Singleton (Baltimore: The Johns Hopkins Press, 1968).

† *Ibid.*, p. 285.

‡ M. Clagett, *The Science of Mechanics in the Middle Ages* (Madison: University of Wisconsin Press, 1959), p. 262.

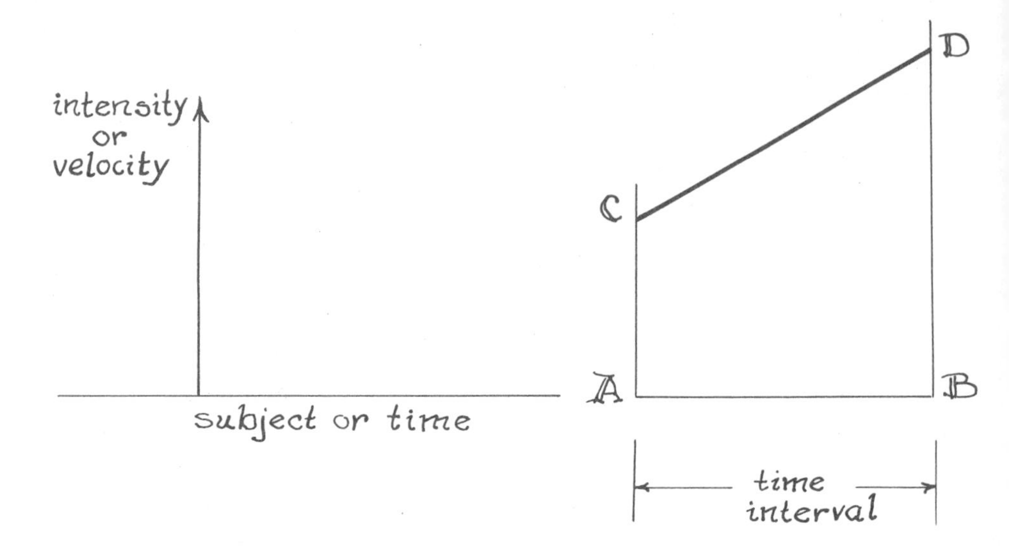

In his graphs of intensity (i.e., velocity) versus time, he clearly related the area under the curve to the distance traveled. As an example, for a uniformly difform quality from a certain degree (i.e., uniform acceleration beginning from a nonzero velocity), his form is a trapezoid whose area is the distance traveled. These works and the later works of Galileo indicate the need for an integral calculus (or a method for obtaining the area under a specified curve).

In his analysis of the Merton kinematics Oresme utilized ratios of quantities taken to powers which were also ratios. This notation was applied generally in his *Algorismus proportionum*, where fractional exponents were displayed as pure numbers: the square root of two was written as $(1/2)2^P$, which implies $2^{1/2}$. In the same manner, $9^{1/3}$ was exhibited as $(1/3)9^P$, and $\boxed{1^P 1/2}\,4$ represented $4^{3/2}$—he gave the result correctly as 8.

Like Galileo, Oresme arrived at the odd number theorem associated with the problem of constant acceleration. Both the Merton group and Oresme became concerned with the summation of infinite series. There is a mention of such sums in the works of Aristotle, and much later Thomas Aquinas gives the specific example of the infinite geometric series which, for reasons which will become obvious, we shall call Zeno's series:

$$1/2 + 1/4 + 1/8 + \cdots + (1/2)^n + \cdots = 1.$$

Aquinas describes the series as the division of the unit line divided first into halves, then a half into halves, one-quarter into halves, and so on to infinity. The proof of the sum is contained in the initial statement; therefore the method is not made explicit.

Because Zeno's series is a special case of the geometric series, we will digress momentarily to evaluate the general geometric sequence, S_N, of $N + 1$ terms. S_N is defined in terms of an arbitrary parameter x as:

$$S_N = 1 + x + x^2 + x^3 + \cdots + x^{N-1} + x^N.$$

By subtracting 1 and multiplying by x, we find that the next higher sequence S_{N+1} terminating in x^{N+1} is defined by the equality:

$$S_{N+1} = xS_N + 1 = S_N + x^{N+1}.$$

Solving for S_N we obtain:

$$S_N = \sum_{n=0}^{N} x^n = \frac{x^{N+1} - 1}{x - 1}.$$

If x is less than 1 in magnitude, and if N is allowed to take on arbitrarily large values, the term x^{N+1} tends to zero, giving:

$$S(x) = \underset{\substack{N \to \infty \\ |x| < 1}}{\text{limit}}\, S_N = \frac{1}{1 - x}.$$

Zeno's series is then this infinite series minus unity evaluated for $x = 1/2$. By the fourteenth century scholars were uncritically assuming the summation of Zeno's series and other more complicated series, such as:

$$1 + 2\left(\frac{1}{2}\right) + 3\left(\frac{1}{4}\right) + 4\left(\frac{1}{8}\right) + \cdots + n\left(\frac{1}{2}\right)^{n-1} + \cdots = 4.$$

Oresme obtained this sum by a geometric argument in which two similar squares are divided to produce the equivalent of the sum in the form:

$$1 + 1 + \left\{1 + \frac{1}{2} + \frac{1}{4} + \cdots + \frac{1}{2^n} + \cdots\right\} = 4.$$

All of these efforts were motivated by an interest in qualities and velocities; thus for the first time we observe the important role physics was to play in stimulating interest in novel mathematical concepts.

The mathematical fame of Nicolas of Cusa (1401–1464), a cardinal of the Church, rests primarily on his work on the quadrature of the circle. Although he achieved an accurate value of π, this cannot be accounted too highly, in view of the earlier Arab calculations, though it is by no means clear that he was aware of this earlier work. Soon after, Georg Puerbach (1423–1461), a Viennese teacher interested in astronomy and trigonometry, provided a translation of Ptolemy's *Almagest*. His student, Johann Müller, known as Regiomontanus, made lasting contributions to trigonometry, both plane and spherical. Like his Hindu and Arab predecessors, he produced a table of tangents which stimulated a popular interest in trigonometric tables. By reducing geometric problems to a stylized rhetorical algebra, he anticipated analytic geometry. From these early efforts we perceive the seeds of a symbolic algebra. Denoting the unknown by *res* and the unknown squared as *census*, a quadratic equation of the type $4x^2 + 170x = 500$ might appear as "4 census et 170 res aequales 500."

In the earlier oriental mathematics some attempts had been made to provide symbols for subtraction. Luca Pacioli, who was born about 1450 and died about 1520, produced his major work in 1494. A Franciscan friar who taught in Rome, Pacioli traveled extensively, possibly even in the East,* after which he wrote a summary of the mathematical knowledge of his time. Although there was little originality in this work, he did introduce symbols for addition, subtraction, and the square root. These were some improvement over the symbols of Regiomontanus. Addition was denoted as *p* (*piu*), subtraction as *m* (*meno*), and the square root as *R* (*radix*). A typical example, then, is "4 p.R.6 4 m.R.6 Productum 16 m.6 10," implying $(4 + \sqrt{6})(4 - \sqrt{6}) = 16 - 6 = 10$.† On the other hand, Pacioli considered only positive roots of quadratics and decided that equations of higher degree could not be solved.

Truly this was an age of invention and creativity. Typical of the period was Leonardo da Vinci. Although historians are not unanimous regarding the importance of his contributions to mechanics and the study of geometric curves, these niceties do not affect his significance. In his work we observe the predominant characteristic of the age, accurate observation. In a drawing of a waterfall the perception of a true experimentalist is evidenced. European scholars began to test every new idea with regard to its accurate portrayal of reality. Because of this, science and painting followed parallel paths for a short time, which accounts in part for the interest which science has shown in Leonardo.

The teachings and writings of Niklas Koppernigk, or Copernicus, provide a fitting climax to this period. Born at Thorn on the Vistula in 1473, Copernicus received his formal education at the University of Cracow and later studied law, medicine, and astronomy at Padua and Bologna. After taking holy orders, he returned to Poland, where, in 1530 he completed his theory of the heliocentric universe. This concept could hardly be viewed as unique because of the earlier Greek theories which assumed the same position; however, considering the tone and dogma of the age, it was truly revolutionary. With the invention of printing, writings such as those of Copernicus could reach a very large audience, and his theory was widely disseminated. The subsequent controversy and the efforts to suppress this heretical idea endowed it with an importance which could not be ignored. In effect, the closed mind of medieval Europe was forced open, and in the process, the stage was set for the new science.

Equations of degree higher than two had been a subject of inquiry by mathematicians earlier than Diophantus, invariably with special cubics. Omar Khayyám had provided a systematic list of the possible cubics with positive roots. The first attempts at a general solution of the cubic equation $x^3 + bx = c$ were made by Scipio del Ferro, a professor of mathematics in Venice. It is reported that he found a root, but the results were not published. Nicola Fontana (1500–1557) of Brescia in the same period effected a general solution. In a contest with del Ferro, Fontana,

* Smith, *History of Mathematics*, 1:253.
† Scott, *A History of Mathematics*, p. 87.

known as "Tartaglia" (the stammerer), came out an easy victor. There are no records of his method—only statements concerning his success. The final published solution of the general solution of the cubic by Girolamo Cardano (Jerome Cardan) (1501–1576) has been clouded by controversy for some four hundred years. Cardano was without question one of the most gifted mathematicians of his day, and serious doubts may be raised concerning the charge of plagiarism.* Often in physics and mathematics the contribution of one creative mind has been undermined by the questioning of others. Mathematics and physics develop in a continuous stream, and jealousy over acknowledgment is always present. This is somewhat inevitable because the number of contributors to science is always more than can be properly acknowledged. Often an original man is neglected, but one must be cautious in assessing the author with only one claim to fame, as compared with an obvious giant of his age. When one considers the facts in Cardano's case, one discovers that he provided a historical account of the problem, giving credit to Scipio del Ferro and Nicola Fontana for the first solution of the cubic; his published solution was that of Fontana.† He thus acknowledged Fontana in the modern tradition. The criticism against Cardano suggests that he should have omitted any reference to a solution of the cubic, but this solution was vital to the whole of his *Ars magna*; moreover, since Fontana could not write Latin, the scientific language of his time, Cardano could hardly be blamed for not waiting for publication of his solution.

Cardano, primarily a physician, was also a prolific writer. His books ranged from ethics and religion to popular works on science. In 1539 he published *Practica arithmeticae generalis*, a thorough treatment of arithmetic. Here the negative numbers and square roots were admitted on the same level as the positives. *Ars magna* was his tenth book on mathematics; it appeared in 1545 and was a milestone in the history of algebra. It presented not only the general solution of the cubic equation but also the quartic, solved by reducing it in general to a cubic equation. This solution of the quartic is attributed to Ludovico Ferrari, a student of Cardano's. The algebraic style is similar to that of Regiomontanus. For instance, $x^3 + 26x = 12x^2 + 12$ is written: "cubus & 26 res, aequantur 12 quadratis p:12."‡ In this form one also observes the *p* (piu) notation of Pacioli. In modern symbolic notation the solution of the cubic equation is compact and ingenious. Consider the general form $y^3 + ay^2 + by + c = 0$. This equation is taken to the form first shown by Scipio del Ferro by the substitution $y = x - a/3$, resulting in $x^3 + px + q = 0$. By converting this equation in one unknown, x, to an equation in two unknowns, u and v, one can force the cubic to assume the form of a quadratic. Let $x = u + v$; then $(u + v)^3 + p(u + v) + q = 0$. By setting the product uv to the

* Henry Morely, *The Life of Girolamo Cardano of Milan*, 2 vols. (London: Chapman and Hall, 1854).

† Oystein Ore, *Cardan the Gambling Scholar* (Princeton, N.J.: Princeton University Press, 1953), p. 84.

‡ Scott, *A History of Mathematics*, p. 90.

value $-p/3$ and substituting, one obtains $u^6 + qu^3 - p^3/27 = 0$; resulting in the roots

$$u_1 = \sqrt[3]{-\frac{q}{2} + \sqrt{\frac{q^2}{4} + \frac{p^3}{27}}},$$

and

$$u_2 = \sqrt[3]{-\frac{q}{2} - \sqrt{\frac{q^2}{4} + \frac{p^3}{27}}}.$$

According to his rule then, $y = x - a/3 = u + v - a/3 = u_1 + u_2 - a/3$. In practice there are two more roots which can be generated from the parameters u_1 and u_2. These are $x = \eta u_1 + \eta^2 u_2$, and $x = \eta^2 u_1 + \eta u_2$, where $\eta = -1/2 + (1/2)\sqrt{-3}$.

These final two roots, of course, are later developments, appearing after 1700. Cardano admitted negative roots, which was in itself an advance, and he recognized the existence of a new type of number—the complex number. He was also aware that the cubic could have no more than three roots. Cardano's rule had this stimulating effect on later mathematicians: they realized that from a real cubic three real roots cannot be extracted without passing through the domain of complex numbers.

In *Ars magna* the fourth-degree polynomial (quartic) was solved by a method of completing the square. Starting with $x^4 + ax^3 + bx^2 + cx + d = 0$, one rearranges terms and completes the square, giving $(x^2 + ax/2)^2 = (a^2/4 - b)x^2 - cx - d$. Now add $(x^2 + ax/2)y + y^2/4$ to both sides to give:

$$\left[x^2 + \frac{ax}{2} + \frac{y}{2}\right]^2 = \left[\frac{a^2}{4} - b + y\right]x^2 + \left[\frac{ay}{2} - c\right]x + \frac{y^2}{4} - d.$$

The term on the right is a quadratic in x. If we set y in such a manner as to make this a square, then we will be able to take the square root of both sides. Since antiquity it was known that a quadratic $Ax^2 + Bx + C$ is a perfect square if the discriminant $(B^2 - 4AC)$ is zero. In such a case both roots are $-B/2A$, and the quadratic may be written as $Ax^2 + Bx + C = (\sqrt{A}\,x + \sqrt{C})^2$. Thus in the original expression one sets the discriminant of the quadratic in x (on the right) equal to zero, forcing y to take on values given by the roots of the cubic, $y^3 - by^2 + (ac - 4d)y - [d(a^2 - 4b) + c^2] = 0$. Call the three roots of this equation y_j; then the values of x can be obtained by taking the square root, giving $x^2 + ax/2 + y_j/2 \pm (\sqrt{A}\,x + \sqrt{C}) = 0$. The major problem then is to solve the cubic equation for y.

Concerning the controversy over the discovery of the solution to the cubic equation, it is known that ninth-century Persians used the method of Menaechmus to solve the cubics in terms of the intersection of a parabola and a hyperbola. Omar Khayyám refers to the method of Alhazen and provides a list of forms of the cubic which have positive roots.

This list represented a decided advance in the general theory. Our knowledge of the extent to which the Arabs advanced this problem is minimal; however, suggestions of the method may have been available in the literature from the Orient. The interest in the cubic was of long standing by Cardano's time. It is significant that in *Ars magna* he provided a list of cubics with positive roots which was the same set of equations given some four hundred years earlier by Omar Khayyám.

Polynomials provided a source of great stimulus to mathematics for three centuries after Cardano. Many mathematicians suspected that the quartic was the equation of the highest degree which could be solved in rational radicals. Attempts to provide a general proof of this statement were to lead to the creation of group theory by Lagrange, Abel, and Galois. The complex roots of polynomials presented mathematics with the complex numbers, which in turn suggested the field of complex variables.

The last of the Italian algebraists of this period was Rafaello Bombelli. His algebra, published in 1527 and 1579, dealt with the reality of the roots of the irreducible case of the cubic. One of his more noteworthy contributions was an improved notation in which the unknown, called *tanto*, was designated by $\overset{1}{\smile}$, the square (*potenza*) by $\overset{2}{\smile}$, the cube (*cubo*) by $\overset{3}{\smile}$, up to the twelfth power (*cubo di potenza di potenza*), written as $\overset{12}{\smile}$. Roots were still designated by R, followed by letters to denote the degree of the root. The power designation is important, for in it we perceive the primitive form of an integral power.

Often a breakthrough in mathematics is the result of nothing more or less than a sudden improvement in notation. Examination may reveal that many hundreds of years of evolution have preceded such events. A former monk and follower of Martin Luther, Michael Stifel (1486–1567), published in Nuremberg his *Arithmetica integra* (1554). He introduced the symbols used today for plus and minus. In 1553 he gave the symbol $\surd$ as the representation of the square root. The cube root was written $\wedge\!\wedge\!\!\surd$, and the fourth root as $\wedge\!\surd$. Thus modern symbolism was slowly being forged.

Little known but of great significance in the development of mathematics was Albert Girard (1595–1632). Although French, he spent most of his life in Holland. He followed Stifel in using the $+$ and $-$ signs, and his studies of algebraic polynomials led him to employ trigonometric functions to solve the cubic. He acknowledged the negative roots of polynomials. Girard asserted that a polynomial of nth degree necessarily has n roots.

By showing that certain roots involved the square root of a negative number, he pointed out the occurrence of complex roots, which he called inexplicable. He employed integral indices for powers and suggested the fractional indices. One also observes in his work the beginnings of number theory. A theorem which appears somewhat later with Fermat was given by Girard: "p, a prime number, is the sum of two squares if and only if $p - 1$ is divisible by 4."

Preceding Girard, François Viète (1540–1603), trained as a lawyer, pursued mathematics in his leisure time while a member of the privy council of the king of France. Among his many contributions to algebra, he is remembered best for Viète's formulas, which relate the coefficients of an algebraic polynomial to its roots. If the polynomial

$$f(x) = \sum_{n=0}^{N} a_n x^n = 0 \quad (\text{with } a_N = 1),$$

$$= x^N + a_{N-1}x^{N-1} + \cdots + a_1 x + a_0 = 0;$$

and if the roots of $f(x)$ are designated by $-x_j$, where

$$f(x) = (x + x_N)(x + x_{N-1}) \cdots (x + x_1) = 0;$$

then the coefficients a_j and the roots x_j are related by the Viète formulas;

$$a_{N-1} = \sum_{n=1}^{N} x_n = x_1 + x_2 + \cdots + x_{N-1} + x_N,$$

$$a_{N-2} = \sum_{k>j}^{N} \sum_{j=1}^{N} x_k x_j = x_1 x_2 + x_1 x_3 + \cdots + x_1 x_N + x_2 x_3 + \cdots + x_{N-1} x_N,$$

$$\vdots$$

$$a_0 = \prod_{n=1}^{N} x_n = x_1 x_2 x_3 \cdots x_{N-1} x_N.$$

In words, these equations state that the second coefficient a_{N-1} is the sum of the negatives of the roots of the polynomial. The next coefficient a_{N-2} is the sum of all of the triple products of separate roots, and so on. Thus a_{N-k} is the sum of the k-tuple products of separate roots. Finally, since there is only one N-tupled product, a_0 is just the product of all of the negatives of the roots (the roots being $-x_j$). To see this more clearly, we expand the cubic:

$$(x + x_1)(x + x_2)(x + x_3)$$
$$= x^3 + (x_1 + x_2 + x_3)x^2 + (x_1 x_2 + x_1 x_3 + x_2 x_3)x + x_1 x_2 x_3.$$

Here one observes that the coefficient of $x^2(a_{3-1})$ is the sum of the roots. The coefficient of $x(a_{3-2} = a_1)$ is the sum of the double products $(x_1 x_2 + x_1 x_3 + x_2 x_3)$. Finally, the last term a_0 is the product of the three roots.

Decimal fractions were invented in very early times by the Chinese and Hindus; they were later introduced into Europe by the Arabs. Simon Stevin of Bruges (1548–1620) was one of the first to provide Europe with a systematic treatment of decimal fractions. Like Girard and Bombelli, he rejected the cumbersome notation for the power of a quantity and employed numerical indices, again considering the use of fractional powers for square and cube roots. At the time of Stevin, the sine of an angle was still considered by some as a length and not a ratio: the influence of Ptolemy was still present some 1400 years later. Stevin provided improved methods for determining sines of angles and showed the fundamental relationship between angles and half angles. His method is equivalent to the modern relation: $\sin(1/2)\alpha = \sqrt{(1/2)(1 - \cos\alpha)}$. The geometric proof can be taken from the diagram $a^2 + f^2 = e^2 = 4c^2$. Because $f^2 = (1 - b)^2$,

$$a^2 + f^2 = a^2 + 1 - 2b + b^2 = 2 - 2b = 4c^2,$$

giving

$$c = \sin(1/2)\alpha = \sqrt{(1/2)(1 - b)} = \sqrt{(1/2)(1 - \cos\alpha)}.$$

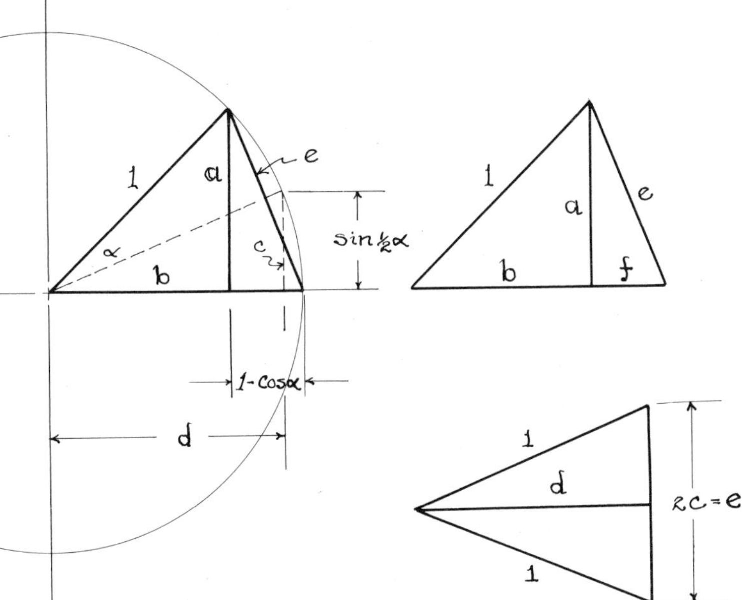

Stevin enunciated the principle of the triangulation of forces, or the law of the vector addition of forces in problems of statics. Addition of vector quantities by triangulation was not new—it appeared in the writings of Apollonius and Heron, to cite an early application. The principle for forces states that the total force on a body acted upon by two vector forces (i.e., with magnitude and direction) is equal to the resultant diagonal force of the parallelogram formed with the two component forces acting as sides of the figure. Regard two forces, $\vec{\mathbf{F}}_1$ and $\vec{\mathbf{F}}_2$, acting at a point P; the resulting total force, $\vec{\mathbf{F}}_{\text{total}}$, is determined from the parallelogram, as shown. If we regard the individual forces as directed line segments or vectors, the resultant is formed by adding the vectors head to tail and by joining the tail of the first vector to the head of the last in the order of their addition. Static equilibrium of a body is achieved when the sum of all individual forces adds to zero. The parallelogram rule must be applied between forces taken two at a time. The equivalent vector sum can be done without the intermediate step of diagonalization.

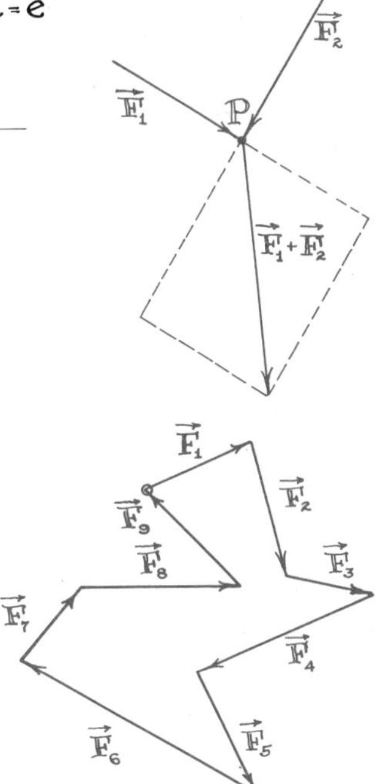

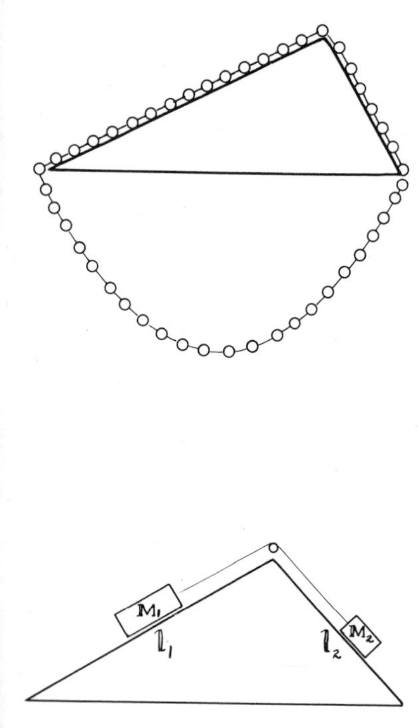

In his *Premier livre de la statique* (1624) Stevin gave an account of the static equilibrium of weights on an inclined plane. Perhaps he performed some experiments to substantiate his results; on the other hand, he introduced an idealization of the problem in order to solve it *a priori* in his book. He envisaged two inclined planes, one twice the length of the other. Solution of the problem of equilibrium of masses suspended from the apex and resting on the planes was achieved by considering a uniform chain or necklace hanging about the triangle formed by the two planes.

Arguing that such a system is necessarily in equilibrium because any displacement of the chain leaves the system in its initial condition, he demonstrated that the two masses required to produce equilibrium when each was resting on one of the planes and connected by a continuous weightless string supported by a frictionless peg at the apex are to one another as the ratio of the corresponding lengths of the planes: $M_1/M_2 = l_1/l_2$, for equilibrium. When the gravitational force components parallel to the plane are introduced, this result is equivalent to $M_1 \sin \alpha_1 = M_1 h/l_1 = M_2 \sin \alpha_2 = M_2 h/l_2$.

Here and later one encounters the concept of idealization, frictionless planes, weightless strings, frictionless pulleys, etc. The idealization of a real problem in order to obtain a meaningful generalized result is the key to scientific advancement. Without the principle of idealization, mathematical physics would be extremely unwieldly. Starting with the scientists of the early Renaissance, this philosophy became a well-established tradition and by the time of Galileo was clearly spelled out in his writings.

In the tradition of Archimedes, Stevin incorporated in his work many problems in hydrostatics. He proved that the pressure of a fluid on the walls of a container was independent of the shape of the container. Included were studies of the equilibrium conditions for floating bodies: here he showed that in static equilibrium the center of gravity of the displaced fluid and the center of gravity of the body were on the same vertical line.

The first law of motion was anticipated by the Merton College group and by Buridan and Oresme in the thirteenth century. Giovanni Battista Benedetti (1530–1590) suggested the first law of undisturbed motion in the form in which it now appears: the undisturbed motion of a body is uniform and rectilinear. He also realized that when the constraint holding a body in uniform circular motion is removed, the detached body moves away in a straight-line path which is tangent to the circle at the position from which it was detached.

Computational facility alone was hardly sufficient to cope with problems which were generated by the discoveries and ideas of this period. As navigational techniques improved and astronomical observations became more precise, common multiplication and division presented a pressing inconvenience. The Arabic number notation and the institution of decimal fractions greatly facilitated the newer computations, and another great discovery was the logarithm. John Napier, Baron Merchiston (1550–1617), is credited with this invention; his method was based upon kinematic constructions which obscured the underlying principles. Napier

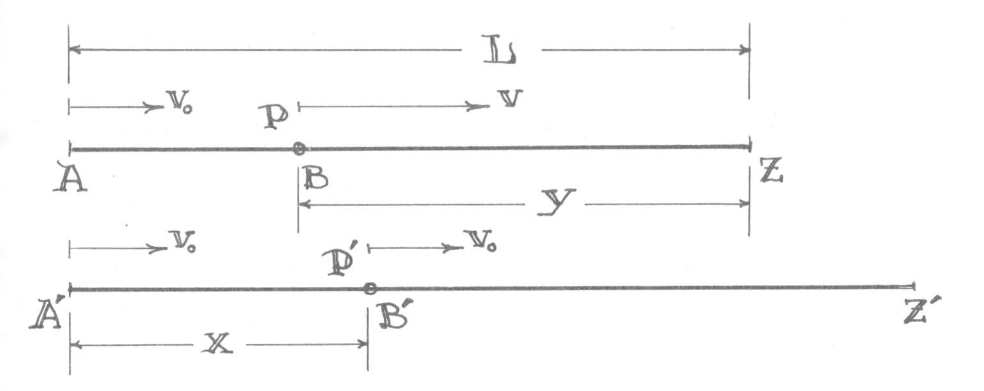

considered two points, P and P', moving respectively along two lines, AZ and $A'Z'$. Assuming that both points begin with the same velocity, we constrain P' to move at constant velocity along $A'Z'$, while P moves with a velocity proportional to the distance remaining. Thus, when P' is at B', the P will be at B; the length $A'B'$ is defined as the logarithm of BZ. Carrying out the instructions using the calculus, one finds that $x/L = \log_e L/y$.

The implied base of Napier's system is the natural base e, whose value is obtained via the calculus as 2.71828183. Logarithms with a base 10 were suggested soon after by Henry Briggs (1561–1631), who called attention to the advantage of a system within which the logarithm of 1 would be 0 and the log of 10 would be 1. The principle of the logarithm was grasped by several others shortly afterward, with the result that many useful tables were produced. These aided significantly the computations of Johannes Kepler.

During this period, a second major field of science was attracting attention. The ancients were familiar with the peculiar properties of electrified amber and magnetic iron, and in the thirteenth century Pierre de Maricourt (Peregrinus), using a magnetized needle as an indicator, plotted the magnetic lines of force on the surface of a globe-shaped lodestone. As a result, the magnetic poles were identified, and because of the directionality of the needle they were identified as being perhaps different—Peregrinus realized that the orientation of a magnetized needle depended upon its relation to the poles of the lodestone.

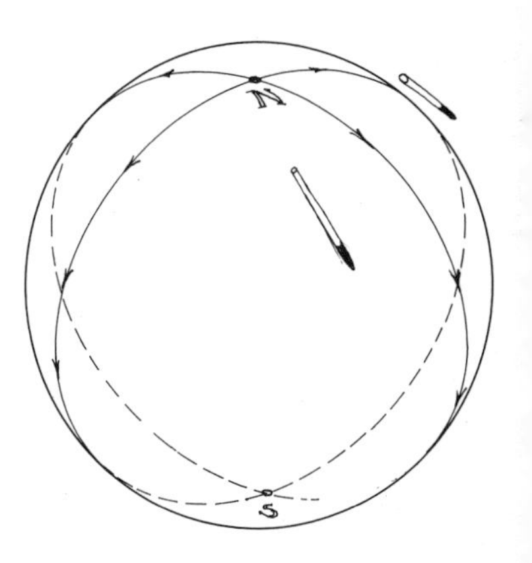

William Gilbert (1540–1603) received his early education in medicine and possessed all of the scientific curiosity of a true scholar. He postulated that the earth was a spherical magnet with two unlike poles; in other words, it was similar to the spherical lodestone of Peregrinus. Gilbert incorrectly conjectured that the gravitational attraction of the earth and the planetary motion of the solar system were products of this magnetic attraction. By forcing a hypothesis on one branch of science because of a new discovery in quite a different field, he uncovered the weakness of scientific investigation when it is stripped of supporting experimentation.

For over two thousand years amber had been known to have attractive powers under certain circumstances. Gilbert demonstrated that similar frictional electrical effects could be produced in sulphur and wax, and he gave the forces set up in frictional electrification the name "electric force." Further experiments convinced him that the forces exhibited by the magnetic lodestone and by the electrified bodies were different: the lodestone needed no friction. More important, he found that while electrical forces could be screened from a given region by conducting bodies, these same conducting bodies were ineffective in screening the magnetic force of the lodestone. In the *Dialogue Concerning the Two Chief World Systems*, Galileo speaks with admiration of the experiments of Gilbert with the lodestone, although he disagreed with several of his conclusions.

Galileo Galilei (1564–1642) was born in Pisa. His life and works have become the subject of innumerable plays, poems, and biographical studies. Many have asked, "Why Galileo; why not a number of others?" In view of the extensive study of the mean velocity theorem and the laws of constant acceleration conducted by the Merton College group and by Oresme, it is questionable to credit Galileo with the full development of these concepts. On the other hand, there is no evidence that Galileo was directly aware of even the Oresme manuscripts, particularly Oresme's *Questions on the Geometry of Euclid*, which was not published until the twentieth century. Although there is no obvious connection, there is no doubt that Galileo lived and developed in an atmosphere in which these concepts already had substance.

During his formative years Galileo began his studies in medicine and then became interested in mathematics and physics, which led to his appointment to the chair of scientific studies at Pisa at the age of 26. His well-known espousal of the Copernican system in 1591 led to a forced resignation and to a subsequent appointment to a professorship at Padua, where he carried out his investigations without interference from the Church.

Much of Galileo's early fame was derived from his improvement and use of the telescope. It has been claimed that he invented it, but the telescope was actually discovered somewhat earlier (1608) by Hans Lippershef in Holland. Galileo may be credited with the development and refinement of a workable instrument, which he employed to advance the astronomical knowledge of the time, discovering among other things the satellites of Jupiter and the craters of the moon.

In 1610 he was invited to Florence to serve as court mathematician and philosopher to the grand duke of Tuscany. He continued his bold support of the Copernican theory which had brought him before the Inquisition, and after the appearance of the *Dialogo*, now an old man of seventy, he was subjected to a second trial, during which he was forced to renounce his "heresy of the movement of the earth." His greatest work, *Discourse on Two New Sciences*, published in 1638, was written while he was under confinement by the Church, and it had to be smuggled out of Italy and published in Holland. From it one sees that Galileo was not crushed in spirit by the Inquisition. Despite his many hardships, he produced probably the greatest physics book of all time in content, literary style, and clarity. Some critics have pointed out that its contents were not wholly original, but this is a small defect in a great work. Much of the *Discourse* is based on his experiments, and the remainder serves to make the manuscript as complete as possible a report of the level of physics at that time.

The *Discourse* begins with a discussion of the strength of materials, including an analysis of the optimum size of mammals according to the type of bone structure they possess. Following this is a rather interesting account of the concept of infinity, in which we find a subtle understanding of the paradox of infinity: the relationship between points and lines, the order of an infinity, and the concept of the continuum.

Through the speech of his three characters, Salviati, Sagredo, and Simplicio, Galileo described the method of exhaustion and the paradoxes which are uncovered when the limiting case is considered. In the First Day, Salviati remarks, "Let us remember that we are dealing with infinities and indivisibles, both of which transcend our finite understanding, the former on account of their magnitude, the latter because of their smallness. In spite of this, men cannot refrain from discussing them, even though *it must be done in a roundabout way.*"

Salviati then proceeds to demonstrate that the volume of a point equals the volume of a line. In this particularly charming construction, a cone and a hemispherical bowl are inscribed in a cylinder of height R and radius R. A classical Greek proof shows that the volumes of the bowl and cone above an intersecting plane passed parallel to the base are equal. By considering the limiting case when the plane approaches the top face of the cylinder, Salviati implies continuity of the volume elements and concludes that the resulting point and circle are equal.

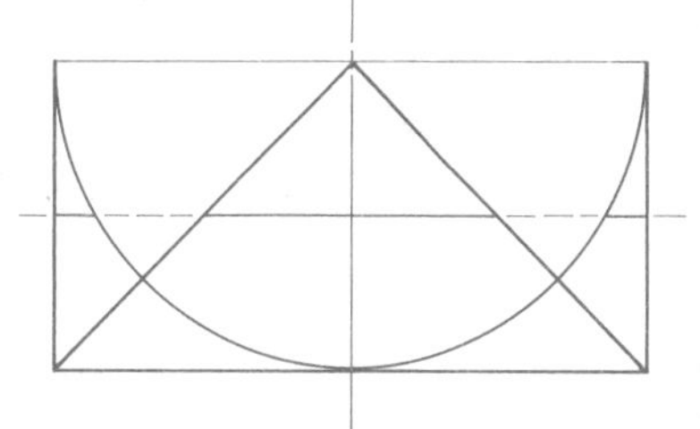

Because of his influence upon his student Cavalieri, it is instructive to observe Galileo's attempts to deal with the concept of infinity. In a further statement, Salviati summarizes with the remark, "And thus from your ingenious argument we are led to conclude that the attributes 'larger,' 'smaller,' and 'equal' have no place either in comparing infinite quantities with each other or in comparing infinite with finite quantities." All in all, his remarks represent a marked appreciation of this difficult concept, an appreciation which was to become of fundamental importance.

Galileo attempted to measure the speed of light by successive shuttering of two lanterns placed a large distance apart. The experiment failed, but Galileo conjectured that the speed was finite. He also considered an atomic view of matter in which there was empty space containing particles of small size. His summary of dynamics represents the investigations for which he is best remembered. On falling bodies, he concludes, "In a medium totally devoid of resistance, all bodies would fall with the same speed." As an example of a perfect demonstration of the principle of levers, consider Galileo's homogeneous block of marble, of length L, suspended from a weightless support rod.

If the block is cut into two pieces of lengths fL and $(1 - f)L$, where f is

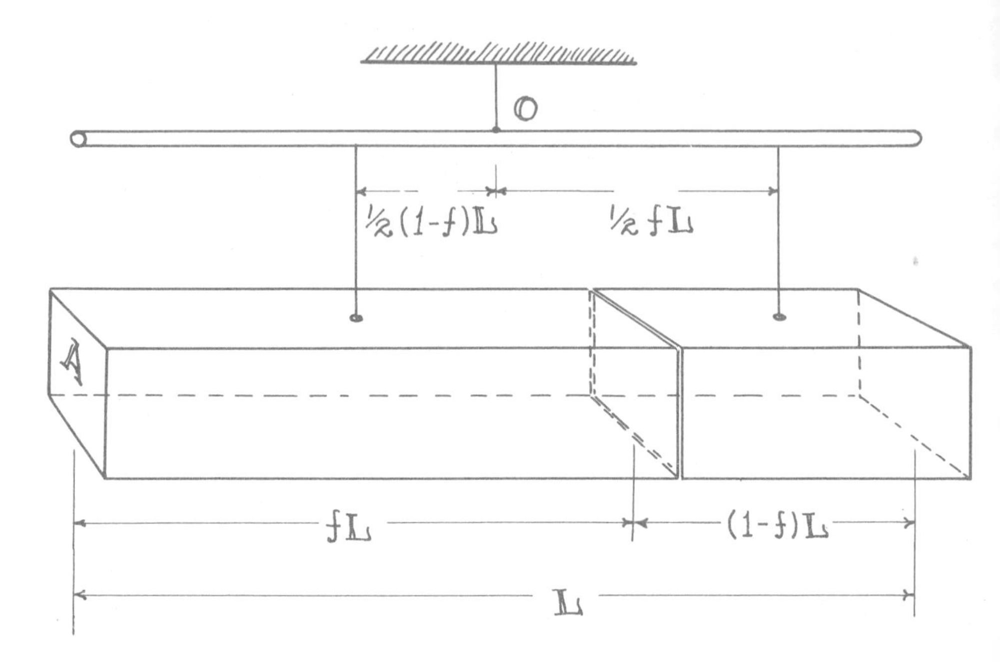

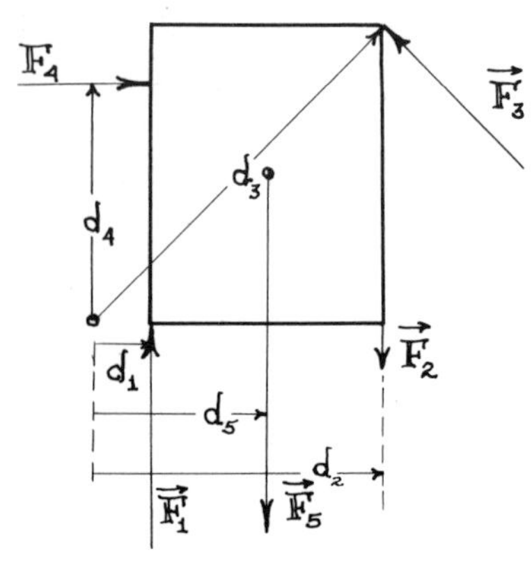

a fraction less than 1 and greater than 0, then to achieve equilibrium the parts must be suspended by strings passing through the centers of gravity of each piece. We therefore observe that the center of gravity of the portion of length fL is a distance $(1/2)(1 - f)L$ from the center 0, while that of length $(1 - f)L$ is a distance $(1/2)fL$ from the center. The weight of the pieces is proportional to their lengths, and therefore the products of the respective moment arm and weight are equal for the two parts:

$$(1/2)(1 - f)L \times fL\rho A = (1/2)fL \times (1 - f)L\rho A,$$

where A is the cross-sectional area and ρ is the weight per unit volume or density of the marble. Earlier Stevin had introduced the law that the sum of the vector forces on a body in static equilibrium is zero. Here we encounter the law of moments given by Archimedes, which stated that, when a body is in static equilibrium, the sum of the moments (a moment is a force times the perpendicular distance from the center of rotation) must be zero. The moment arm is the distance from the center of rotation measured perpendicular to the line of action of the associated force.

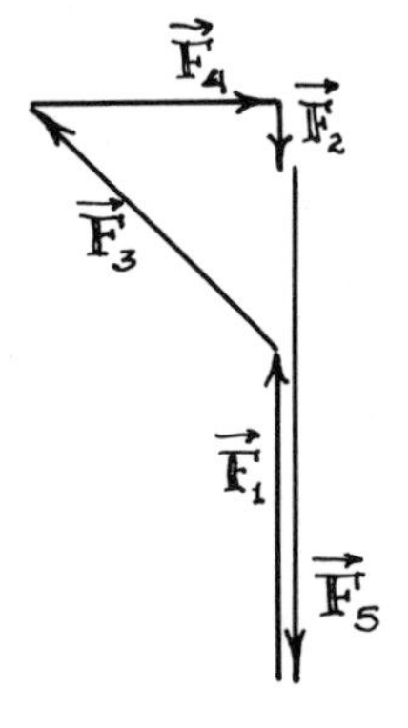

In the example shown, a rectangular object is in static equilibrium under the action of forces $\vec{\mathbf{F}}_1$, $\vec{\mathbf{F}}_2$, $\vec{\mathbf{F}}_3$, $\vec{\mathbf{F}}_4$, and $\vec{\mathbf{F}}_5$, and the vector sum of the forces must be zero:

$$\sum_{j=1}^{5} \vec{\mathbf{F}}_j = \vec{\mathbf{F}}_1 + \vec{\mathbf{F}}_2 + \vec{\mathbf{F}}_3 + \vec{\mathbf{F}}_4 + \vec{\mathbf{F}}_5 = 0.$$

The sum of the moments must also be zero, taking counterclockwise rotations as positive:

$$F_1d_1 + F_2d_2 + F_3d_3 - F_4d_4 - F_5d_5 = 0.$$

Liberties have been taken in these examples in order to illustrate the principles in modern symbolic notation.

In his discussion of the motion of bodies under constant acceleration, Galileo reached the climax of his work in dynamics. As pointed out in earlier chapters, these were not new concepts, but they were treated systematically for the first time; moreover, the derivation which showed that the path of a projectile was a parabola seems to be original with Galileo. His discussion entitled The Third Day puts forth the definition of constant acceleration: "A motion is said to be uniformly accelerated when, starting from rest, it acquires, during equal time intervals, equal increments of speed."

To derive the distance traveled, Galileo employed an ingenious velocity time diagram, reminiscent of that of the Paris group. This diagram defines the distance traveled as the area enclosed in the triangle formed by plotting the velocity at any instant against the time. First he demonstrated the

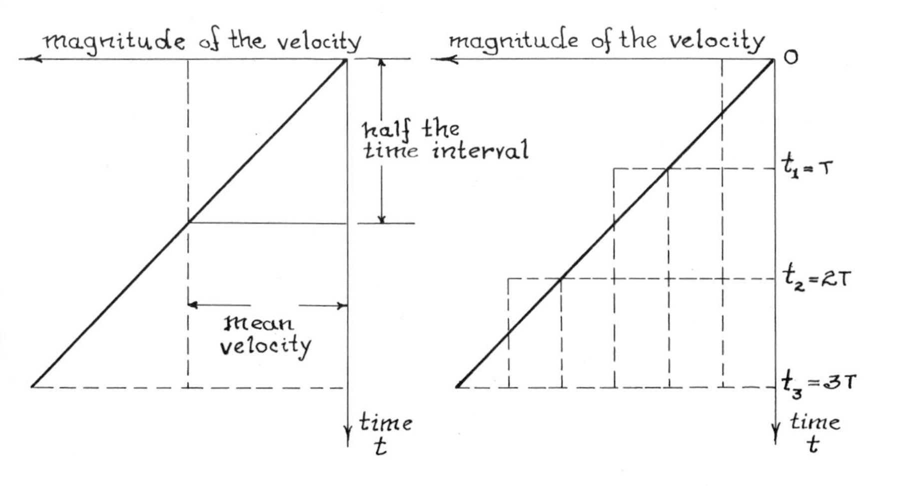

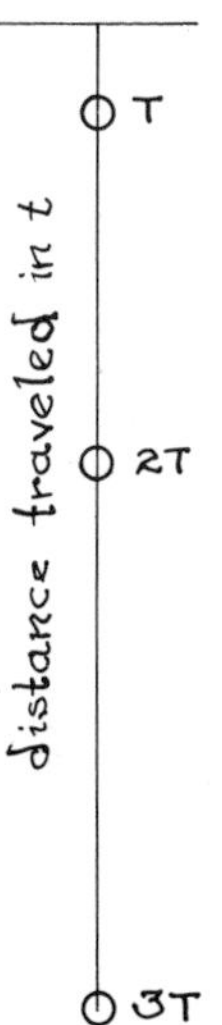

mean velocity theorem: the area of the triangle is equal to the area of the rectangle of equal height, t_f, of base width, $(1/2)v_{\text{final}}$; i.e., distance = $(1/2)v_{\text{final}} \cdot t_{\text{final}}$. This result was known in the thirteenth century. Galileo then stated that if the total time interval is split into equal subintervals of time T, the area in successive time intervals goes as the *sum of the odd integers*. Here knowledge of the Pythagorean square numbers is assumed. The distance or area increases with time as $(1 + 3 + 5 + 7 \cdots)T^2$. Consequently, because this sum is a square, the distance is proportional to the square of the total time:

$$T^2 \sum_{n=1}^{N} (2n - 1) = (NT)^2 = t_f^2.$$

The constant of proportionality is $(1/2)a$, where a is the acceleration, and, therefore, starting from rest the distance d covered in a time t is $d = (1/2)at_f^2$. This may be extended to show that if a body starts off with an initial magnitude of velocity v_0, and undergoes a constant acceleration for a time t, the total distance covered in t is the sum of the rectangle v_0t plus the triangle $(1/2)at^2$: $d = v_0t + (1/2)at^2$. Further, one can relate a

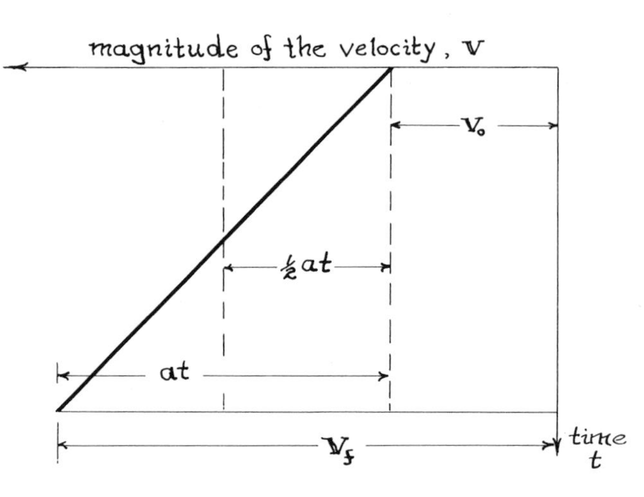

and the final velocity v_f. The triangle has a base equal to at because the mean velocity [measured at $(1/2)t$] is $(1/2)at$, and therefore $v_{\text{final}} = v_0 + at$. Galileo checked these results experimentally using inclined planes and water stopclocks. His descriptions of the pendulum problem, for instance, was a sound experimental demonstration of the conservation of mechanical energy.

Finally, in the discussion The Fourth Day, he utilized the vector resolution of velocities along two perpendicular axes to derive the orbit of a projectile. The axes were taken along the vertical and horizontal, and he showed that free fall along the vertical, coupled with a motion at constant velocity along the horizontal, produces a parabola.

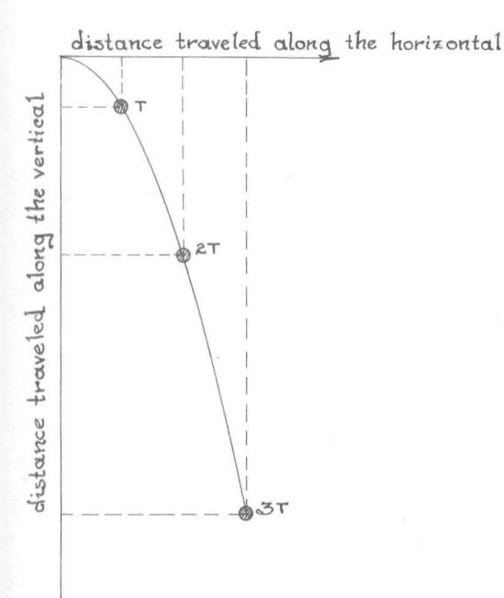

With the invention of the telescope and its development into a useful astronomical tool, the data on planetary motion became sufficiently precise for astronomers to perform detailed checks upon the shape of the orbits. Originally, the Copernican scheme had suggested that the orbits of the planets about the sun were circular. This has a striking analogy with the invention of the Bohr atom in the early twentieth century. Let us make a small digression to acknowledge that all motion of the planets is relative. It is not proper to say, for instance, that the heliocentric model is correct, as opposed to a system referred to the earth as a center. The heliocentric system is the *superior representation* because the orbits have a much simpler functional form when referred to a coordinate system located at the center of the sun. Thus, correctness in this controversy merely means a more convenient choice of coordinates: the orbits indeed have a realizable form when referred to the earth. On the other hand, this form is much too complex to be useful. Johannes Kepler was perhaps one of the first theoretical astronomers. He was not an experimenter and had no obvious talent for such work, but fortunately he began his great work as an assistant to the great Danish experimental astronomer Tycho Brahe. Although Brahe clung to the pre-Copernican views of the solar system, the accuracy of his collection of data on planetary positions was not affected thereby.

Attacking the planetary problem in terms of a heliocentric model, Kepler initially attempted to use circular orbits. When this failed, he resorted to the epicycles of Ptolemy; the epicycle can be constructed as a first-order approximation of an ellipse. Because of the accuracy of Brahe's data, Kepler found that the best epicyclic representation of the planetary orbits was off by eight minutes of arc, a sizable inaccuracy!

In the process of analyzing the orbits, Kepler divided them into a sum of small triangular-shaped segments and discovered that the areas, swept out in equal time intervals, were in fact equal. This observation is known as Kepler's Second Law, and it became one of the crucial tests of any model of planetary motion in a gravitational field. After Newton's solution to the problem of planetary motion, it was found that Kepler's Second Law was true for any orbital motion under the influence of a central force field and that, in fact, it represented a statement of the law of conservation of angular momentum.

During his early work, Kepler shared the Pythagorean dream of fitting the planetary orbits to the five regular solids as inscribed and circumscribed circles (the octahedron, icosahedron, dodecahedron, tetrahedron, and cube), but this hypothesis proved to be a factitious one.

In all his work Kepler stressed the importance of mathematics in the study of nature and himself made significant contributions to the early techniques of integration. While his methods were basically those of Archimedes and Eudoxus, his success with them stimulated interest in this exciting area. After spending most of his life on the problem of the planetary orbits, Kepler discovered that the ellipses of Apollonius gave relatively accurate fits to his data, whereupon he postulated his three empirical laws, which were to form the basic test for all future theories of gravitation.

The First Law: the planets move in elliptical orbits with the sun at one focus
The Second Law: a planet sweeps out equal areas in equal times
The Third Law: the ratio of the cube of the semimajor axis to the square of the period is the same constant for all planets

Newton demonstrated that the first and third laws are a consequence of the inverse square law of gravitation, while the second is a general result of the fact that the gravitational force is central. By central we mean that the attractive force between two point masses acts along the line of centers of the two points. To accommodate the real problem of attracting spheres, Newton showed that a central force which is an inverse square law—namely, varying as one over the square of the distance between centers—causes two spheres to interact as if they were point masses located at the centers of the spheres.

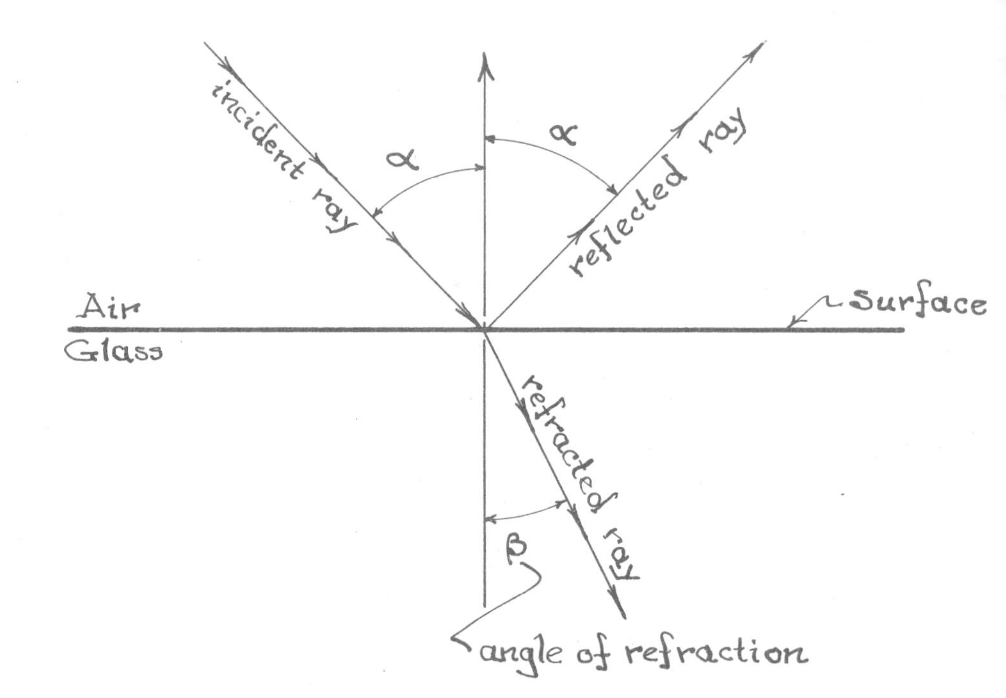

Kepler was intrigued by other physical phenomena, including optics. In an attempt to compute the focus of a double convex lens, he discovered that when light is incident upon glass near the perpendicular, the ratio of the angle of incidence to the angle of refraction is approximately 3/2. He did not discover the general law of refraction, which was stated by Willebrord Snell (1591–1626). Snell found that the sine of the angle of incidence (α in the diagram) was proportional to the sine of the angle of refraction, β, and that the constant of proportionality was approximately 3/2 for glass. The constant, however, varied somewhat from one type of glass to another. Although the scientific explanation of this law was not yet propounded, it had been used in one empirical approximation or another for a thousand years, its general nature having been recognized by Ptolemy and later by Alhazen. The proper law forms the basis of the

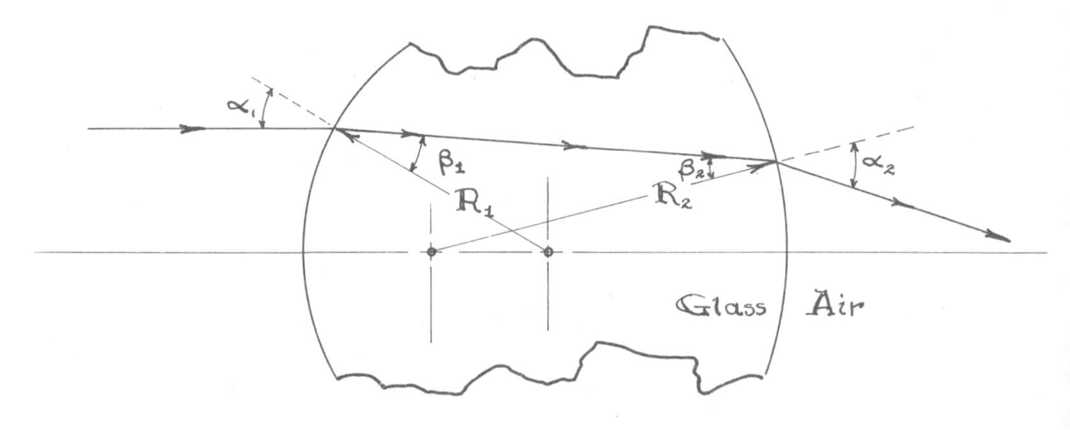

analysis of the focusing properties of optical lenses. One can observe roughly in the diagram how a convex lens focuses parallel light rays into a point.

In the following figure, assume that the angle of incidence, α, is such that sin α is approximately equal to α (α is less than 10°). Calling the constant of proportionality n (where n is approximately 3/2 for glass) the angle of refraction is approximately α/n. Here we repeat the approximation that the sine of the small angle of refraction is approximately equal to the angle of refraction. Assuming that the lens is thin, implying that the distance h of the ray from the horizontal axis at the point of emergence is essentially unchanged during transit through the lens, the emerging ray makes an angle $(n - 1)(\alpha + \gamma)$ with the horizontal. Again,

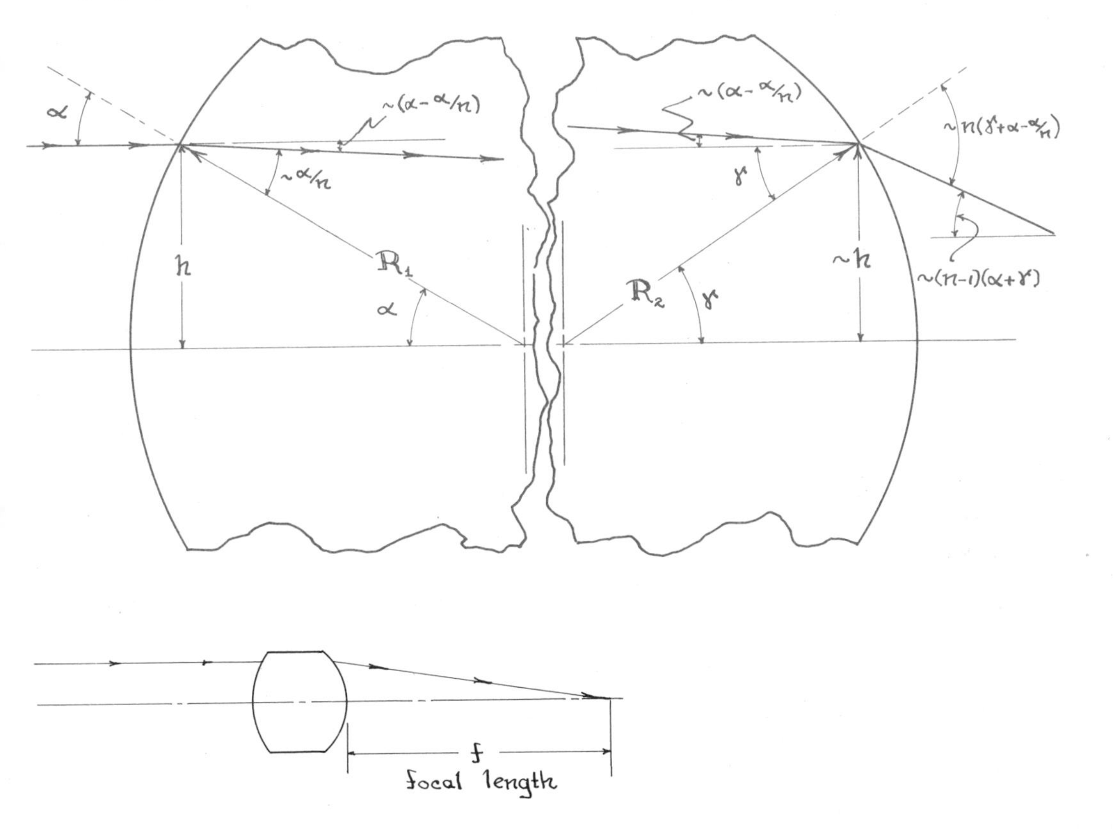

if α and γ are small angles, one can write $\alpha \simeq h/R_1$, and $\gamma \simeq h/R_2$, with f the distance from the lens to the focal point given by $h/f \simeq (n - 1)(\alpha + \gamma)$. Substituting for α and γ, one obtains the famous thin lens formula, which approximates the focal length f in terms of the radii of the two surfaces, R_1 and R_2: $1/f \simeq (n - 1)(1/R_1 + 1/R_2)$. Finally then, the focal length of a thin convex lens for rays near the axis of symmetry is determined to first order by the radius of curvature of the two surfaces and the constant of proportionality n (the index of refraction). Because of the symmetry of this result in terms of the ordering of R_1 and R_2, the lens has the same focal length for light incident from either the left or the right.

At the turn of the sixteenth century, the failures and triumphs of science were in the hitherto unexplored area of the analysis of functions. The leading minds of the time were concentrated upon two major topics: the problem of computing the tangent to a given curve at a specified point, and the challenge of developing a general technique for obtaining the area under a curve. While establishing his laws of planetary motion, Kepler was confronted with the task of determining the area of a sector of an ellipse and the length of a segment of arc of an ellipse. In place of the tedious method of exhaustion, he introduced the concepts of the infinitely large and the infinitesimally small. He regarded the circle as composed of an infinite number of triangles, each with its vertex at the center, the cone as an infinite number of pyramids, and the cylinder as an infinite number of prisms the sum of whose heights were that of the cylinder.

Many fresh problems were suggested by Kepler's endeavors, and the hope of solving them inspired many of his contemporaries. Bonaventura Cavalieri (1598–1647) made an early attempt to develop the infinitesimal geometry. His approach, though successful in many instances, was riddled with faults of rigor. By itself the work could have been a step forward; unfortunately, his basic assumptions were carried into calculus books for several hundred years to come, and consequently Cavalieri has been blamed not only for the mistakes natural to his age but for the uncritical use of his methods by later teachers of mathematics as well.

Cavalieri assumed that a line could be viewed as the assemblage of an infinite number of points, a plane as an infinite set of lines, and a solid as an infinite set of planes. Viewing these assumptions in the light of the rigorous arithmetization of the calculus during the nineteenth century, they are disastrous, and the influence of his methods, as carried into the textbooks of the nineteenth century, appalled mathematicians. Put into practice, Cavalieri's methods resemble those of ancient Babylon and Greece. The area of a triangle was made up of the sum of an infinite set of lines which increase by one at each step; we notice here that Cavalieri avoided undefined problems by normalizing his series with respect to a rectangle, which in itself is made up of an infinite set of lines of equal length.

$$\frac{\text{area of an isosceles triangle}}{\text{area of a rectangle with the same base and altitude}} = \frac{0 + 1 + 2 + 3 + \cdots + (n-1) + n}{n + n + n + n + \cdots + n + n}$$

$$= \frac{(1/2)n(n+1)}{n(n+1)} = \frac{1}{2}.$$

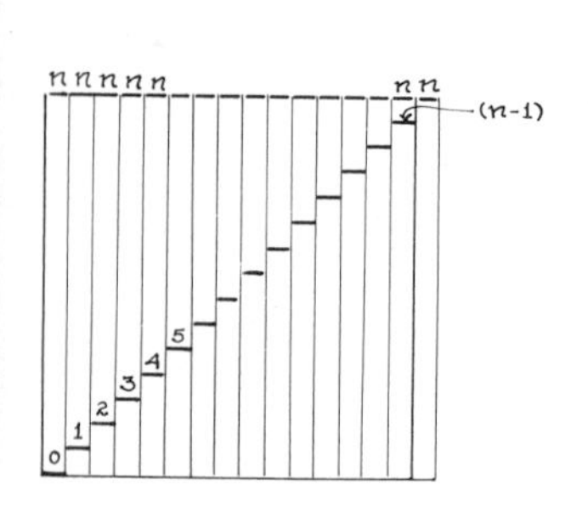

The volume of a cone as compared to the volume of a cylinder of the same base and height was found by using the sum of the squares of the integers. From the Babylonian series, this becomes 1/3 for large n. He likewise employed the sum of the cubes, normalized to give the limit 1/4.

From these series, of ancient lineage, Cavalieri generalized the method without proof to conclude that

$$\frac{0^m + 1^m + 2^m + \cdots + (n-1)^m + n^m}{n^m + n^m + \cdots + n^m} = \frac{1}{m+1}.$$

Although these results were obtained with rather shaky assumptions, they do illustrate the beginning of a method.

In the midst of the extensive activity centering about the creation of the calculus and analytic geometry, a quite separate field of mathematics came into being—projective geometry. Gérard Desargues (1593–1662) published his *Brouillon projet d'une atteinte aux evénemens des recontres d'un cone avec un plan*, a study of the geometry of projected shapes, in 1639. Aside from its mathematical invention, this work concentrates upon the concept of mathematical invariance, an idea which came to dominate the mathematics and physics of the nineteenth and twentieth centuries. It has become a philosophical principle that the significant quantities of a problem are those which are unaffected and remain unchanged under the action of certain transformations. Our fascination with invariance in a sense represents an abstract search for those things which do not age and are not destroyed. Desargues had a remarkable ability to generalize and saw quite clearly those properties of a given construction which remained unchanged under a given set of transformations. Certain properties of geometric figures are so fundamental to the figure that they persist even after the figures undergo rather drastic deformations.

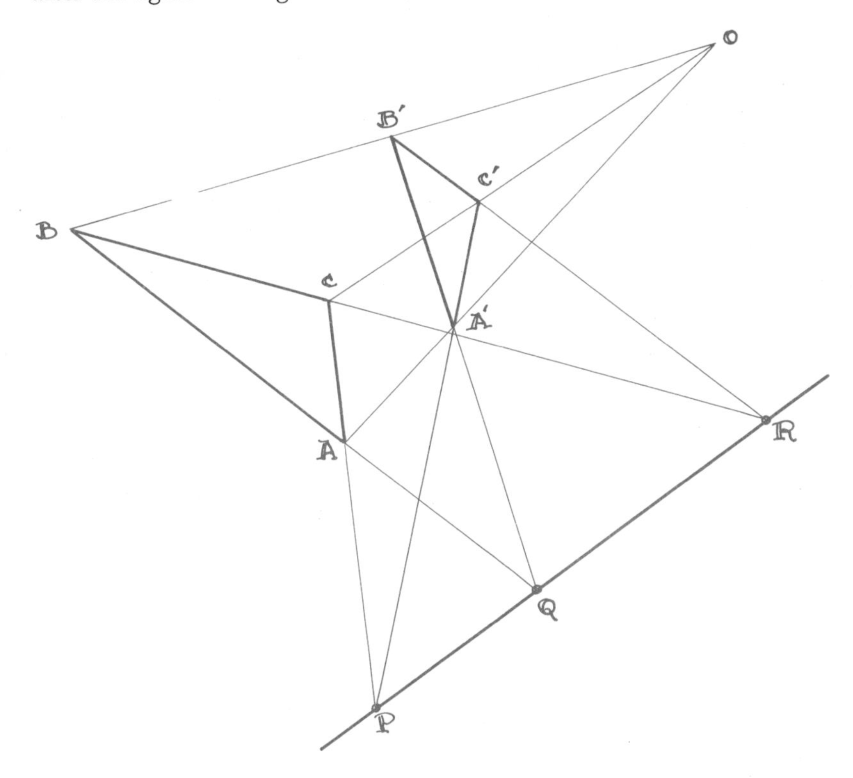

Perspective drawing on a plane is the projection of three-dimensional objects onto two dimensions. Although angles and lengths may be distorted, the objects are still recognizable, partly because certain properties are invariant under projection. Desargues' theorem is one of the earliest and most famous of the theorems of projective geometry: if in a plane two triangles are so situated that straight lines joining pairs of corresponding vertices are concurrent in a point, then the extension of corresponding sides will intersect in three colinear points, as illustrated in the diagram above. If the two triangles are not coplanar, this theorem may be demonstrated in a rather simple fashion. Some judgment must be exercised because this demonstration, although suggestive, is not rigorous.

In the diagram it will be noted that ABC and $A'B'C'$ form two planes which intersect in the line PQR. The extensions of the sides AB and $A'B'$, for example, now become the intersections of the plane AOB

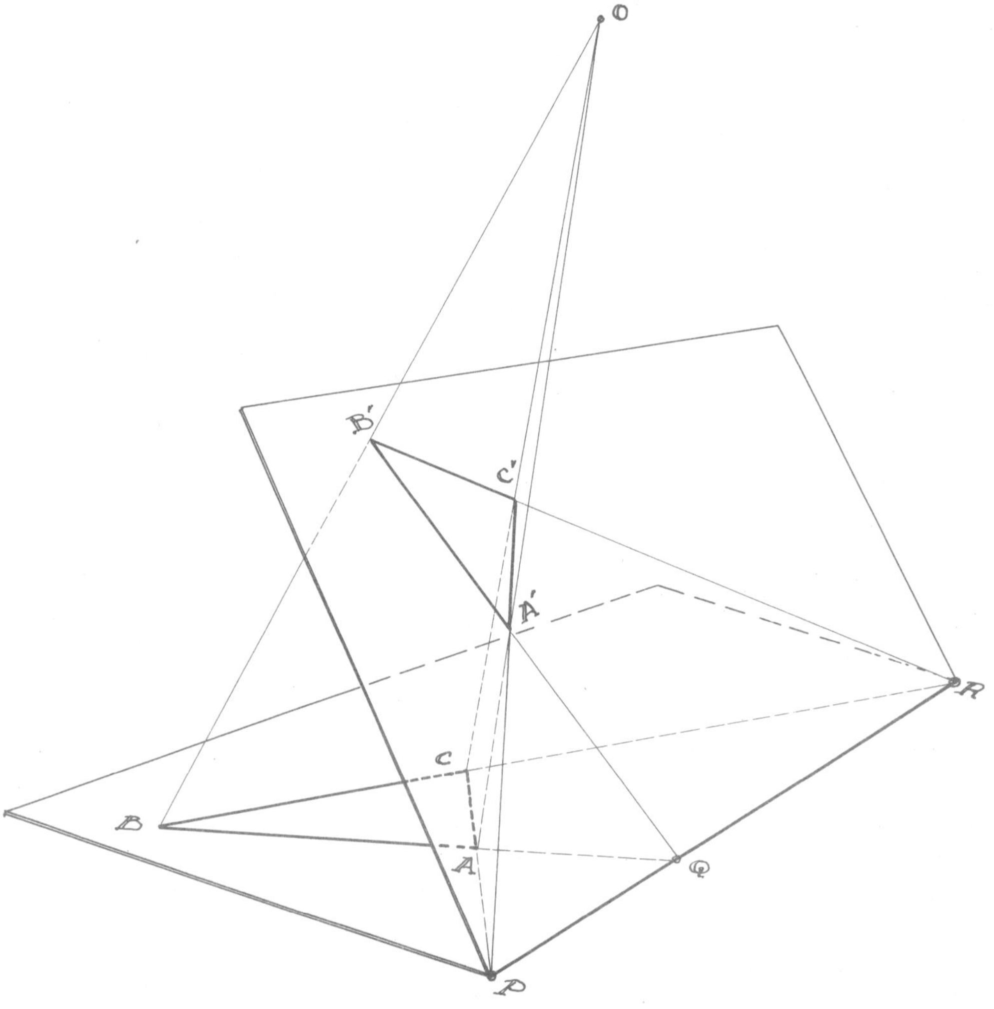

O
B'
C'
A'
R
C
B
A
Q
P

with the two planes. Naturally, the lines of intersection of this third plane with the other two meet in the line of intersection PR. The same argument holds for the extensions of BC, $B'C'$, and CA, $C'A'$. The diagram suggests that when planes ABC and $A'B'C'$ coincide, in the limit, the three-dimensional problem collapses into a two-dimensional one. Although this seems obvious, the line of intersection PR is not uniquely determined. The proof can actually be carried out with triangles ABC and $A'B'C'$ in the same plane but constructed to give the special case of AB parallel to $A'B'$, BC parallel to $B'C'$, and CA parallel to $C'A'$. As often occurs in mathematics, the general solution can be proved by establishing one special case.

Desargues' brilliant contributions were overshadowed by the epochal achievements in mathematics which mark this age. René Descartes' (1596–1650) invention of the analytic geometry was a triumph of notation. *La Géométrie* appeared in 1637; in it Descartes demonstrated that a geometric problem could be put into algebraic form, thus allowing all the analytic tools of algebra to be applied to the solution of geometric problems.

Many innovations in mathematical notation were introduced in *La Géométrie*. Exponents were written in Hindu-Arabic numerals, although this improvement was not extended to fractional powers. Descartes created the convention of writing unknown quantities as the last letters of the alphabet, x, y, and z, and known quantities as the first letters, a, b, c, etc. Lower-case letters were used, a convention which obtains to this day. Negative roots of equations were assigned equal status relative to positive roots, and Descartes noted that the maximum number of roots of a polynomial equals the degree of the polynomial, clearly suggesting the possible presence of complex roots; for the first time the factor theorem appears, which states that if a is a root of $f(x)$, then $f(x)$ is divisible by $(x - a)$.

La Géométrie appeared in three volumes, the first two on analytic geometry and the third pertaining to algebra. In Book I Descartes proclaimed that "any problem of geometry can easily be reduced to such terms that a knowledge of the lengths of certain straight lines is sufficient for its construction." He proceeded to illustrate various algebraic operations in terms of geometry and stated that a plane curve is defined by some specific property which holds for every point on the curve. Further, a correspondence was established between plane curves and equations in two variables, i.e., the correspondence between the algebraic and analytic properties of the equation $f(x, y) = 0$ and the geometric properties of the curve. Axes both rectangular and oblique were admitted.

As simple examples in present-day notation, several plots of simple algebraic functions are now shown. The coordinate axes have been taken perpendicular to one another, and the curves are obtained by assuming values for one variable, substituting, and then computing the resultant value of the second variable. For example, in computing $x^2 + y^2 = 1$, if we take $x = 3/5$, then the corresponding value of y can be obtained from $y^2 = 1 - 9/25 = 16/25$, etc.

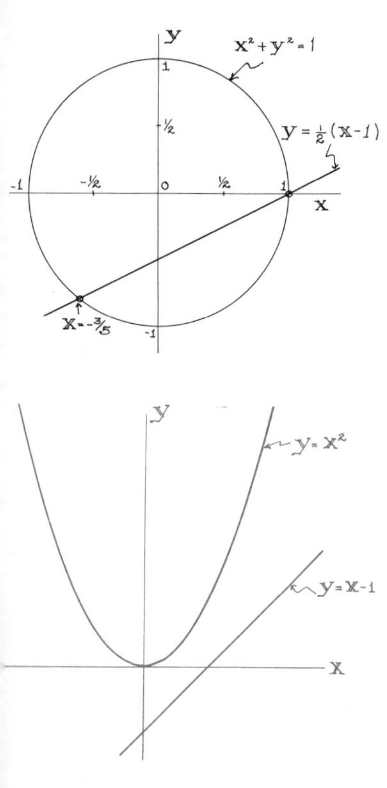

One of the striking consequences of Descartes' method was that the intersection of two curves $f(x, y) = 0$ and $g(x, y) = 0$ represented the simultaneous algebraic solution of the two equations. First consider the intersection of a straight line $y = (1/2)(x - 1)$ [i.e., $f(x, y) = y - (1/2)(x - 1)$], with the circle $x^2 + y^2 = 1$ [i.e., $g(x, y) = x^2 + y^2 - 1$]. Algebraically the solution may be obtained by substituting for either x or y to give, for example, $x^2 + (1/4)(x - 1)^2 = 2$, which can be solved for the two roots representing the values of x, which correspond to the two points of intersection. These are $x = 1$ and $x = -3/5$.

Because complex numbers had not yet been admitted as respectable solutions to algebraic problems, there were a few dangers inherent in the unjustified assumption that all simultaneous solutions of algebraic equations in two variables could be exhibited in a simple two-dimensional diagram. For instance, if one regards the curves formed by the two equations $y = x^2$ (a parabola) and $y = x - 1$ (a straight line), there appears to be no intersection when the curves are plotted.

If one substitutes for y to give the quadratic in x, $x^2 - x + 1 = 0$. There are two solutions for x (i.e., $x = 1/2 \pm (\sqrt{-1})\sqrt{3/2}$), indicating a type of intersection which does not appear on the real plane. The intersections are complex and consequently would not have been accepted as solutions of the Cartesian type.

Polynomials with real roots have an appealing representation when graphed onto the real plane. Consider $y = x^2 - 2x + 1/4$. This expression exhibits roots on the real xy plane at the position $y = 0$ [i.e., at $x = 1 \pm (1/2)\sqrt{3}$]. If in the figure the x axis is displaced downward by one, the parabola no longer zeros in the real plane. On the other hand, the new expression, $y = (y' - 1) = x^2 - 2x + 1/4$, giving $y' = x^2 - 2x + 5/4$, has roots $[x = 1 \pm (1/2)\sqrt{-1}]$ corresponding to $y' = 0$, but these roots do not show up as an intersection of the curve with the x axis. Therefore, despite the wide application of the analytic geometry of real numbers, it was quite apparent that it represented only a limited domain of problems.

The variables of the analytic geometry can be extended to surfaces and curves in three real dimensions; for instance, the function $x^2 + y^2 + z^2 - a^2 = 0$ represents a sphere of radius a in three dimensions, where the corresponding values of x, y, and z are taken along three mutually perpendicular axes. The generalization of these functions to higher-dimensional spaces is apparent here. The complex numbers and higher-dimensional spaces were not fully appreciated until the early part of the nineteenth century.

Descartes approached physics in the classical Greek spirit, and as a result his achievements in physics were less impressive than those in mathematics. A metaphysician, he confidently inferred a large array of conclusions from the minimal amoùnt of physical data. Any discrepancy between his conclusions and observations failed to disturb him. His method is well illustrated by his development of Snell's law of refraction. Here he deduced the law on the basis of three postulates, two of which were invalid. Concurring with Aristotle, Descartes proclaimed that a vacuum

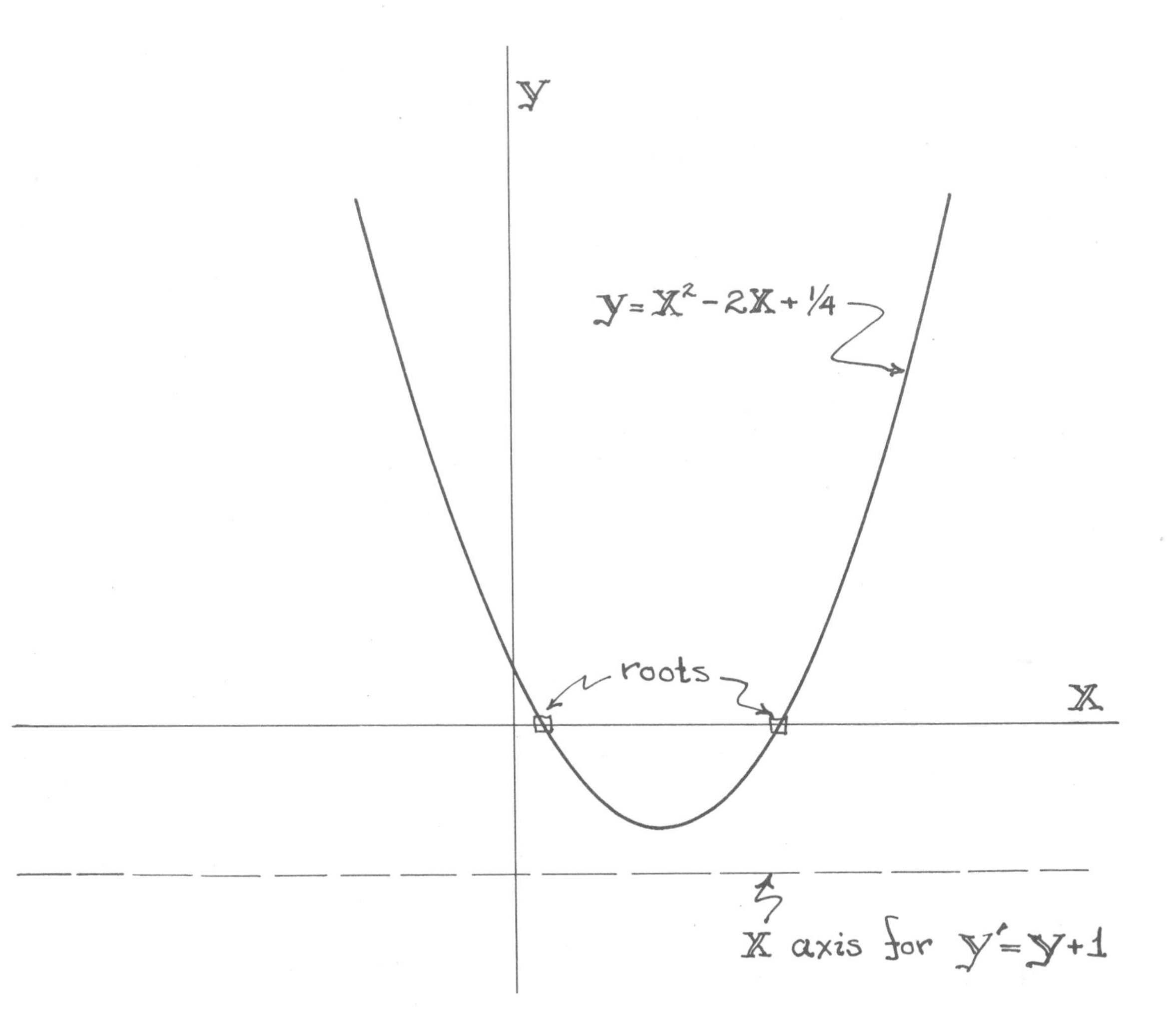

could not exist and, in a more decisive blow to the scientific method, stipulated that all successful physical theory must encompass all of experience. This grandiose posture led to unfortunate criticism of many local theories, such as Newton's law of universal gravitation. Merely because Newton had failed to incorporate a mechanical system in his theory which explained "action at a distance," Cartesian philosophers attempted to discredit an entirely successful postulate.

Where Descartes had failed as a mathematical physicist, Christian Huygens (1629–1695) prospered. A native of The Hague, Huygens was educated at the University of Leyden. With much greater precision than Galileo he was able to apply mathematical methods to the puzzles of physics. An interesting case in point is the pendulum. A pendulum clock, calibrated in Paris, was taken in 1671 to Cayenne, French Guiana, by Jean Richer, who observed that there it lost two and one-half minutes a day in mean solar time. By decreasing the length of the pendulum, he successfully calibrated this clock to maintain the correct time in Cayenne.

Back in Paris, however, the clock gained several minutes daily, indicating that, conversely, the length of the pendulum was now too short. To what extent did oscillation time depend upon the acceleration of gravity? Galileo had drawn attention to the possibilities of the use of the pendulum as a clock, in an essentially experimental rather than mathematical analysis, but the theory of the pendulum was first developed by Huygens and grew out of his interest in astronomy.

In 1666 Huygens was persuaded to move to Paris, and by 1671 he was in residence at the Royal Observatory. His earliest contribution was an improved telescope, with which he identified the rings of Saturn, and his interest turned to seeking an improved method for measuring time. In 1668 he invented a pendulum clock driven by weights. His famous *Horologium oscillatorium*, a masterful discussion of dynamics, contains an extensive treatment of pendulums.

The basic problem of pendulums is isochronism, that is, the constancy of the period when there are small variations in amplitude. Of all pendulum problems, that of the simple pendulum or small-amplitude pendulum is the most important. If a point mass m is suspended by a rigid weightless rod of length l from a pivot O, the motion is described either by Newton's Second Law or its equivalent (in the case of the gravitational attraction), the conservation of energy. Because Huygens is responsible for an early postulate of the conservation of energy, we may analyze the problem in the light of this principle. In the presence of the gravitational field, this conservation relation can be stated thus: "The sum of the kinetic energy and the potential energy is a constant known as the total energy."

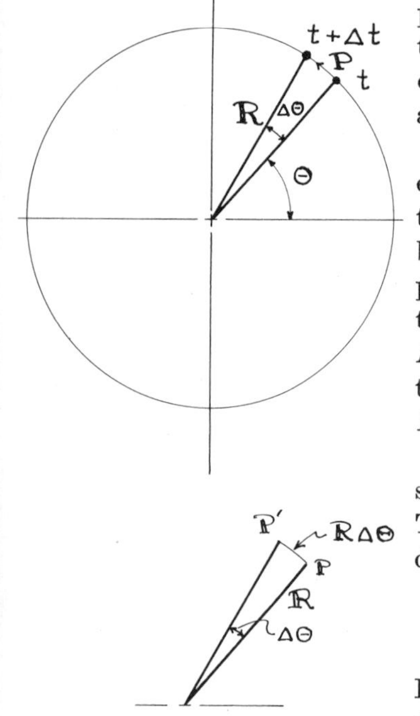

For the problem of the rotation of a simple pendulum, the kinetic energy is one-half the mass of the pendulum bob times the square of the tangential velocity. The potential energy is the product of the mass of the bob, the acceleration of gravity, and the height measured from the lowest point of the circle of rotation of the pendulum bob. The notion of a tangential velocity of rotation appeared as early as the age of Heron of Alexandria and has a very simple relationship to the angular velocity or time rate of change of angular position. In the diagram shown, a point, P, moves on the circumference of a circle of radius R.

If in a time interval Δt the position of the point moves from P to P', sweeping out an angular interval $\Delta\theta$, the length of arc swept out is $R\,\Delta\theta$. The linear velocity on the circumference of the circle is then just the distance covered in Δt divided by the time interval Δt.

$$\text{tangential velocity} = \frac{R\,\Delta\theta}{\Delta t} = R\left(\frac{\Delta\theta}{\Delta t}\right)$$

In the expression above, $\Delta\theta/\Delta t$, the change in angle expressed in radians divided by the change in time is the angular velocity.

We may now consider the simple pendulum as previously described. This motion is in a plane and is described by a rotation of the bob of mass m about the pivot O. At any instant, then, the angle as measured from the vertical is changing in time at a rate $\Delta\theta/\Delta t$, while θ provides a measure of the instantaneous height h. The sum of the kinetic energy plus potential energy is

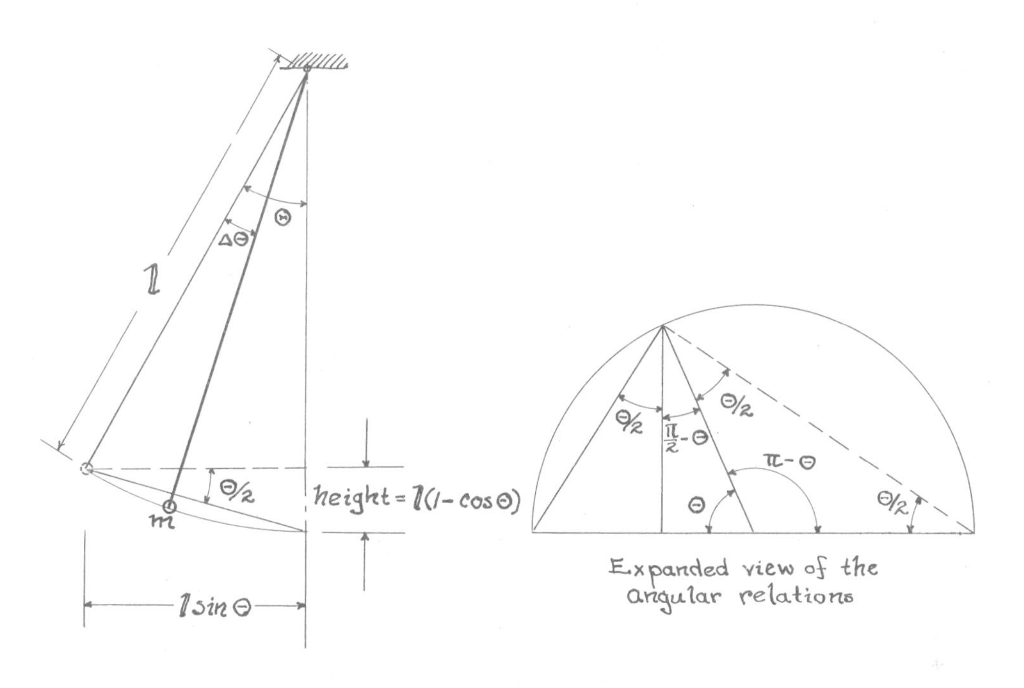

$$\frac{1}{2} m\left\{l \frac{\Delta\theta}{\Delta t}\right\}^2 + 2mgl \sin^2 \frac{\theta}{2} = \text{a constant} = 2mgl \sin^2 \frac{\theta_0}{2}.$$

The constant can be evaluated by examining the expression on the left at the maximum point of the swing, where the angular velocity is zero (i.e., $\Delta\theta/\Delta t = 0$ when $\theta = \theta_0$). The simple pendulum expression is obtainable when θ_0 (the maximum angular deflection) is sufficiently small that $\sin \theta_0/2 \simeq \theta_0/2$. This condition is satisfied when θ_0 is less than about 10 degrees (or 0.1745 radians). Because θ is always less than or equal to the maximum value θ_0, the same approximation holds: $\sin \theta/2 \simeq \theta/2$. Dividing by $(1/2)mgl\theta_0{}^2$ the original expression can be presented in a Pythagorean form:

$$\left[\frac{\Delta(\theta/\theta_0)}{\Delta(t\sqrt{g/l})}\right]^2 + \left(\frac{\theta}{\theta_0}\right)^2 = 1.$$

By defining new variables, $\Omega = \theta/\theta_0$, and $T = t\sqrt{g/l}$, this equation becomes

$$\left(\frac{\Delta\Omega}{\Delta T}\right)^2 + \Omega^2 = 1.$$

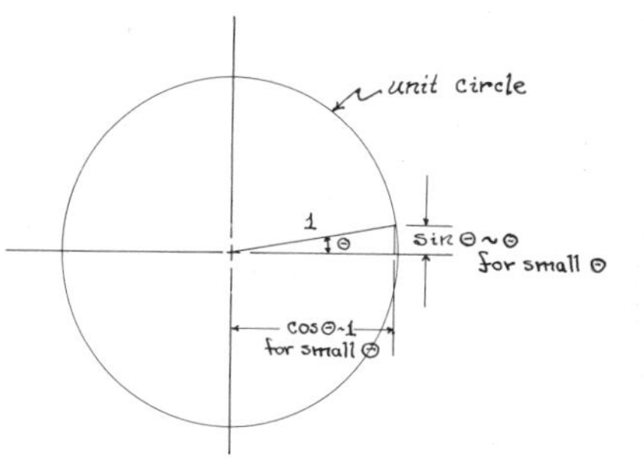

From the Pythagorean form of the trigonometric relations, we recognize that $\Omega = \sin T$ satisfies this relation if $\Delta\Omega/\Delta T = \cos T$. To show this, the trigonometric relation for the sin of the sum of two terms may be applied, namely, $\sin(A + B) = \sin A \cos B + \sin B \cos A$. Using this we can test the assumed solution $\Omega = \sin T$. As T goes from T to $T + \Delta T$, Ω changes to $\Omega + \Delta\Omega$. Therefore $\Omega + \Delta\Omega = \sin(T + \Delta T)$.

Expanding this by the $\sin(A + B)$ law, we obtain

$$\Omega + \Delta\Omega = \sin T \cos \Delta T + \cos T \sin \Delta T.$$

If ΔT is very small, $\cos \Delta T$ is approximately unity, and $\sin \Delta T$ is approximately equal to ΔT. Substituting these relations for small ΔT and employing the original assumption that $\Omega = \sin T$ (substituting for Ω on the left-hand side), we find that $\sin T + \Delta\Omega = 1 \cdot \sin T + \Delta T \cos T$, giving $\Delta\Omega/\Delta T = \cos T$. This result vindicates the assumed solution $\Omega = \sin T$. Because $\Omega = \theta/\theta_0$, and $T = t\sqrt{g/l}$, the solution takes the form

$$\theta = \theta_0 \sin t\sqrt{g/l}.$$

From the final expression above the period is obtained. Starting from $t = 0$, the pendulum completes one period when $T = 2\pi$, or $t = 2\pi\sqrt{l/g}$. Therefore, the period of the simple pendulum is $2\pi\sqrt{l/g}$. The solution of this problem for θ as a function of time introduces a smooth function of the time t. By the seventeenth century the concept of a continuous function of an independent variable became a serious consideration for mathematicians and scientists. A plot of the solution, $\theta = \theta_0 \sin t\sqrt{g/l}$, provides an excellent example of a continuous function, for which every value of the independent variable t has a corresponding value of the function $\theta_0 \sin t\sqrt{g/l}$. Earlier the sine was defined in terms of the side of a right triangle of hypotenuse 1. This triangle was referred to a unit circle, and consequently the function $\theta_0 \sin t\sqrt{g/l}$ can be referred to the vertical projection of the moving radius of a circle. Furthermore, the radius is of length θ_0 and rotates with a constant angular velocity $\sqrt{g/l}$. A plot of the correspondence between $\theta_0 \sin t\sqrt{g/l}$ and the rotary motion is shown.

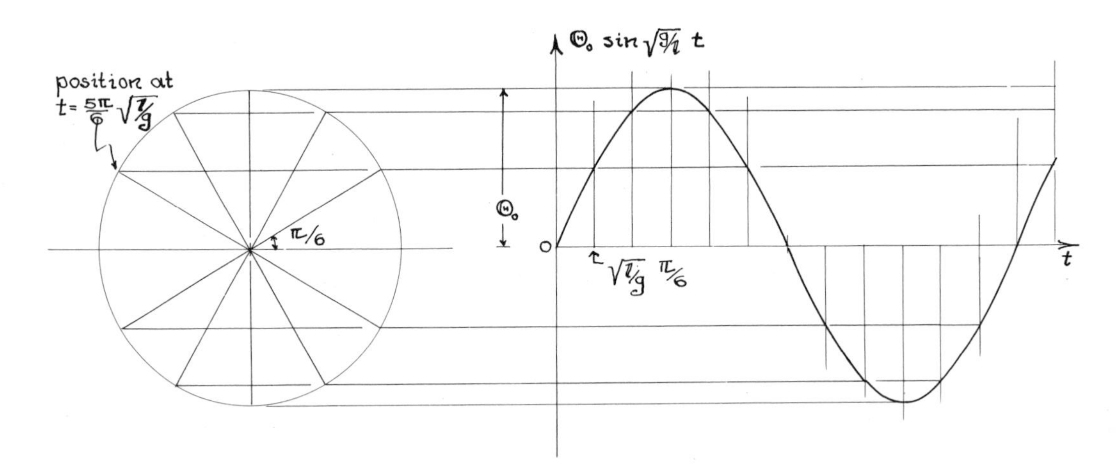

From the derivation it may be concluded that the simple pendulum is only approximately isochronous, with a sinusoidal motion. Huygens realized that to make an isochronous pendulum its length, l, must be shortened slightly as the angular displacement from the vertical increases. To be truly isochronous the pendulum must swing in a cycloidal arc. Huygens achieved this by introducing curved forms of a cycloidal shape to shorten the length of the pendulum as it swings away from the vertical. In the diagram the effect of the constraining forms is illustrated for two different deflections of the pendulum.

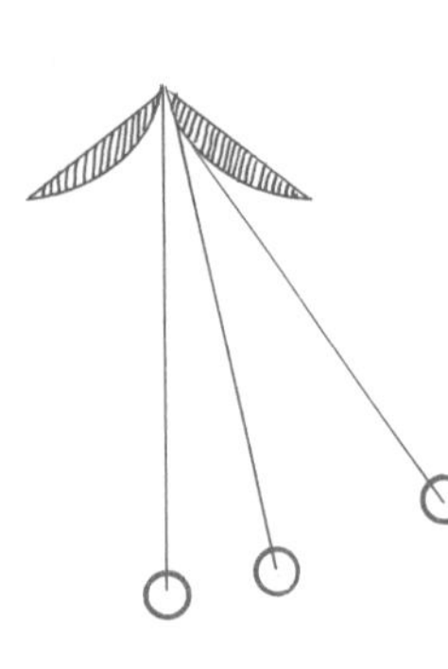

Huygens with others formulated the first law of dynamics (commonly known as Newton's First Law): if gravity did not exist and the atmosphere did not obstruct the motions of bodies, a body would maintain forever, with constant velocity in a straight line, the motion once impressed on it. The formula for the centripetal acceleration of a mass point moving on the circumference of a circle with constant angular velocity (or tangential velocity) was derived by Huygens.

The acceleration of a point is determined by the time rate of change of velocity. Thus if in a short time interval Δt the velocity $\vec{v}$ of a body changes by $\Delta\vec{v}$, the acceleration is given by $\Delta\vec{v}/\Delta t$. Velocity and acceleration are both quantities having direction and magnitude; therefore, a change implies an alteration of either quantity or of both. In the notation used here, for each example the directions associated with the quantities derived must be explained in words. The tangential velocity of a point moving on a circle is always tangent to the curve and thus constitutes a rotating vector. Consider a point moving with a constant tangential velocity on a circle. At a time t the vector has the direction $\vec{v}$. After an interval Δt the vector changes direction to have the form $\vec{v}$ plus $\Delta\vec{v}$. During Δt the angle θ changes, and consequently the tangential velocity changes by $\Delta\vec{v}$, a directed quantity. Using the parallelogram method or its equivalent, adding the tail of one vector to the head of another, $\vec{v} + \Delta\vec{v}$ can be graphed as one leg of an isosceles triangle. The triangle is isosceles because the tangential velocity is constant in magnitude when the rotation is uniform. Since the radius vector rotates through $\Delta\theta$ in Δt, the angle between $\vec{v}$ and $\vec{v} + \Delta\vec{v}$ is $\Delta\theta$. Finally, the change $\Delta\vec{v}$ divided by Δt has a magnitude of approximately $v\,\Delta\theta$, if $\Delta\theta$ is small and is directed radially inward toward the center of the circle. This is the centripetal acceleration $\vec{a}_c$ of the point during uniform rotary motion: centripetal acceleration (magnitude) $= a_c = v(\Delta\theta/\Delta t)$. We have previously shown that the magnitude of the tangential velocity, $\vec{v}$, was given by $R(\Delta\theta/\Delta t)$. Substituting for $\Delta\theta/\Delta t$, $a_c = v^2/R$. This expression indicates that the acceleration required to maintain uniform circular motion is equal to the square of the tangential velocity divided by the distance from the center of rotation, and further, it must be specified that $\vec{a}_c$ is directed inward toward the center of rotation.

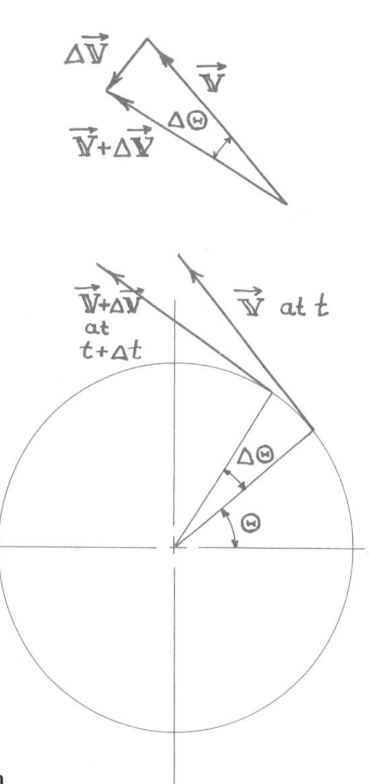

In 1681 Huygens returned to Holland and to his early interest in optics. For solving optical problems he created the principle which carries his name, a technique of enduring validity, which is still employed in many introductory texts. All space was treated as an invisible elastic medium, the "aether," and light was regarded as undulatory, propagating through the aether in spherical waves, somewhat as diverging circular waves are produced when a small stone is dropped into still water. The propagation of a wave in water may be traced by observing the motion of a maximum (crest) or a minimum (trough). Likewise when Huygens' principle is applied, the spherical wave front may be followed (usually a maximum of the undulation) to trace the motion of the light wave. The wave front propagates by considering each element on the front as a radiator. Each element emits a spherical wavelet at t and in a time $t + \Delta t$; the resulting wavelets create an envelope which constitutes the new wave front. The fact that light is propagated at a finite velocity is implied in this construction, and the method is very useful in the analysis of wave fronts which are limited by obstacles.

With the knowledge that in the refraction of light in a dense medium the ray perpendicular to the wave front is bent toward the surface normal, that is, toward the perpendicular, Huygens was able to provide a description of the refraction of light, assuming that waves traveled slower in the dense medium such as water or glass than in the less dense medium of air. In the diagram the incident wave front is plane, and it strikes the

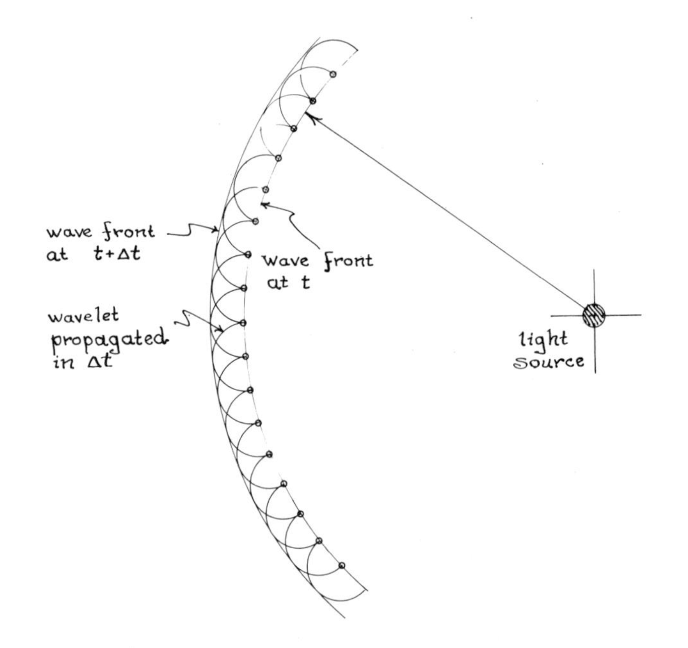

plane surface of a dense refracting medium at point A and at time t_A. Considering only a limited portion of wave front AB, we assume that the velocity of propagation in the less dense medium is v_1; thus the time for point B of the wave front to reach the surface is $t_B - t_A$.

In the time interval $(t_B - t_A)$, B moves to B' [a distance of $v_1(t_B - t_A)$], while A moves in terms of a spherical wave to A' at a lesser velocity, v_2 $(v_2 < v_1)$, covering a distance in the dense refracting medium $v_2(t_B - t_A)$. The original wave front AB is inside the dense medium at t_B and forms a plane wave front $A'B'$, which makes an angle β with the surface which is less than α, the angle of incidence. From the two right triangles ABB' and $AA'B$, with a common hypotenuse AB', one can solve for the relation between $\sin \alpha$ and $\sin \beta$ in terms of v_1 and v_2, the velocities of propagation of the wave fronts in the two media:

$$AB' = \frac{BB'}{\sin \alpha} = \frac{v_1(t_B - t_A)}{\sin \alpha},$$

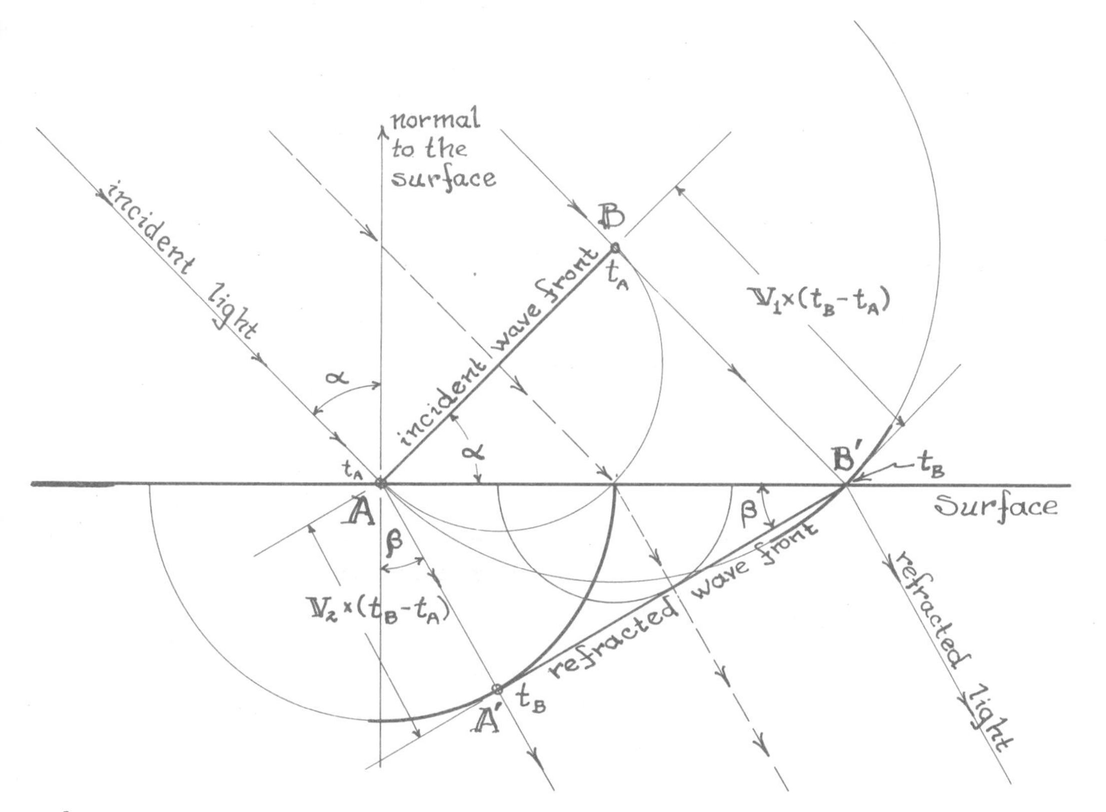

and

$$AB' = \frac{AA'}{\sin\beta} = \frac{v_2(t_B - t_A)}{\sin\beta},$$

giving

$$\frac{1}{v_1}\sin\alpha = \frac{1}{v_2}\sin\beta.$$

If the velocity of propagation in a vacuum is c, we multiply both sides by c and obtain Snell's law in terms of the indices of refraction: $n_1 = c/v_1$ and $n_2 = c/v_2$: $n_1 \sin\alpha = n_2 \sin\beta$.

Even more praiseworthy is the fact that Huygens was able to provide a model to explain the phenomenon of double refraction in calcite, discovered by Erasmus Bartholinus in 1669. Calcite has an intriguing property: when placed over a spot and viewed through the crystal, two images are observed. Huygens postulated that rays are composed of two parts, an ordinary and an extraordinary ray, and that calcite crystals have two different indices of refraction, each acting on one of the two rays. Upon refraction at the entrance surface, the two rays are bent toward the surface normal by different angles; after being transmitted through the calcite crystal, they emerge at different points, creating two different images.

Primitive as this explanation was, it became the basis of the more modern treatment of the phenomenon. When polarization of light was discovered, it developed that the two rays merely corresponded to two polarized components of the light. The two polarizations are perpendicular to the ray and to one another; the three quantities form an orthogonal coordinate system.

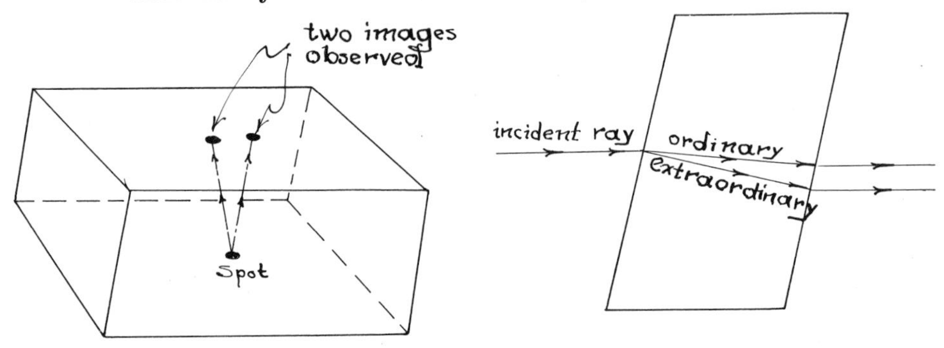

During this period of intense speculation concerning the laws of the physical world, the nature of the atmosphere and the behavior of gases were subjects of special interest. Galileo's pupil and secretary, Evangelista Torricelli (1608–1647), adduced the first explanation for the failure of a suction pump to lift water higher than thirty-two feet. Though aware that atmosphere had weight, Galileo failed to utilize this hypothesis; however, Torricelli reasoned that the height of a column of mercury in an evacuated tube should be to the height of a water column as the inverse ratios of the densities. With this method he invented the mercury barometer and was at once confronted with the fact that the atmospheric pressure may vary slightly from day to day.

With the invention of the air pump by Otto von Guerricke (1602–1686) convincing evidence of air pressure was offered, and following this line, Robert Boyle (1627–1691) demonstrated that at a fixed temperature the volume of a gas varied in inverse proportion to the pressure. Boyle's assertions on the atomic structure of matter were vastly important in setting forth a very old idea in straightforward terms. He created the concept of an element as "certain primitive and simple bodies, which not being made up of any other bodies . . . are the ingredients of which all those called perfectly mixt bodies are immediately compounded and into which they are ultimately resolved." A universal matter common to all bodies was assumed. Certainly Boyle's views coincided with those of the French philosopher Pierre Gassendi (1592–1655) and no doubt in some measure derived from them.

Even though he was unable to measure it, Galileo felt that the velocity of light was finite. In 1675 Olaus Roemer (1644–1710) offered the first evidence that light has a finite velocity. Of the eleven satellites of Jupiter, four were visible through the available telescopes. By observing that intervals between eclipses of a given satellite were shorter than the average when Jupiter and the earth approached each other and were longer when the planets were receding, Roemer provided a rough estimate of the velocity of light. Because of the motion of the earth about the sun during the interval of a pair of eclipses, the earth could move either toward or away from the second light signal, making the total distance of travel for the second signal less or greater than the average. Roemer reckoned that it took light 22 minutes to cover a distance equal to the diameter of the earth's orbit, and he obtained a value of $2.1(10)^{10}$ centimeters per second for the velocity of light. Although this value is too low by 30 per cent, it was a remarkable achievement. An elementary analysis of this calculation can be made with the diagram shown.

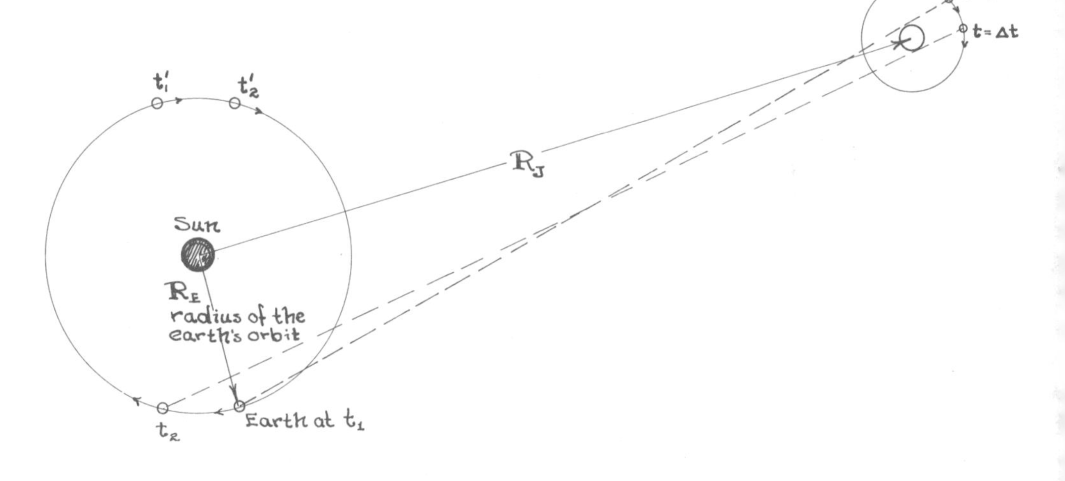

Assume that the eclipse is initiated at $t = 0$ and that the first signal is received at $t_1 = R_J/c$, where R_J is the distance to Jupiter and c is the velocity of light. At a time Δt after $t = 0$ the second signal starts toward the earth. The second signal arrives at t_2 but it covers an additional distance equal to $R_J + R_E\omega_E(t_2 - t_1)$, where ω_E is the angular velocity of the earth about the sun (2π divided by one year) and R_E is the radius of the earth's orbit. The time of arrival, t_2, is then this distance divided by c plus the time between eclipses, Δt:

$$t_2 = \frac{R_J + R_E\omega_E(t_2 - t_1)}{c} + \Delta t.$$

Solving for t_2 we obtain (using $t_1 = R_J/c$)

$$t_2 = \frac{R_J}{c} + \frac{\Delta t}{(1 - R_E\omega_E/c)}.$$

Then

$$t_2 - t_1 = \frac{\Delta t}{(1 - R_E\omega_E/c)}.$$

On the approaching side of the orbit the extra distance of travel is subtracted, giving the apparent time between observations as

$$t'_2 - t'_1 = \frac{\Delta t}{(1 + R_E\omega_E/c)}.$$

Since Δt can be measured when the earth is moving perpendicular to R_J, either measurement, $(t_2 - t_1)$ or $(t_2' - t_1')$, will give a value for c:

$$c = \frac{R_E\omega_E}{[1 - \Delta t/(t_2 - t_1)]} = \frac{R_E\omega_E}{[\Delta t/(t_2' - t_1') - 1]}.$$

The quantities on the right are measured and $R_E\omega_E$ is known. Δt, the eclipse interval, can also be obtained if necessary:

$$\Delta t = \frac{2(t_2' - t_1')(t_2 - t_1)}{(t_2' - t_1') + (t_2 - t_1)}.$$

Jean Dominique Cassini (1625–1712) provided an estimate of R_E, the distance between the earth and sun, in 1673. Good as his value was at the time, it was inaccurate by 7 per cent. The velocity of light was also indicated by the aberration of light from the fixed stars between summer and winter. This was first observed by James Bradley (1692–1762) in 1729. Aberration under these conditions implies that the angle of a telescope for alignment on a fixed star varies between summer and winter. For the star under consideration, the earth was moving perpendicular to the direction of the starlight, and Bradley deduced that the finite velocity of light and the earth's motion combined to determine the angles of sight.

VI

A Man for All Science

In the short space of a hundred years, Cardano, Galileo, Stevin, Gilbert, Kepler, Descartes, and Huygens set the stage for the Age of Newton. Algebra, analytics, kinematics and dynamics, and planetary orbits presented the European with a bold picture of the future. Galileo's analysis of constant acceleration added a further awareness of the importance of an area under a curve and of the slope of a curve. Descartes had made a successful attempt at taking the slope of certain curves, an achievement in which he had, as a matter of fact, been anticipated in the unpublished work of Pierre de Fermat (1601–1665). This "prince of amateurs," as he is sometimes called, trained in the law, pursued mathematics as a hobby. His understanding of the basic principles of analytic geometry is evident in *Ad Locos planos et solidos isagoge*, which he compiled in 1629. Hoping to restore some lost proofs of Apollonius, Fermat employed a system of coordinates. Although his work had a decided influence on other mathematicians, with whom he carried on an extensive correspondence, this manuscript was not made public for half a century and thus did not receive the general acclaim one might expect. Fortunately, much of his correspondence has been preserved. In a letter dated September 22, 1636, a year before Descartes published *La Géométrie*, Fermat wrote to Personne de Roberval (1602–1675) concerning his own work on maxima

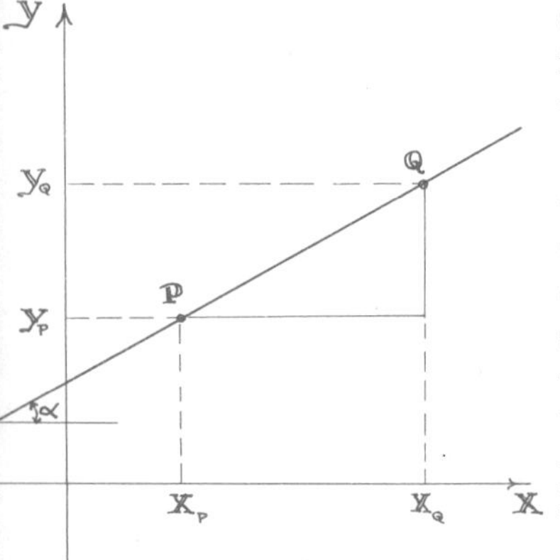

and minima and referred to his theorems on tangents, amicable numbers, integration, loci of plane and solid figures—in short, a broad spectrum of work in analytics.

To find the maxima and minima of certain plotted curves, Fermat created a differential calculus of infinitesimals. To obtain the tangent to a curve at a given point, the slope of the curve at that point must be calculated. The slope of a line is defined as the ratio of increase or decrease along the vertical axis to the corresponding length interval taken along the horizontal axis. In the accompanying figure, the slope of the line L can be calculated by taking any two points, P and Q, on L, measuring the projection of the line segment PG on the vertical y axis, and dividing by the horizontal projection:

$$\text{slope of } L = \frac{y_Q - y_P}{x_Q - x_P} = \tan \alpha .$$

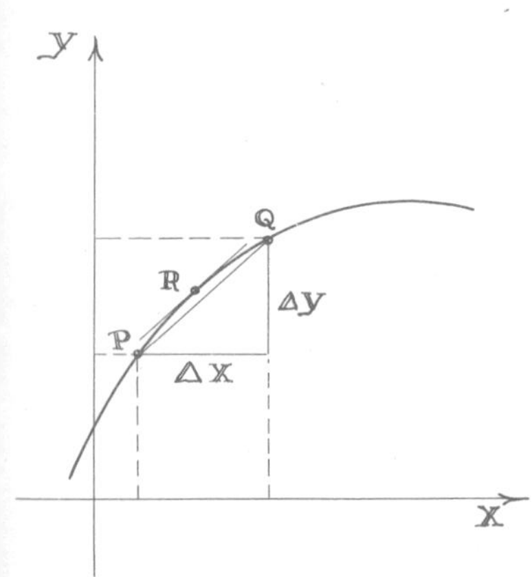

It should be apparent that the slope is positive if the y projection increases as the point moves from left to right and negative if the y projection decreases; therefore, the slope is a simple quantity to measure if one deals only with straight lines. The slope of a general curve varies from point to point, and consequently a separate computation must be performed, in some sense, at each point in question. Consider the curve $y = f(x)$, shown here: an approximate value of the slope at a point R can be achieved by measuring the slope of the straight line connecting two points, P and Q, where P and Q fall on opposite sides of R and are very close to R.

By designating the y projection of PQ as Δy and the x projection as Δx, we obtain the approximate value: slope at $R \simeq \Delta y/\Delta x$ (here the sign $\simeq$ means "approximately equal to"). The major question here is how close is "very close"? If we say that Δx is taken as small as we please (to borrow from Eudoxus), one can obtain values for the slope as close to the true value as may be wished. Whenever the slope at a point is zero, the point represents a maximum, a minimum, or an inflexion point of the curve, as illustrated.

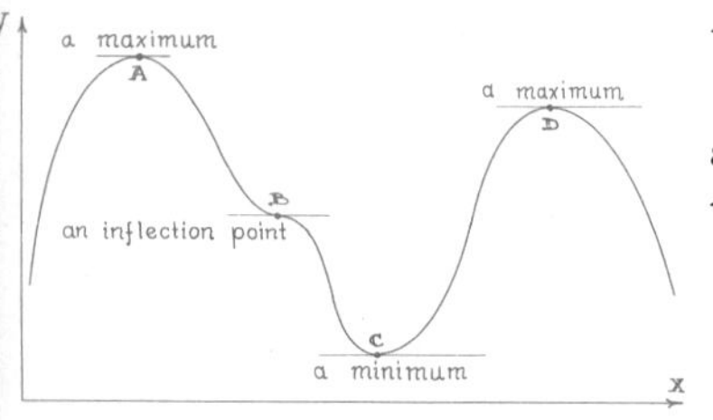

Since there are three different ways in which a zero slope can occur, a further criterion is required to differentiate between them. To systematize this classification, we first notice that slopes have a sign associated with

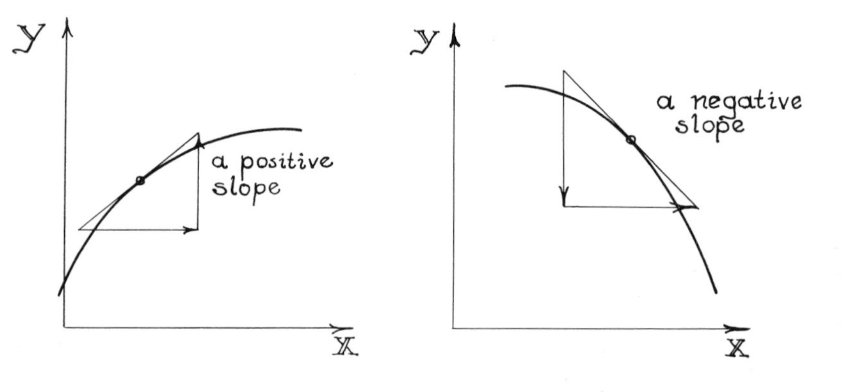

them. If y increases as x increases, the slope is positive, while a decrease in y for a corresponding increase in x signifies that the slope is negative.

In the vicinity of a *maximum* of a curve, the slope is progressively positive, zero, and, finally, negative. Therefore, if one plots the slope of a curve $f(x)$ as a function of x, the slope at a maximum passes through zero going from positive to negative values as x increases. Minima, then, are characterized by a slope progressing from negative to positive. The anomalous zero derivative point which can occur is the inflection point, and in this case the slope on either side of the inflection point retains the same sign.

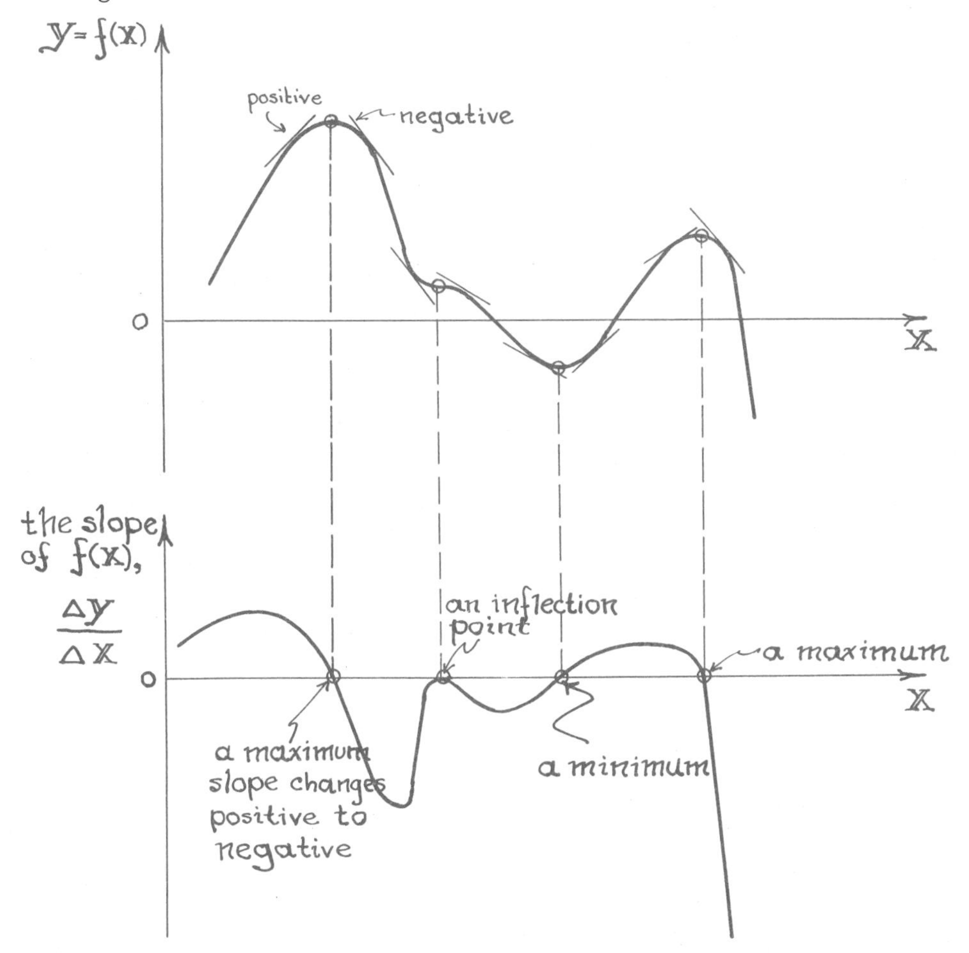

Fermat's interest in mathematics extended far beyond analytics and the calculus. The theory of numbers, indeterminate equations, and the theory of probability were major areas to which he made lasting contributions. In letters exchanged with Pascal he set forth the classical problem of the division of the stake between two dice players whose game is interrupted; the ideas discussed in these letters were fundamental to the development of the modern theory of probability, and mingled with the argument on probability one also discovers a lively exchange on the theory of numbers.

Although he considered mathematics his hobby, the "prince of amateurs" did not neglect physics, and one of the most beautiful of his physical hypotheses is his Least Time principle. Attempting to derive Snell's law of refraction, Fermat, like Heron of Alexandria in his analysis of reflection, suggested that a light ray in passing from a point A to a point B in a medium of variable refractive index (or in a medium wherein the velocity of light varied) would make the passage in the least possible time.

Regard the diagram illustrating the refraction of a light ray at a plane boundary. Assume a velocity of v_1 in the upper medium and a lesser velocity of v_2 in the lower medium. With the dimensions given in the diagram, the only variable is x, the point along the horizontal where the light ray strikes the surface of medium 2.

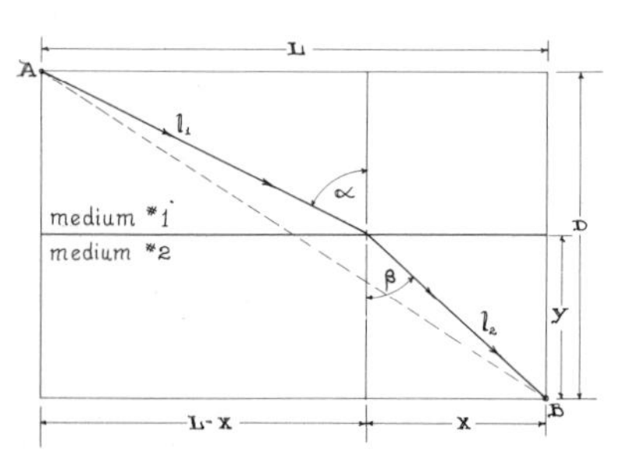

$$\text{distance traveled in medium } 1 = l_1 = \sqrt{(L-x)^2 + (D-y)^2};$$

$$\text{distance traveled in medium } 2 = l_2 = \sqrt{x^2 + y^2};$$

$$\text{total traveling time from } A \text{ to } B = T = \frac{l_1}{v_1} + \frac{l_2}{v_2}.$$

Using Fermat's minimum slope principle, if we vary x by Δx, then the corresponding variation in T, that is, ΔT, should be zero at the minimum. With the definition of T given above, a value $x + \Delta x$ should produce $T + \Delta T$, of the following form:

$$T + \Delta T = \frac{1}{v_1}\sqrt{(L - x - \Delta x)^2 + (D-y)^2} + \frac{1}{v_2}\sqrt{(x + \Delta x)^2 + y^2}.$$

Extracting $\sqrt{(L-x)^2 + (D-y)^2} = l_1$ from the first term on the right, and $l_2 = \sqrt{x^2 + y^2}$ from the second term:

$$T + \Delta T = \frac{l_1}{v_1}\sqrt{1 - \left[\frac{2(L-x)\,\Delta x - (\Delta x)^2}{l_1^2}\right]} + \frac{l_2}{v_2}\sqrt{1 + \left[\frac{2x\,\Delta x + (\Delta x)^2}{l_2^2}\right]}.$$

If Δx is small compared to l_1 and l_2, we use the approximation that $\sqrt{1 \pm \epsilon} \simeq 1 \pm \epsilon/2$ when ϵ is small. Using this approximation, dropping the very small terms $(\Delta x)^2$,

$$T + \Delta T \simeq -\frac{(L-x)\,\Delta x}{v_1 l_1} + \frac{x\,\Delta x}{v_2 l_2} + \left\{\frac{l_1}{v_1} + \frac{l_2}{v_2}\right\}.$$

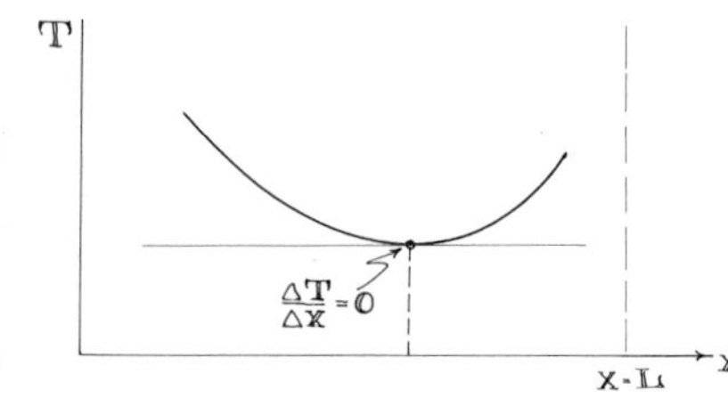

The sum of the third and fourth terms above is just T; therefore, canceling T and dividing by Δx,

$$\frac{\Delta T}{\Delta x} \simeq -\frac{(L-x)}{v_1 l_1} + \frac{x}{v_2 l_2}.$$

Set $\Delta T/\Delta x = 0$ to find the condition for the minimum time; i.e.,

$$\frac{1}{v_1}\frac{(L-x)}{l_1} \simeq \frac{1}{v_2}\frac{x}{l_2},$$

or $(1/v_1) \sin \alpha = (1/v_2) \sin \beta$. This calculation is equivalent to finding the value of x, which gives the minimum value of T in the curve of T versus x.

Roberval, John Wallis, and Isaac Barrow all turned their attention to the absorbing problem of finding the tangent to a curve. Roberval regarded the tangent to a curve at a point as the direction of motion of a point moving on the curve. He considered the velocity to be resolved vectorially into a vertical component and a horizontal component. Consequently, the tangent to the point would lie along the diagonal of the rectangle formed by the two velocity components. As we shall observe, Newton used this same approach to differentiation, in denoting the y component of velocity along the curve as $\dot{y}$ and the x component of velocity as $\dot{x}$. Thus the slope is just $\dot{y}/\dot{x}$.

Several mathematicians after Cavalieri had attempted to free his integration technique of its apparent flaws. Both Roberval and Pascal addressed themselves to this exercise. They considered a line to be composed of infinitesimal line segments, an area to be formed of infinitesimally narrow rectangles, and a solid to consist of infinitesimally thin slices. These assumptions were well founded. Roberval computed the area of a cycloid and found it to be three times the area of the generating circle. He also investigated the areas under the simple curves $x^{1/2}$, $x^{1/3}$, $x^{1/4}$, and $x^{1/5}$. Fractional powers were not used at this time, so that his examples were stated as curves in which the abscissas are as the squares, cubes, fourth, and fifth powers of the ordinate (the vertical). He compared each area with the area of the circumscribing rectangle and found them to be 2/3, 3/4, 4/5, and 5/6, respectively. To view a simpler exercise, let us first obtain the areas under curves for which the ordinate behaves as the square and the cube. Although Cavalieri accomplished this, it is instructive to picture these problems using the rectangular partitioning of Roberval. This method approximates the area under a curve $y = f(x)$ by filling the space in question with narrow vertical rectangles of width ϵ and a height equal to the mean value of the function within each respective interval.

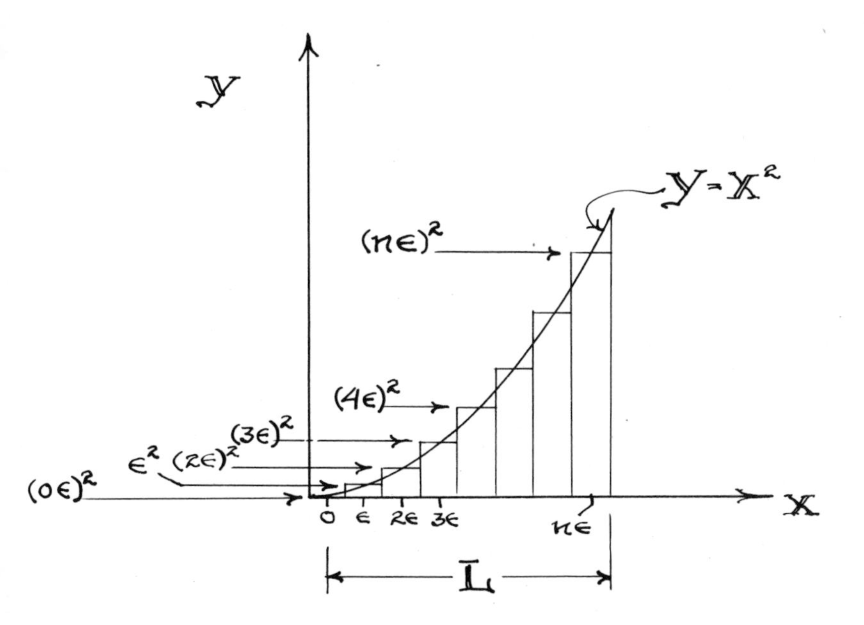

The sum of the rectangles in the figure behaves as

$$\sum_{m=0}^{n} \epsilon\cdot(m\epsilon)^2 = \epsilon\cdot(0\epsilon)^2 + \epsilon\cdot(1\epsilon)^2 + \epsilon\cdot(2\epsilon)^2 + \cdots$$
$$+ \epsilon\cdot([n-1]\epsilon)^2 + \epsilon\cdot(n\epsilon)^2,$$
$$= \epsilon^3[0^2 + 1^2 + 2^2 + \cdots + (n-1)^2 + n^2],$$
$$= \epsilon^3\cdot\frac{1}{6}\,n(n+1)(2n+1).$$

Once again we have used the result for the squares of the integers, first mentioned in connection with the Babylonians and Pythagoreans. If the length of the base of this figure is $L = (n + 1/2)\epsilon$, then the sum can be written in terms of L and the circumscribing rectangle $L\cdot L^2$ as

$$\frac{\epsilon^3}{6}\,n(n+1)(2n+1) = \frac{1}{3}\,L\left(L^2 - \frac{1}{4}\,\epsilon^2\right) \xrightarrow[\epsilon\ll L]{} \frac{1}{3}\,L^3.$$

Thus, when ϵ is infinitesimally small compared to L, we may neglect $(1/4)\epsilon^2$ relative to L^2. If the area under the curve is compared with the rectangle of area $L\cdot L^2$, one obtains 1/3. The area under $y = x^3$ can be computed in the same fashion: in the figure the sum $\sum$ of the rectangles is obtained from the ancient relation for the sum of the cubes of the integers:

$$\sum_{m=0}^{n} \epsilon\cdot(m\epsilon)^3 = \epsilon\cdot(0\epsilon)^3 + \epsilon\cdot(1\epsilon)^3 + \cdots + \epsilon\cdot[(n-1)\epsilon]^3 + \epsilon\cdot(n\epsilon)^3,$$
$$= \epsilon^4[0^3 + 1^3 + 2^3 + 3^3 + \cdots + (n-1)^3 + n^3],$$
$$= \epsilon^4\cdot\frac{1}{4}\,n^2(n+1)^2.$$

Once again $L = (n + 1/2)\epsilon$; therefore the sum gives

$$\frac{1}{4}\,(n\epsilon)^2(n\epsilon + \epsilon)^2 = \frac{1}{4}\left[L^2 - \frac{1}{4}\,\epsilon^2\right]^2 \xrightarrow[\epsilon\ll L]{} \frac{1}{4}\,L^4;$$

a comparison of this with the rectangle of area $L\cdot L^3$ gives 1/4. In general, the area under a curve $y = x^r$ from $x = 0$ to $x = L$ is the sum

$$\sum_{m=0}^{n} \epsilon\cdot(m\epsilon)^r \xrightarrow[\epsilon\ll L]{} \frac{(n\epsilon)^{r+1}}{r+1} \longrightarrow \frac{L^{r+1}}{r+1},$$

and the ratio of this area to that of the circumscribing rectangles is $1/(r+1)$.

In the light of these elementary results, we observe that Roberval's solutions for the areas under the curves (in modern notation), $x^{1/2}$, $x^{1/3}$, $x^{1/4}$, and $x^{1/5}$, are quite significant. The values of these areas when compared with the values of the circumscribed rectangles are also $1/(r+1)$, where r is the power of x. In these examples, however, r is no longer

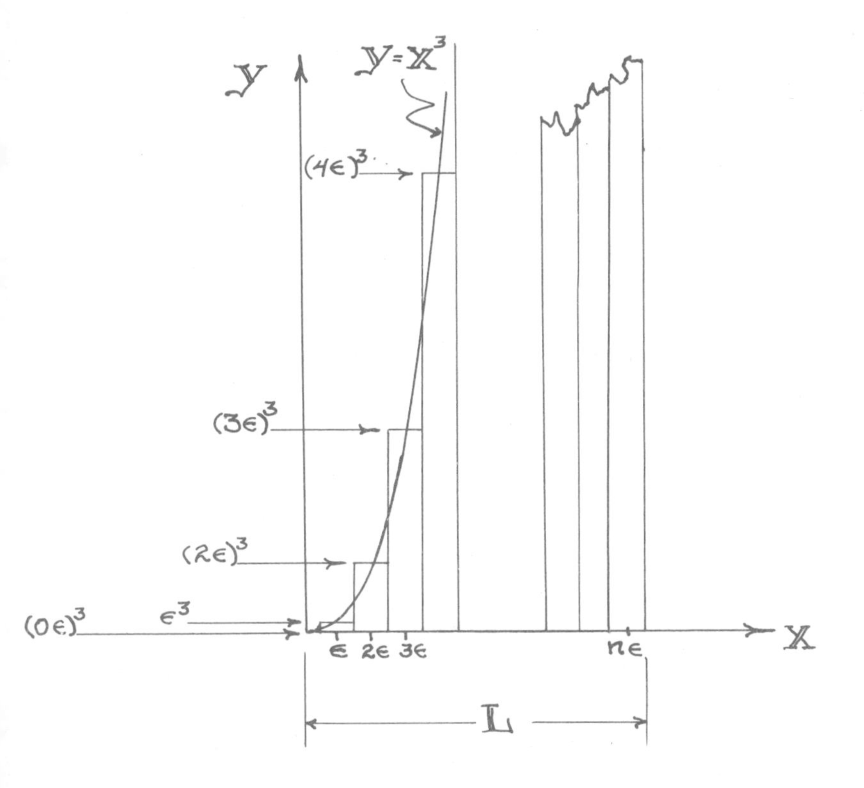

restricted to integral values, which suggests that the fractional powers are a proper representation for roots. Torricelli (1608–1647) accomplished the same result and in addition showed that the area under the curve, $x^{2/3}$, is equal to 3/5 of the circumscribing rectangle.

John Wallis (1616–1703) advanced the knowledge of integration in his *Arithmetica infinitorum* of 1656. He suggested there that the general rule for areas under curves, which was indicated as the normalized value $1/(n+1)$, was valid for nonintegral powers as well as for integral powers. He found that the nonintegral powers encompassed all of the roots; i.e., that $\sqrt{x}$ could be written as $x^{1/2}$, $\sqrt[3]{x}$ as $x^{1/3}$, etc. Wallis then investigated the integrals (or areas) of a particular set of algebraic polynomials. By this time it was customary (from Descartes) to label the unknown variable as x or y, etc.

The instruction "to sum" was idealized to the infinite sum of infinitesimals and was designated by an elongated S symbol, $\int$ (from Leibniz, 1686). To maintain modern notation, we shall introduce the infinitesimal width of such a rectangle as would be employed in rectangular partitioning. This width was denoted as dx by Leibniz in 1684. With this notation the area under a curve x^m between $x = 0$ and $x = 1$ would be the infinite sum of rectangles $x^m\,dx$, written as:

$$\int_0^1 x^m\,dx = \frac{1}{m+1}\cdot$$

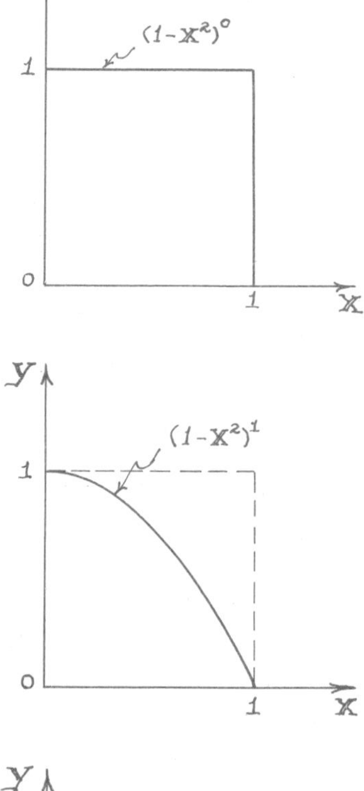

Wallis presented the areas of the polynomials $(1 - x \cdot x)^n$, where n is an integer in his work; some of his results are:

$$\int_0^1 (1 - xx)^0\, dx = \int_0^1 (1 - x^2)^0\, dx = \int_0^1 dx = 1,$$

$$\int_0^1 (1 - x^2)^1\, dx = \int_0^1 dx - \int_0^1 x^2\, dx = 1 - \frac{1}{3} = \frac{2}{3},$$

and

$$\int_0^1 (1 - x^2)^2\, dx = \int_0^1 dx - 2\int_0^1 x^2\, dx + \int_0^1 x^4\, dx = 1 - \frac{2}{3} + \frac{1}{5} = \frac{8}{15}.$$

These forms are quite suggestive; for instance, the circular quadrant is represented by $y = \sqrt{1 - x^2}$; thus to obtain the numerical value of $\pi/4$ we need only compute:

$$\int_0^1 \sqrt{1 - x^2}\, dx = \int_0^1 (1 - x^2)^{1/2}\, dx.$$

The challenge of this particular integral (an infinite sum) led Newton to establish an infinite series expansion for the quantity $(1 - x^2)^{1/2}$. From the known binomial expansions of $(1 - x^2)^n$ (n being an integer), Newton inferred that a binomial expansion for $(1 - x^2)^{1/2}$ also existed. Wallis obtained a solution for this integral in terms of an infinite product rather than an infinite series: he found that

$$\frac{\pi}{4} = \int_0^1 (1 - x^2)^{1/2}\, dx = \frac{2}{3}\cdot\frac{4}{3}\cdot\frac{4}{5}\cdot\frac{6}{5}\cdot\frac{6}{7}\cdot\frac{8}{7}\cdot\frac{8}{9}\cdots.$$

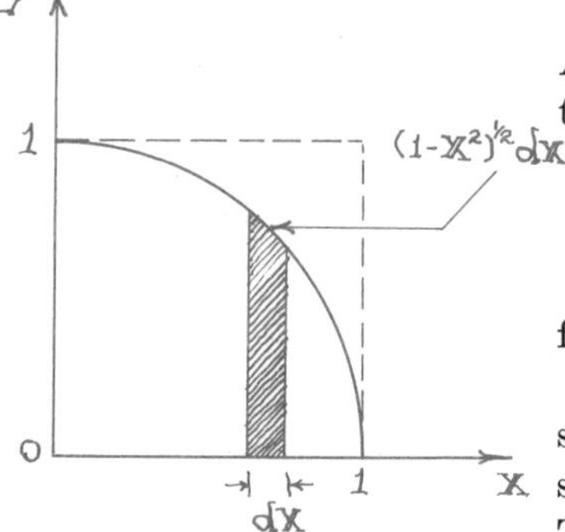

Although the area under the quantity x^m was known, Wallis formalized this result by stating the rule for the general indefinite integral

$$\int x^m\, dx = \frac{x^{m+1}}{m + 1}$$

for all m except the transition value $m = -1$.

The value of the integral of x^{-1} is the natural logarithm of x, and this solution in series form is readily obtained from the geometric series. This series was described earlier in the discussion of the Merton College group. The sequence S_N was shown to have the form

$$S_N = 1 + x + x^2 + \cdots + x^{N-1} + x^N = \frac{1 - x^{N+1}}{1 - x}.$$

For values of x less than 1, greater than -1, and for arbitrarily large N, x^{N+1} goes to zero, and

$$S_N \xrightarrow[|x|<1]{N\to\infty} S = \frac{1}{1 - x} = 1 + x + x^2 + x^3 + \cdots.$$

Because of Wallis' rule for the integral of x^m, the integral of $1/(1 - x)$ can be obtained by expanding in the power series given above:

$$\int \frac{dx}{1 - x} = \int [1 + x + x^2 + x^3 + \cdots] \, dx$$

$$= x + \frac{1}{2}x^2 + \frac{1}{3}x^3 + \frac{1}{4}x^4 + \cdots.$$

Here we have assumed that the values of x are less than $+1$ and greater than -1. The natural logarithm of $(1 - x)$ is defined by this indefinite integral:

$$\log_e (1 - x) = -\int \frac{dx}{1 - x} = -\left[x + \frac{1}{2}x^2 + \frac{1}{3}x^3 + \frac{1}{4}x^4 + \cdots\right].$$

By replacing x with $-y$ in the original expression, one obtains the famous Mercator's series for

$$\log_e (1 + y) = y - \frac{1}{2}y^2 + \frac{1}{3}y^2 - \cdots.$$

The other subdivision of the calculus involved the calculation of the slope of a curve. Newton's preceptor at Cambridge, Isaac Barrow (1630–1677), also obtained methods for determining the tangents to curves and was close to discovery of the general methods of the differential calculus. Born in 1642 in Lincolnshire, Isaac Newton was educated in public schools. Evidently he was a good student, but there is little in his early life to suggest his capacity for invention. In 1660 he entered Trinity College, Cambridge, and as an undergraduate studied Kepler's *Optics*, Barrow's *Lectures*, and Wallis' *Arithmetic infinitorum*. Within nine years he had made major advances in optics, created his first tests of the inverse square law of gravitation, and had communicated his first ideas of the unified calculus. In addition to these memorable achievements, he assembled the three laws of mechanics and generalized the binomial expansion.

These remarkable works should be kept in perspective. Newton himself acknowledged his debt to the giants of his age, who had provided the framework for many of these advances. There is a tendency to cast the events of scientific history into a simple pattern, in which as many ideas as possible are associated with a single individual. However, it must be evident from what has been presented in these pages that most major advances were the culmination of a recognizable process of evolution, in which brilliant suggestions and contributions were made by an impressive array of minds. But because it is usually a long and tedious process to trace the origins of a subject, one tends to favor the "one man, one try, one success" description of scientific discovery. It is commonly accepted that Newton and Leibniz invented the calculus, but methods for taking the slope of a curve (differentiation) and the techniques of integration (or the measurement of areas under curves) were clearly understood by 1660, the year in which Newton entered college as an undergraduate. Newton and Leibniz supplied the missing concept: differentiation and integration were connected, and one operation was the inverse of the other.

Newton was stimulated by his early interest in the calculations of John Wallis and freely acknowledged the influence of Wallis' book. He obtained the infinite series for the value of $\pi/4$ in an indirect fashion. Rather than attempt the expansion of $\sqrt{1 - x^2}$, he considered the integrals of the known polynomials and interpolated between them. As Wallis had done, he began with the integrals of $(1 - x^2)^n$:

$$\int (1 - x^2)^0 \, dx = x,$$

$$\int (1 - x^2)^{1/2} \, dx = ?$$

$$\int (1 - x^2)^{2/2} \, dx = x + \frac{1}{-3} x^3,$$

$$\int (1 - x^2)^{3/2} \, dx = ?$$

$$\int (1 - x^2)^{4/2} \, dx = x + \frac{2}{-3} x^3 + \frac{1}{5} x^5,$$

and so on.

With the array of the integrals of $(1 - x^2)^m$ before him, Newton reasoned that the series representation for the integral of $(1 - x^2)^{1/2}$ would be infinite and odd. He further assumed from the pattern that the terms -3, 5, -7, etc., in the denominators of succeeding coefficients would be the same in the case of the integral of $(1 - x^2)^{1/2}$. He next observed that the coefficient (involving $1/-3$) of Wallis' series went as 0, 1, 2, 3, etc.; thus the term in the numerator of the second term of the second integral above should be 1/2. In this manner of interpolation he guessed the other coefficients of the series, giving

$$\int (1 - x^2)^{1/2} \, dx = x + \frac{1/2}{-3} x^3 + \frac{-1/8}{5} x^5 + \frac{1/16}{-7} x^7 + \cdots.$$

From this result he obtained the series expansion of $\pi/4$ by integrating from $x = 0$ to $x = 1$:

$$\frac{\pi}{4} = \int_0^1 (1 - x^2)^{1/2} \, dx = 1 - \frac{1}{6} - \frac{1}{40} - \frac{1}{112} + \cdots.$$

Knowing the Wallis rule for integrating x^m, Newton was then able to extract the series expansion of $(1 - x^2)^{1/2}$ from this result. After a lengthy examination of these interpolated series, he produced a binomial expansion for nonintegral as well as integral powers. Even though his results were obtained without rigorous proofs, they were correct, and they placed an emphasis on infinite series. Such series have dominated the field of analysis from that time onward. These early studies were devoid of such obvious tests of a series as a test of convergence.

Newton's first work on the calculus, *De analysi*, was compiled in 1669, though it did not appear until 1711. Like his predecessors he adhered to the infinitesimals; he perceived the connection between integration and differentiation, however, by relating the change in the area under the curve to small changes in the ordinate. *De analysi* opens with a statement found in Wallis' work: if a curve y is equal to $ax^{m/n}$, then the area bounded by the curve is $[an/(m + n)]x^{(m+n)/n}$.

His second book on the calculus, *Methodus fluxionum*, was written in 1671 and appeared in 1736. The notations of fluent (the growing quantity) and the fluxion (rate of generation) are historically interesting, but they were superceded by the more appealing notation of Leibniz. Moreover, the approach of Newton was essentially that of Roberval, in that the variation of the two variables is taken with respect to a third parameter, the time, and derivatives are computed by taking the ratios of the time rates of change of the two major variables.

As an example of Newton's method, consider the curve $x^2 = ay$. If a point on the curve moves smoothly in time such that the x and the y components of the velocity are $\dot{x}$ and $\dot{y}$,* then in a very small interval of time o (nowadays we would use Δt) the abscissa x would change by $\dot{x}$o, while the ordinate y is changed by $\dot{y}$o. At the new point, $ay = x^2$ can be written:

$$a(y + \dot{y}\text{o}) = (x + \dot{x}\text{o})^2 = x^2 + 2x\dot{x}\text{o} + (\dot{x}\text{o})^2.$$

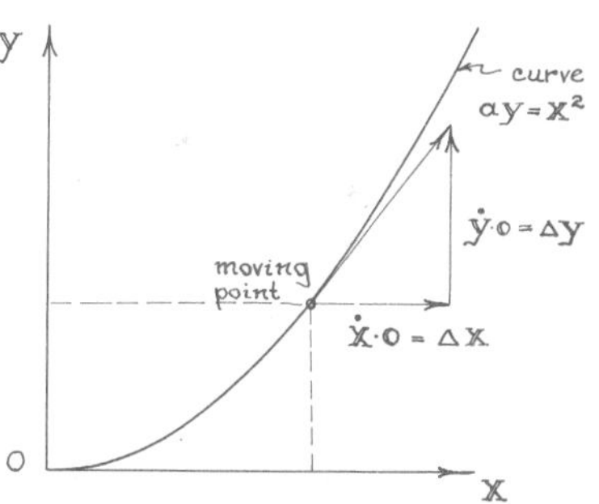

Assuming that the terms in o^2 can be neglected, and using the fact $ay = x^2$, we find $a\dot{y}\text{o} = 2x\dot{x}\text{o}$, or the slope of the curve $= \dot{y}/\dot{x} = 2x/a$. Since (in the notation used previously) $\dot{x} = \Delta x/\Delta t$ and $\dot{y} = \Delta y/\Delta t$, the ratio $\dot{y}/\dot{x}$ becomes $\Delta y/\Delta x$.

This notation is representative of Newton's method, and with it he developed the relationship between the operation of differentiation and that of integration. Incorporated in these studies were the determinations of the maxima and minima of curves.

The paradox of the discovery of the fundamental theorem of the calculus was that two minds arrived independently at the same conclusion within a few years of one another. To all intents an amateur, Gottfried Wilhelm von Leibniz (1646–1716) published his first paper on the calculus in 1684, some twenty years before Newton's. However, Newton developed the calculus between 1665 and 1666, while Leibniz did not arrive at his version until the period 1673 to 1676. The calculus of both men was hampered by a vague treatment of the infinitesimal quantity. Newton's "moments of fluxions," such as $\dot{x}$o and $\dot{y}$o, were the equivalents of Leibniz's dx and dy. In both treatments the small zero or the differential element was thought of as *infinitely small* quantities and was not clearly

* In terms of the notation used in previous sections, $\dot{x}$ and $\dot{y}$ are shorthand representations of dx/dt and dy/dt when the increments become exceedingly small.

defined. The ensuing confusion lasted until the beginning of the nineteenth century, when the modern concept of a limit was firmly established.

Great controversy and polemics were indulged in by the followers of Leibniz and Newton over the priority of invention of what was believed to be the calculus. In fact, the two men had merely provided the connection between the operations of differentiation and integration. The earlier investigations of Huygens, Roberval, Fermat, and Wallis had developed the practical techniques for these operations by 1660, and it is therefore quite conceivable that both men should make the connection independently. The amusing aspect of the controversy is that Newton and Leibniz communicated with each other freely and apparently without antagonism. The argument over priority was evidently chauvinistic and developed only among the followers of these two giants, not an uncommon experience in history.

Leibniz was a man of varied talents—a philosopher, a lawyer, and ultimately a mathematician. It is clear that he had studied the works of the earlier masters, including John Wallis. In his 1684 paper, he introduced the symbols for the differential calculus which were to remain the standard notation in that subject: dx became the infinitesimal length along the abscissa, and dy was the corresponding increment of length along the ordinate. Using the example given above, the curve $ay = x^2$, Leibniz expansion given without proof is:

$$a(y + dy) = (x + dx)^2 = x^2 + 2x\,dx + (dx)^2,$$
$$a\,dy = 2x\,dx,$$

or $dy/dx = 2x/a$.

Included in this early paper was the statement of the chain rule of differentiation of the product of two functions. If u and v are two functions of x, the differential of uv is $d(uv) = u\,dv + v\,du$. With this start the reader should make an attempt to develop the next rule for the ratio of two functions: $d(u/v) = -(u/v^2)\,dv + (1/v)\,du$.

Leibniz stated the rules of integration in a book review written in 1686. Here the elongated S symbol was introduced, $\int$. The equation of the cycloid was expressed as:

$$y = \sqrt{2x - x^2} + \int \frac{dx}{\sqrt{2x - x^2}}.$$

The inverse relationship between differentiation and integration is well demonstrated with the graphs of a function, its derivative at every point, and the integral of the derivative curve.

In the figures shown integration of dy/dx to give y is indicated by the indefinite integral:*

$$\int \frac{dy}{dx}\,dx = \int dy = y.$$

* Remember that the definite integral is a statement of taking the area under a curve between two specified values of the independent variable, x.

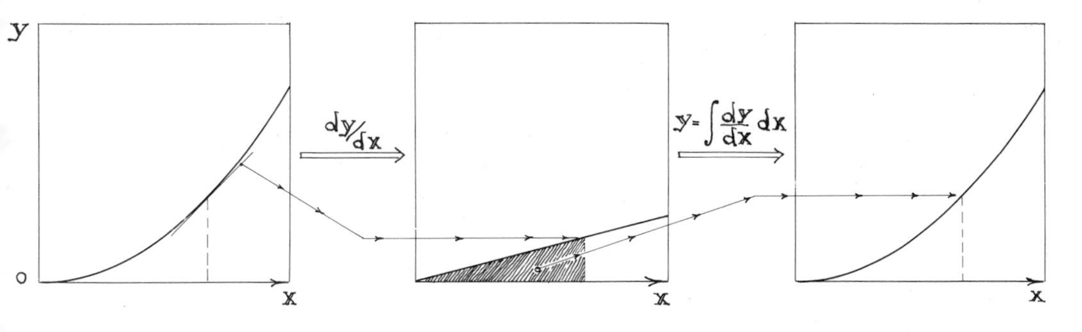

Leibniz showed the connection between integration and differentiation by a method of rectangular partitioning:

$$\frac{y(x + \Delta x) - y(x)}{\Delta x} \xrightarrow[\Delta x \to 0]{} \frac{dy}{dx},$$

and $\int y(x)\,dx$ = the sum of the rectangles under the curve $y(x)$ in the limit as $\Delta x \to 0$. This is not exactly the original notation, but it serves to illustrate the method.

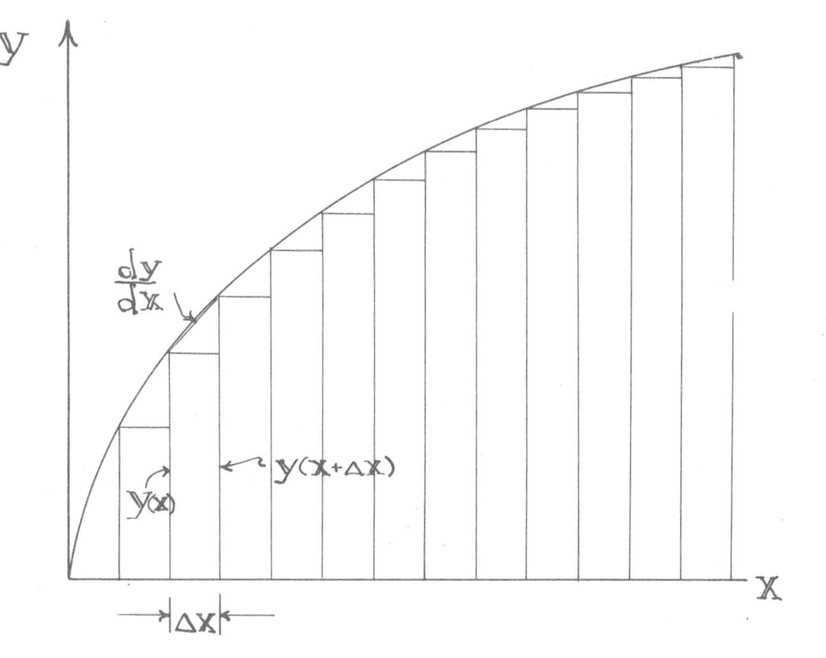

The inverse relationship between integration and differentiation may be demonstrated with the function $f(x)$, and the associated primitive function $F(x)$. Two types of integrals were involved, the definite and the indefinite. The definite integral represents the area under $f(x)$ between two fixed points on the x axis, say a and b. The indefinite integral is related but does not exhibit the end points a and b: the indefinite integral = $F(x) = \int f(x)\,dx$. If the integral exists, every indefinite integral $F(x)$ of the function $f(x)$ is the primitive of $f(x)$; *further,* f(x) *is equal to the*

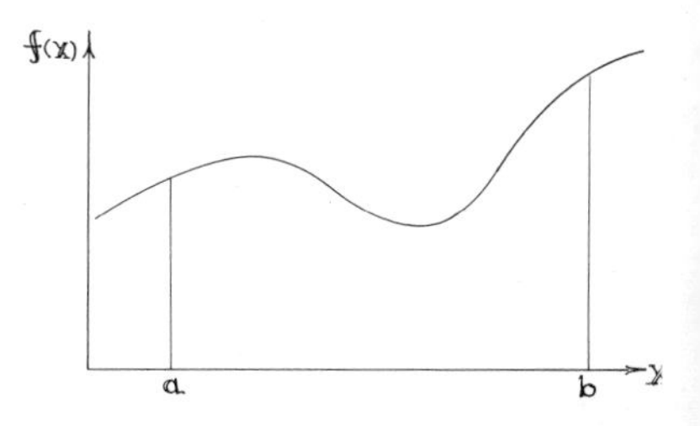

derivative of the primitive function F(x). This last statement can be demonstrated from the expression for the definite integral taken over an infinitesimal, say between x' and $x' + \Delta x'$; then

$$F(x' + \Delta x') - F(x') = \int_{x'}^{x'+\Delta x'} f(x)\,dx \xrightarrow[\text{for small } \Delta x]{} f(x')\,\Delta x'.$$

Divide by $\Delta x'$, and

$$\underset{\Delta x' \to 0}{\text{limit}} \frac{F(x' + \Delta x') - F(x')}{\Delta x'} = \frac{dF(x')}{dx'} = f(x').$$

The statement relating $f(x)$ and its primitive $F(x)$ is essentially the fundamental theorem of calculus. With this definition, the definite integral is also given in terms of the primitive function:

$$\int_a^b f(x)\,dx = \int_a^b \frac{dF(x)}{dx}\,dx = \int_a^b dF(x) = F(b) - F(a).$$

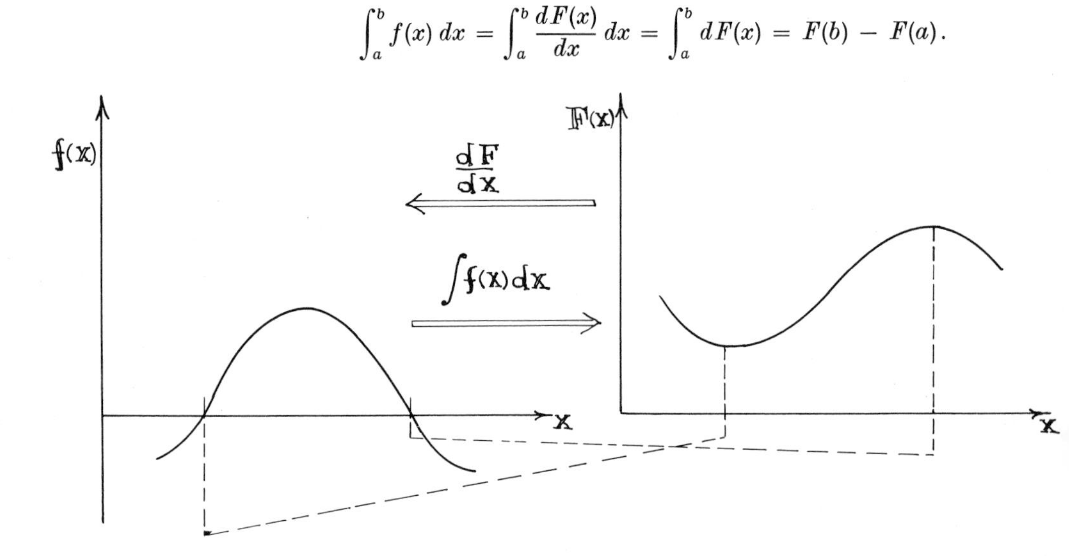

The superiority of Leibniz' method is borne out by the test of time. As an example of this notation, consider his evaluation of $\pi/4$. Newton approximated this number by extrapolation between the series provided by known integrals of polynomials. Leibniz developed a very simple integral* for $\pi/4$ by using the derivatives of the trigonometric functions and investigating the derivatives of $\tan x = \sin x/\cos x$:

$$\frac{\pi}{4} = \int_0^1 \frac{dx}{1 + x^2}.$$

* The derivatives of the trigonometric functions can be demonstrated in a number of ways. On page 97 above the derivative of $\sin x$ was derived using the law for the sine of the sum of two angles. Briefly, we showed that

$$\sin(\alpha + \Delta\alpha) = \sin\alpha\cos\Delta\alpha + \cos\alpha\sin\Delta\alpha;$$

and that when $\Delta\alpha$ was very small,

$$\cos\Delta\alpha \xrightarrow[\Delta\alpha \text{ small}]{} 1,$$

Recall the agony of extrapolation with the method of Newton. This integral of Leibniz is easy to expand as a geometric series:

$$\frac{1}{1+x^2} = 1 - (x^2) + (x^2)^2 - (x^2)^3 + (x^2)^4 + \cdots,$$
$$= 1 - x^2 + x^4 - x^6 + x^8 + \cdots.$$

Substituting and integrating term by term,

$$\frac{\pi}{4} = \int_0^1 \frac{dx}{1+x^2} = \int_0^1 (1 - x^2 + x^4 - x^6 + x^8 - + \cdots)\, dx,$$

and

$$\frac{\pi}{4} = \left(x - \frac{x^3}{3} + \frac{x^5}{5} - \frac{x^7}{7} + \cdots\right)_0^1 = 1 - \frac{1}{3} + \frac{1}{5} - \frac{1}{7} + \frac{1}{9} - \cdots.$$

At first glance this has a form similar to the series obtained by Newton, but the two are quite different. By comparison, the latter series converges more rapidly.

Newton's influence on the world of science is marked by much more than just the calculus. In the field of algebra he recognized the complex roots of a polynomial, and he stated in his *Arithmetica universalis* that

---

and

$$\sin \Delta\alpha \xrightarrow[\Delta\alpha \text{ small}]{} \Delta\alpha.$$

Then

$$\sin(\alpha + \Delta\alpha) \xrightarrow[\Delta\alpha \text{ small}]{} \sin\alpha + \Delta\alpha \cos\alpha.$$

The derivative of $\sin\alpha$ can then be developed from

$$\frac{\sin(\alpha + \Delta\alpha) - \sin\alpha}{\Delta\alpha} \xrightarrow[\Delta\alpha \text{ small}]{} \cos\alpha;$$

therefore, $d \sin\alpha/d\alpha = \cos\alpha$. In the same manner one can show that $d\cos\alpha/d\alpha = -\sin\alpha$.

The derivative of the tangent of α is just the derivative of the ratio of the sine to the cosine. Using the chain rule for ratios,

$$\frac{d\tan\alpha}{d\alpha} = \frac{d}{d\alpha}\left(\frac{\sin\alpha}{\cos\alpha}\right) = -\frac{\sin\alpha}{\cos^2\alpha}\frac{d\cos\alpha}{d\alpha} + \frac{1}{\cos\alpha}\frac{d\sin\alpha}{d\alpha},$$
$$= \frac{\sin^2\alpha + \cos^2\alpha}{\cos^2\alpha} = \sec^2\alpha = \frac{1}{\cos^2\alpha}.$$

We further employ the identity that

$$1 + \tan^2\alpha = 1 + \frac{\sin^2\alpha}{\cos^2\alpha} = \frac{\cos^2\alpha + \sin^2\alpha}{\cos^2\alpha} = \frac{1}{\cos^2\alpha} = \sec^2\alpha.$$

Therefore, in the integral of $1/(1 + x^2)$ we let $x = \tan\alpha$ and $dx = d\tan\alpha = \sec^2\alpha\, d\alpha$. Then

$$\int_0^1 \frac{dx}{1+x^2} = \int_0^{\pi/4} \frac{\sec^2\alpha}{\sec^2\alpha}\, d\alpha = \int_0^{\pi/4} d\alpha = \frac{\pi}{4}.$$

Here the limits on the integral over α were found by noticing that when $x = \tan\alpha = 0$, $\alpha = 0$; and when $x = \tan\alpha = 1$, $\alpha = \pi/4$.

when the coefficients of a polynomial are real, the complex roots occur in pairs.

Newton's research on the dispersion of light in a prism led him to the conclusion that white light was a mixture of all colors. The formation of colors from white light was known in antiquity; Seneca (2–66 C.E.)* referred to the identity of the colors of the rainbow and to those formed by the edges of a piece of glass. Descartes, Hooke, and others thought that the refractory power of a medium modified white light and so produced color. Newton proceeded by first noting that when white light from a slit S fell upon a prism at an angle of incidence, ϕ, the refracted light emerged from the opposite face of the prism in a spread of angles, δ_j, where δ_j is measured relative to the axis of the entering ray.

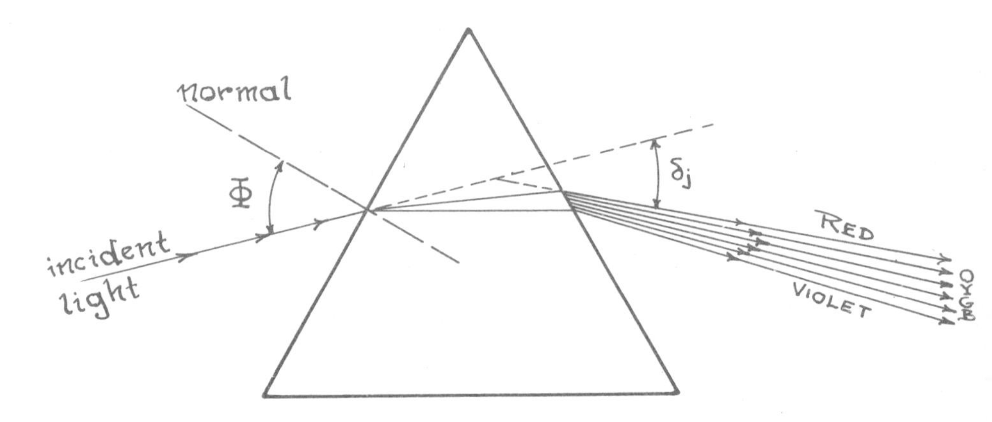

The emerging light was spread in a range of colors which varied in succession from red to violet, red having the least and violet the greatest deviation. To show that it was not the prism which modified the light, Newton separated part of the emerging beam of a specific color range by masking, and this he refracted in a second prism. The light emerged from the second prism the same color as it entered, indicating no modification.

Newton correctly concluded that white light was composed of light of different colors, that the dispersion observed in the first prism indicated

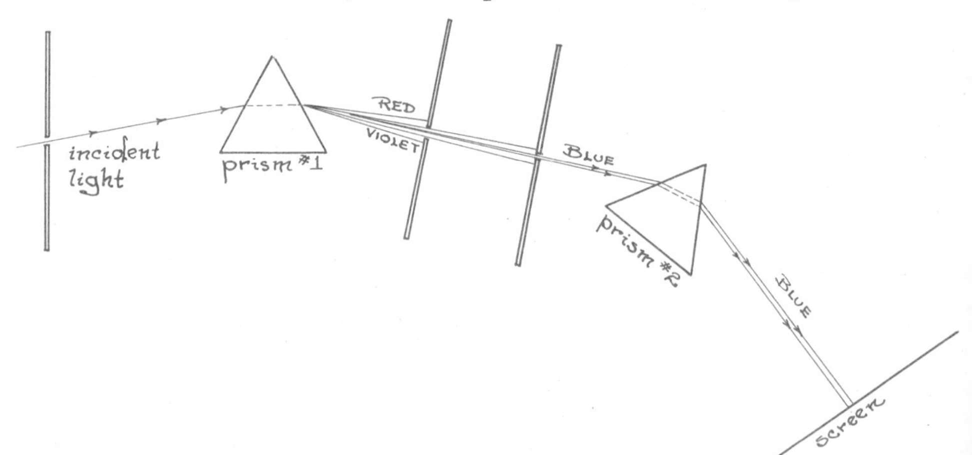

a varying refractive index for each color in glass, and that this, in turn, produced an array of refractions of the different colors: the effect is known as dispersion. In the visible region the refractive index in glass is smallest for red light and largest for violet, ranging from about 1.61 to 1.66 (in silicate flint glass).

Because the wave theory of light, supported by Huygens, Hooke, and others of the day, did not appear to explain the rectilinear path of light rays, Newton rejected this description. It is to be regretted that, because of the weight of his scientific authority, the wave theory of light was suppressed for almost a century. Ironically enough, he employed part of the wave theory in an explanation of his own. After becoming intrigued with the colors produced in thin plates, he empirically obtained the laws of interference—understanding that these effects required multiple reflection. By using the analogue of pitch from the theory of sound to explain color, and enunciating the conviction that homogeneous light was periodic, Newton in a sense incorporated part of the wave theory of light.

Dynamics was the "new frontier" of science at this time. Descartes' earlier influence upon mechanics was unfortunate: to him the fundamental test of truth lay in the clarity with which we apprehend it. Consequently, experimental evidence was not much emphasized; moreover, Descartes demanded that a successful theory of mechanics be universal. As a result, he created a model of the universe devoid of any empty space and filled with closely packed matter, of endless variety, whirling in an endless vortex motion.

Although the theory of vortices failed to account for such known phenomena as Kepler's laws, the influence of the Cartesian school of thought was sufficiently powerful to overwhelm the Newtonian version of mechanics in Europe for quite some time. A fundamental difference o philosophy existed between Newton and Descartes. Newton held that investigations in theoretical physics are concerned primarily with observed events, these events being called "phenomena." A theory would perforce contain components which could not be detected in themselves but which would be assumed in order to establish continuity between separate events which were in fact observed. These hidden phenomena were called "inter-phenomena" and were inherent in any isolated theory. This principle, first enunciated by Newton, was later expanded by Roger Cotes (1682–1716) in his defense of Newton, and is known as Cotes' principle.

Before examining the law of universal gravitation, the laws of motion must be considered. Originally there were two points of view on the fundamental quantities of dynamics, the Cartesian view that the basic quantity of motion was what is now called momentum, mv (the mass times the velocity), and the Leibnizian view that the basic quantity was proportional to the square of the velocity. The latter view was held intuitively by Galileo and was clearly enunciated by Huygens. Within the Cartesian view, $ft = mv$ was the basic equation of motion, implying that a force, f, acting for a time, t, over a distance, s, gives rise to a change in momentum of mv. Both formulations apply to motions which start from rest. Both points of view were correct, but the controversy lasted fifty years (until 1743).

One may initiate a study of mechanics by starting with Newton's second law, a force law, and then deriving the conservation of energy. Equally valid are formulations which begin with the total energy (or even the difference between kinetic and potential energy), and from these derive the force equation. Newton's three laws of motion were, in fact, laws which had been proposed by various earlier investigators and were later combined and unified in Newton's formulation. We write the three laws of motion in the more modern form as:

1. every material body continues in its state of rest or of uniform motion in a straight line unless it is compelled to change that state by the action of a force impressed upon it;
2. the time rate of change of the momentum (a vector quantity) of a body is equal to the (vector) sum of the forces exerted upon it;
3. whenever one body exerts a force on another body, the second always exerts on the first a force which is equal in magnitude, opposite in direction, and has the same line of action.

With the advent of the calculus the first law is, in a sense, a direct consequence of the second. Using vector notation to indicate that each vector equation is literally three equations, one for each coordinate, the second law can be written:

$$\vec{\mathbf{F}} = \sum_{\text{all } j} \vec{\mathbf{F}}_j = \frac{d}{dt}(m\vec{\mathbf{v}}) \xrightarrow[\substack{\text{For constant}\\ \text{mass and in}\\ \text{Cartesian}\\ \text{coordinates}}]{} \begin{cases} F_x = m\dfrac{d^2x}{dt^2}, \\ F_y = m\dfrac{d^2y}{dt^2}, \\ F_z = m\dfrac{d^2z}{dt^2}; \end{cases}$$

where F_x, F_y, and F_z are the x, y, and z components of the resultant vector force, $\vec{\mathbf{F}}$. The quantity m is the mass of the body, while $\vec{\mathbf{v}}$ is its vector velocity.

The derivative $d\vec{\mathbf{v}}/dt = d^2\vec{\mathbf{r}}/dt^2$ is the vector acceleration. When the total force exerted upon a mass m is zero, the time rate of change of the velocity is zero, and therefore the velocity is a constant. This is the first law, and it implies that the mass is constant in time.

To understand the connection between the Cartesian and Leibnizian approach to mechanics, consider the motion of a mass, m, in one dimension along the x axis, under the action of a force which can be a function of the displacement x. For instance, a stretched spring exerts a force which is proportional to the displacement. The equation of motion is

$$\frac{d}{dt}\left[m\frac{dx}{dt}\right] = m\frac{d^2x}{dt^2} = F_x(x).$$

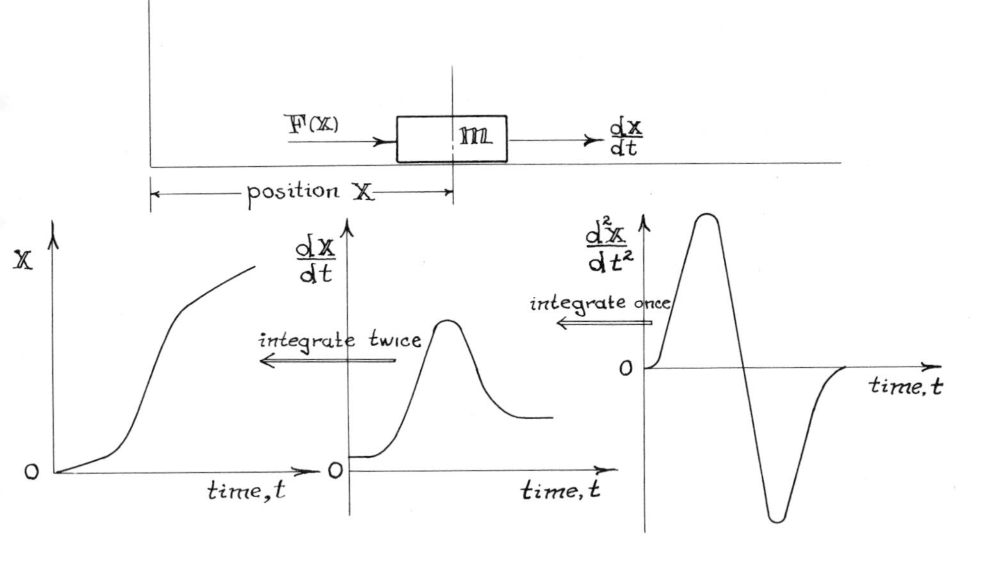

Multiply both sides by dx/dt, and this equation can be reduced to the energy form:

$$m\frac{dx}{dt}\cdot\frac{d}{dt}\left(\frac{dx}{dt}\right) = \frac{d}{dt}\left[\frac{1}{2}m\left(\frac{dx}{dt}\right)^2\right] = \frac{d}{dt}\left(\frac{1}{2}mv^2\right) = F(x)\frac{dx}{dt}.$$

Now integrate with respect to time,

$$\int \frac{d}{dt}\left(\frac{1}{2}mv^2\right) dt = \frac{1}{2}mv^2 = \int F(x)\frac{dx}{dt}\,dt + E = \int F(x)\,dx + E,$$

where E is a constant of integration.

Rearrange, bringing the integral of $F(x)$ to the lefthand side. The quantity $(1/2)m(dx/dt)^2 = (1/2)mv^2$ is called the kinetic energy. The integral of $F(x)$ over the distance x is the work done by the force F, and the constant E is the total energy of the system. Then $(1/2)mv^2 - \int F(x)\,dx = E$. This is the law of conservation of energy only if F is not a function of time, and if F can be derived from a scalar function $V(x)$, according to $F(x) = -dV(x)/dx$. When F is not an explicit function of time, the integral $\int F(x)(dx/dt)\,dt$ can be written as $\int F\,dx$. In either case, this integral is the work. Further, when $F(x) = -dV(x)/dx$ the force is defined as conservative, and the indefinite integral of $(dV/dx)\,dx$ is $V(x)$, giving $(1/2)mv^2 + V(x) = E$.

The law of conservation of energy is a statement that the kinetic energy plus the potential energy is a constant. Here one observes that the Cartesian and Leibnizian approaches to mechanics are essentially the same. We qualify this because forces can be constructed which do not obey conservation energy; on the other hand, the work done by a force is always equal to the change in kinetic energy.

Our concept of mass is ordinarily obtained from Newton's second law. This approach to a definition can lead to circular reasoning in that one defines the mass in terms of the impressed forces and then reverses the procedure and defines force in terms of the mass times the observed acceleration. It is possible to construct a consistent operational definition of mass in terms of an experimental recipe. Since we only deal with relative mass, in that the unit mass is referred to a standard mass (a unit volume

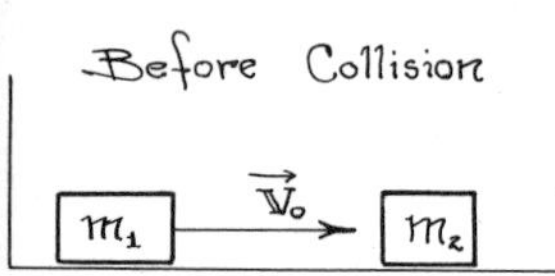

of water at standard temperature and pressure), the relative masses of two bodies can be obtained from elastic collisions in an isolated system. "Elastic" implies that the energy is conserved in the collision; the total kinetic energy before contact equals the total kinetic energy after it. Consider two blocks of mass m_1 and m_2, sliding upon a frictionless horizontal plane. Initially block 1 moves to the right with a velocity v_0; while block 2 is at rest. After collision, block 1 has a velocity v_1 and block 2 a velocity of v_2. The system is isolated (no external forces); therefore the total momentum before collision equals the total momentum after: $m_1 V_0 = m_1 v_1 + m_2 v_2$.

Because the kinematic factors can be measured, a solution appears as the ratio of m_2 to m_1: $m_2/m_1 = (v_0 - v_1)/v_2$. One must recognize that the sign of v_1 can be negative if m_1 is less than m_2; this is accounted for by the appearance of a minus times a minus to give a plus in the term $v_0 - v_1$. Therefore, by elastic scattering all masses m_2 could be determined by a standard m_1. The second law can be written in terms of various coordinate systems; in the presentation just given, it was shown in Cartesian coordinates. For two-dimensional polar coordinates, which were invented by Jacques Bernoulli, the acceleration in terms of the angle of position, ϕ, and the distance from the origin, r, takes on a form different from that of the Cartesian system.

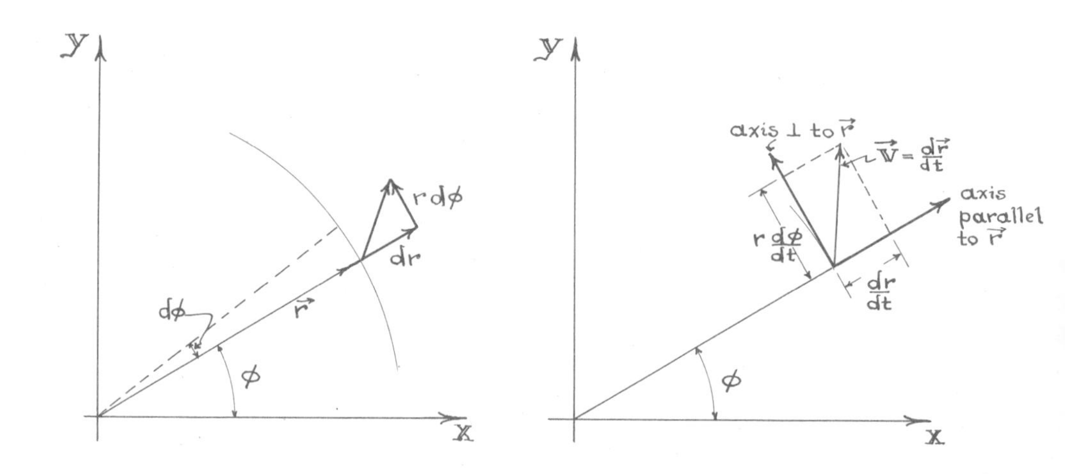

Displacements in a polar diagram are resolved along two mutually perpendicular axes, one parallel to r, the radius, and one perpendicular to r, changing positively in the direction of increasing ϕ. The two velocity components along these axes are dr/dt (parallel to r) and $r(d\phi/dt)$ (perpendicular to r); these are known as the radial velocity and the tangential velocity. When the velocity changes in time to give an acceleration, the components of the acceleration must also be resolved along these two axes. The changes in *each component* of velocity are of two types: the first is a functional change in the main coordinate, and the second is caused by the rotation or by the change in angular orientation of $\vec{v}$. Angular variations of the radial velocity term, dr/dt, appear in the direction perpendicular to the r axis, while the changes in the tangential component $r(d\phi/dt)$ as ϕ changes appear as $r[(d\phi/dt)^2]$ in a direction opposite to the direction of r. Finally, summing up the velocity changes parallel to r, we obtain the radial acceleration as

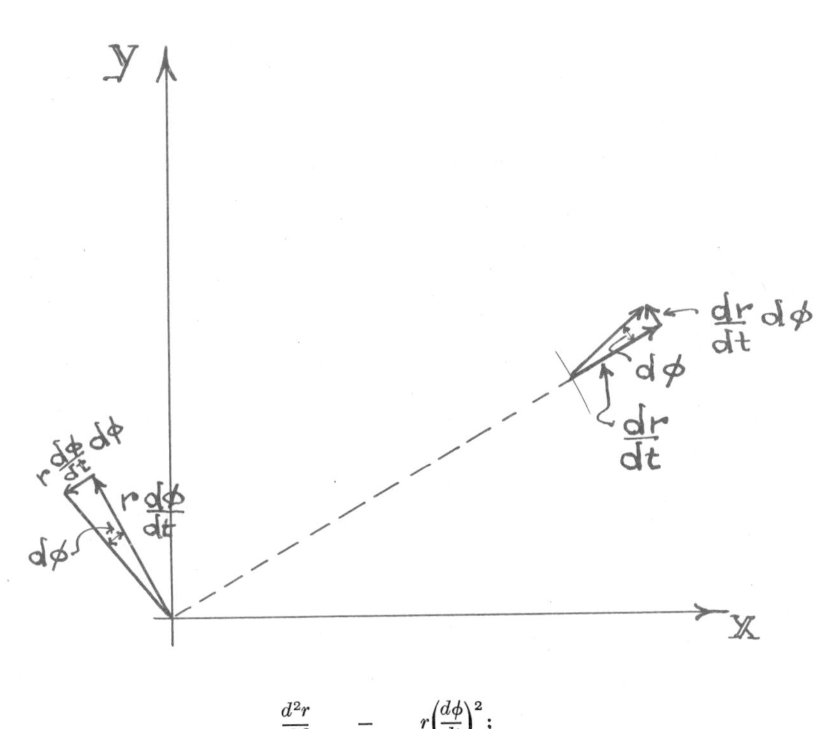

$$\underbrace{\frac{d^2r}{dt^2}}_{\text{functional}} - \underbrace{r\left(\frac{d\phi}{dt}\right)^2}_{\text{the centripetal acceleration, which is caused by directional changes}};$$

and tangential acceleration,

$$\underbrace{2\left(\frac{dr}{dt}\right)\left(\frac{d\phi}{dt}\right)}_{\text{the Coriolis acceleration}} + \underbrace{r\,\frac{d^2\phi}{dt^2}}_{\text{functional}}.$$

In the radial acceleration the second term, $-r[(d\phi/dt)^2]$, is the centripetal acceleration term first derived by Huygens, for if the motion is circular at constant radius and constant angular velocity, $d\phi/dt$, the only acceleration term is

$$-r\left(\frac{d\phi}{dt}\right)^2 = -\frac{v_{\text{tang}}^2}{r}.$$

The first term in the tangential acceleration represents the Coriolis acceleration $[2(dr/dt)(d\phi/dt)]$, showing that a simultaneous variation in r and ϕ gives a mixed term in the two velocities. When Newton's second law is written in terms of the radial and tangential components of the force, some rather simple concepts result:

$$ma_{\text{radial}} = m\left[\frac{d^2r}{dt^2} - r\left(\frac{d\phi}{dt}\right)^2\right] = F_{\text{radial}},$$

and

$$ma_{\text{tangential}} = m\left[2\left(\frac{dr}{dt}\right)\left(\frac{d\phi}{dt}\right) + r\,\frac{d^2\phi}{dt^2}\right] = F_{\text{tangential}}.$$

Kepler's second law can be obtained from the tangential equation. If a force is *central*, by definition, it is directed parallel (or antiparallel) to r and consequently F_{tang} is zero. Thus when there is no tangential component to a force, the tangential acceleration is zero. By multiplying ma_{tang} by r, we obtain a conserved quantity:

$$rma_{\text{tang}} = m\left[2r\left(\frac{dr}{dt}\right)\left(\frac{d\phi}{dt}\right) + r^2\frac{d^2\phi}{dt^2}\right] = \frac{d}{dt}\left[mr^2\left(\frac{d\phi}{dt}\right)\right].$$

Thus the time rate of change of the quantity $mr^2(d\phi/dt)$ is zero for $F_{\text{tang}} = 0$, and therefore $mr^2(d\phi/dt)$ is a constant.*

The quantity $mr^2(d\phi/dt)$ is called the angular momentum, $\mathscr{L}$. In this development it has been demonstrated that whenever the tangential component of the force (F_{tang}) vanishes, the angular momentum, $\mathscr{L}$, is conserved or is constant. Further, if $mr^2(d\phi/dt) = \mathscr{L} =$ a constant, then $d\phi/dt = \mathscr{L}/mr^2$, and the radial acceleration term can be expressed only in terms of r and its time derivatives. Substituting for $d\phi/dt$ in ma_{radial} we obtain

$$m\left(\frac{d^2r}{dt^2} - \frac{\mathscr{L}^2}{m^2r^3}\right) = F_{\text{radial}}.$$

By using the conservation of angular momentum the two equations are reduced to one, leaving a differential equation which has a solution which gives r as a function of the independent variable t. Various systematic methods exist for solving this problem, given F_r. If F_r, the radial component of the force, depends upon r only, this equation can be reduced by multiplying by the radial velocity component dr/dt (we now introduce a more compact notation in which a superior dot indicates a time derivative: $dr/dt = \dot{r}$):

$$m\dot{r}\left(\frac{d\dot{r}}{dt} - \frac{\mathscr{L}^2}{m^2r^3}\right) = \frac{d}{dt}\left(\frac{1}{2}m\dot{r}^2 + \frac{\mathscr{L}^2}{2mr^2}\right) = F_r\frac{dr}{dt}.$$

Integrating this expression with respect to t, one obtains

$$\frac{1}{2}m\dot{r}^2 + \frac{\mathscr{L}^2}{2mr^2} = \int F_r(r)\,dr + \text{a constant}.$$

The first term on the right is the work done by the radial force acting through a distance dr. The term $(1/2)m(dr/dt)^2$ is the radial kinetic energy, while $\mathscr{L}^2/2mr^2$ is called the centripetal barrier energy and represents the kinetic energy of the rotary motion. All of this is mentioned to form some sort of background for the motion under gravitational attraction.

As shown in the preceding development of the motion about a center, the angular momentum $\mathscr{L}$ for a mass point is defined as the mass, m, times the square of the distance, r, to the center of the motion times the angular velocity, $\omega = d\phi/dt$: $\mathscr{L} = mr^2(d\phi/dt) = mr^2\omega$. This is the analogue in angular coordinates to the linear momentum. The quantity $\mathscr{L}$ is constant whenever the forces present are "central"; i.e., the central force acts only parallel, not perpendicular, to r. With this condition satisfied, Kepler's second law can be satisfied whenever $\mathscr{L}$ is conserved. In the diagram we observe that for a small variation of the angle (i.e., $\Delta\phi$) in a time interval, Δt, the area ΔA swept out is $(1/2)r^2\,\Delta\phi$; therefore, if $\mathscr{L}$ is constant, the time rate of change of the area is constant: $\mathscr{L}/2m = dA/dt =$ a constant.

Newton postulated the inverse square law of gravitation in 1666. Assuming that the gravitational attraction of the earth decreased as the

* In this manipulation we have used the "chain rule" for differentiation, i.e.,

$$\frac{d}{dt}(AB) = B\frac{dA}{dt} + A\frac{dB}{dt}.$$

Thus,

$$\frac{d}{dt}\left(r^2\frac{d\phi}{dt}\right) = 2r\left(\frac{dr}{dt}\right)\left(\frac{d\phi}{dt}\right) + r^2\frac{d^2\phi}{dt^2}.$$

square of the distance from the earth's center, he viewed the circular motion of the moon as a fall toward the earth's center. His analysis was presented in the classical geometric form.

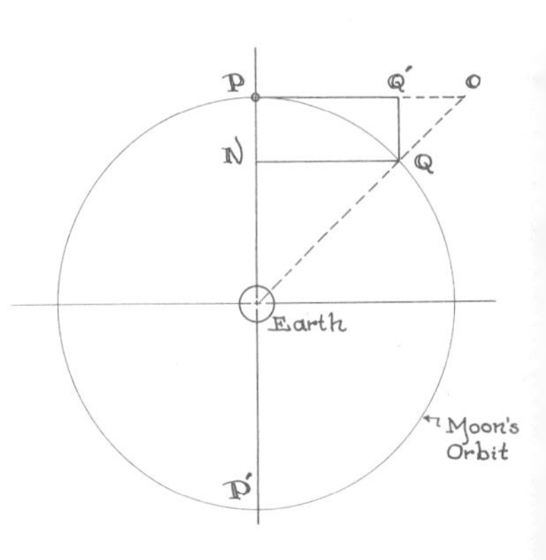

The theory appeared at first invalid because in the calculation he had incorporated the value of the length of a degree of arc on the earth's surface as 60 miles. In 1682 Jean Picard made a more accurate determination of this value, 69.1 miles. The distance of the moon was believed to be $60R$, where R is the radius of the earth, with a mean period of 27 days, 7 hours, 43 minutes. With these values, the circumference of the moon's orbit is 69.1 mi/degree $\times$ 360° $\times$ 60 $\times$ 5280 ft/mi.

Newton assumed that the length of the chord of a small arc was approximately the length of the arc. Using the geometry of a right triangle inscribed in a semicircle: $(PN)(PP') = (PQ)^2 \simeq (\text{arc } PQ)^2$, giving $PN \simeq [(\text{arc } PQ)^2/PP']$. Thus he concluded that the arc PQ traversed in one minute was 200,308 feet. Substituting, the fall of the moon PN in 1 minute is

$$\frac{(69.1 \times 60 \times 360 \times 5{,}280)^2\pi}{(69.1 \times 60 \times 360 \times 5{,}280)(39{,}343 \text{ min})^2} = 16 \text{ ft}.$$

This is the geometric evidence; the inverse square law implies that the moon's fall in one minute referred to the acceleration of gravity at the surface of the earth is

$$\frac{(1/2)32.2 \text{ ft/sec}^2 \times (60 \text{ sec/min})^2}{(60)^2} = 16.1 \text{ ft},$$

demonstrating the first vindication of the inverse square law.

The first volume of Newton's *Principia*, entitled *De motu corporum*, appeared in 1685. Here was presented the gravitational law of force between two bodies, $F = -GM_1GM_2/r^2$, where M_j is the mass of the jth body and r is the distance between the centers. These bodies are considered as points or spheres. In volume 1 Proposition LXXI proved that the mutual attraction between two spheres acts as though the mass of each sphere was concentrated at its respective center. Volume 2, *De motu corporum secundus*, investigated the motion of bodies in a resisting medium. The Cartesian system of vortices was examined and shown to be inconsistent with all observations. *De systemate mundi*, the third volume, provided an account of the motion of the moon and the theory of tides and comets. Although the calculus was employed in the solution of most of the theories presented, Newton, for the sake of familiarity, translated each problem into a classical geometric interpretation, thus making the work more incomprehensible than it need have been. Further, by creating the impression that most problems were solvable by geometric operations, Newton failed to illustrate cohesively the power of the calculus. Moreover, his notation for the calculus was quite unwieldly, and British scholars who clung to a Newtonian formalism through loyalty were severely hampered, in contrast with the Europeans, who accepted the Leibniz formulation.

In the simpler notation of our earlier example of motion in polar coordinates, the third law of Kepler can be easily derived by the simplifying assumption that the orbits of the planets are essentially circular. Denoting the period of one revolution of a given planet as T, then in terms of the angular velocity, $\omega = d\phi/dt$, the period is $T = 2\pi/\omega$ (i.e., ω is in radians per second; thus 2π radians divided by ω radians per second gives seconds). Assume that the radius of the circular orbit is R. For a circular orbit, according to Huygens and later Newton, the centripetal acceleration inward is $-R\omega^2$. From Newton's second law the gravitational force (directed along the radius) is $-(GM_S m/R^2)$. Equating force and mass times acceleration (parallel to R—one must remember that for a circle the magnitude of R does not change; therefore dR/dt is zero):

$$-\mu R\omega^2 = -\mu R\left(\frac{2\pi}{T}\right)^2 = -\frac{GM_S m}{R^2},$$

where μ is the reduced mass $Mm/(M + m)$. When M is much greater than m, μ approaches the value m. Solving for R^3/T^2, we find $G(M_S + m)/(2\pi)^2 =$ a constant. Here G is the gravitational constant, M_S is the mass of the sun, m is the mass of a planet, and R is the length of the semi-major axis (for a circle, the radius). G, the gravitational constant, must be measured independently, and the present value is $6.673(10)^{-8}$ cm^3/gm-sec^2.

To obtain the mass of the earth, the inverse square law of gravity can be employed with the known mean radius of the earth R_E. Because the force of gravity on a mass m at the surface is $mg \simeq GM_E m/R_E^2$, then M_E, the earth's mass, equals gR_E^2/G.* The distance between the sun and the earth can be measured by triangulation. Knowing the radius R_E and the angular difference between the two sightings, one can measure R, the distance from the sun to the earth. With a knowledge of the mass of the earth, one can then determine the mass of the sun from the period of the earth about the sun, T_E and R_E:

$$M_S = \frac{(2\pi)^2 R_E^3}{GT_E^2} - M_E.$$

Other planets may be accounted for once M_S has been calculated.

For almost a hundred years following the death of Newton mathematics made little progress in England, particularly when compared with the brilliant advances made in Europe under the leadership of the Bernoulli family. Among the few English contributions, however, was the introduction of a type of complex variable by Roger Cotes (1682–1716) and Abraham de Moivre (1667–1754). Cotes's formula relates an angle to the natural logarithm of the trigonometric functions:

$$\sqrt{-1}\,\Theta = \log_e(\cos\Theta + \sqrt{-1}\sin\Theta).$$

De Moivre's theorem is equivalent but is stated in the form:

$$\cos n\Theta + \sqrt{-1}\sin n\Theta = (\cos\Theta + \sqrt{-1}\sin\Theta)^n.$$

* The acceleration of gravity had been measured accurately by Huygens and others, using a pendulum.

Both relationships become natural when the exponential function is developed in terms of the trigonometric functions. De Moivre also extended the understanding of the probabilities of points, a problem originally posed by Fermat and Pascal.

The exponential function was the child of Jean Bernoulli, and De Moivre's theorem was demonstrated by Euler. The only two names of much significance in the next hundred years of English mathematics are those of Colin Maclaurin (1698–1746) and Brook Taylor (1685–1731). Maclaurin followed Newton's use of the classical geometry and thus avoided the calculus; however, he did produce a series which was later generalized to the famous power series of Taylor (called the Taylor series). This series allows a function $f(x)$ evaluated at x_0 to be extended to $f(x_0 + h)$ by a power series in h:

$$f(x_0 + h) = f(x_0) + \frac{1}{1!}\left[\frac{df}{dx}\right]_{x=x_0} h + \frac{1}{2!}\left[\frac{d^2f}{dx^2}\right]_{x=x_0} h^2 + \cdots$$

$$+ \frac{1}{n!}\left[\frac{d^nf}{dx^n}\right]_{x=x_0} h^n + \cdots,$$

where $h = x - x_0$. This expansion is less general than it appears at first glance. Because the series is infinite, only functions which have well-defined derivatives to all orders can be represented in this fashion. To understand the limitations of this representation, one need only consider functions which have a cusp at x_0; in such cases the first derivative will exhibit a jump discontinuity at x_0, and the second derivative will be undefined at this point. The difficulties involved in those functions which do not have a Taylor's series representation led to modifications and a reevaluation of the definitions of an integral in the nineteenth century.

To illustrate the use of the Taylor expansion, consider the function $f(x) = \sqrt{1 - x^2}$, which Newton expanded by a technique of interpolation. By successive differentiations this function can be readily expanded in an infinite series about the point $x_0 = 0$. Let us note that

$$\frac{d[g(x)]^{1/2}}{dx} = \frac{1}{2}\frac{1}{(g)^{1/2}}\frac{dg}{dx};$$

therefore

$$\frac{d\sqrt{1 - x^2}}{dx} = \frac{1}{2}\frac{1}{\sqrt{1 - x^2}}(-2x) = \frac{-x}{\sqrt{1 - x^2}}.$$

Evaluating this derivative at $x = 0$ gives zero. The second derivative takes the form $-(1 - 2x^2)/(1 - x^2)^{3/2}$, which, when evaluated at $x = 0$, gives -1. Carrying out successive differentiations, one obtains the series expansion:

$$\sqrt{1 - x^2} = 1 + 0 - (1/2)x^2 - (1/8)x^4 - \cdots$$

for $x \leqslant 1$.

The calculus of Leibniz spread very rapidly throughout Europe. Its superior notation and ease of handling enabled mathematicians to solve many of the problems of application within a brief fifty years. The Bernoullis made astonishing advances in this area. The family came to the free city of Basel as refugees from Antwerp. Jacques (1654–1705) was born in Basel and from 1687 occupied the chair of mathematics there. He and Leibniz carried on a vigorous and profitable correspondence, which greatly influenced the direction of Bernoulli's mathematical interests. His contributions were wide in scope, but his major focus was on analysis and probability. Like his predecessors, Jacques Bernoulli investigated the probabilities involved in games of chance, but his work was the first attempt to give a sound mathematical basis to the subject. In *Ars conjectandi* he relates the probability of occurrences to the binomial coefficients, an old problem which was considered by Omar Khayyám.

In the theory of permutations, the number of arrangements of n objects is $n!$ For instance, the first position can be occupied by any one of n objects. Once this is established, the next position can only be occupied by $n - 1$, and so on to the last, which has only one object remaining. Thus the total number of permutations is the product of these terms n factorial (written $n!$), where $n! = n(n - 1)(n - 2) \cdots 3 \cdot 2 \cdot 1$. The numbers of ordered subsets of n objects containing m objects ($m \leqslant n$; i.e., m less than or equal to n) is

$$n(n - 1)(n - 2) \cdots (n - m + 2)(n - m + 1) = \frac{n!}{(n - m)!}.$$

If we ask for the number of subsets containing m objects out of n, without regard for the order in which they appear, this represents the number of combinations of n objects taken m at a time. To obtain this result we divide the number of ordered subsets by $m!$, since each combination can be arranged internally in $m!$ different ways which must not be counted. Thus the number of n things taken m at a time without regard for the order is the binomial coefficient mentioned earlier: $\binom{n}{m} = \binom{n}{n - m} = n!/m!(n - m)!$. Shown is the symbol $\binom{n}{m}$, which is the accepted representation for this coefficient. Because of the symmetry of $m!$ and $(n - m)!$ in the denominator, either symbol, $\binom{n}{m}$ or $\binom{n}{n - m}$, is acceptable. This calculation may also be viewed in terms of the product of the number of distinguishable and indistinguishable arrangements, which is equal to the total number of arrangements of n objects (i.e., $n!$): (number of distinguishable) $\times$ (number of indistinguishables) = total number of arrangements.

Each subset of m things can be arranged in $m!$ indistinguishable ways, and the subset of $(n - m)$ objects can be arranged in $(n - m)!$ indistinguishable ways. As an example of the indistinguishable permutation, consider one of the arrangements of the second term in the product $(a + b)^4$: i.e., *aaab* or a^3b. In this configuration, any permutation of a pair of as leaves the arrangement in the same form and therefore is considered an indistinguishable permutation. The binomial coefficient, then, is the number of distinguishable arrangements of n things when the number of indistinguishable arrangements is $m!$ times $(n - m)!$.

A typical example of this calculation is the problem of finding the number of different arrangements produced by distributing b balls in a boxes. The total number of arrangements is $(a - 1 + b)!$, while the number of indistinguishable permutations is $b!(a - 1)!$. The number of distinguishable arrangements is $(a - 1 + b)!/b!(a - 1)!$. Examination of this result reveals that this is the binomial coefficient

$$\binom{a - 1 + b}{b} = \binom{a - 1 + b}{a - 1}.$$

When two balls are distributed in three boxes, the number of different arrangements is $4!/2!(3 - 1)! = 6$.

The fundamental laws of probability can be obtained by counting. Call the probability of event A, $P(A)$ and the probabilityof event B, $P(B)$, and let $n_1 =$ the number of outcomes in which A occurs and not B, $n_2 =$ the number of outcomes in which B occurs and not A, $n_3 =$ the number of outcomes in which both A and B occur, $n_4 =$ the number of outcomes in which neither A nor B occurs, and $n = n_1 + n_2 + n_3 + n_4 =$ total number of outcomes. Then, by counting, $P(A) = (n_1 + n_3)/n$, and $P(B) = (n_2 + n_3)/n$. The probability of A or B or $(A + B)$ is $P(A + B) = (n_1 + n_2 + n_3)/n$. The probability of both A and B is $P(AB) = n_3/n$.

Suppose we ask more complex questions: for instance, the conditional probability of A is the probability of A, given that B must occur. Call this $P(A \mid B)$. Then $P(A \mid B) = n_3/(n_2 + n_3)$. Similarly, the probability of B, given that A must occur, is $P(B \mid A) = n_3/(n_1 + n_3)$. From these we can obtain some convenient algebraic connections: $P(A + B) = P(A) + P(B) - P(AB)$, and $P(AB) = P(A)P(B \mid A) = P(B)P(A \mid B)$.

As an example of $P(A + B)$, consider the probability that when one card is drawn from each of two decks, at least one is an ace. This is the probability of A or B, and we use the algebraic expansion of $P(A + B)$. Here $P(A) = 4/52$ and $P(B) = 4/52$, giving $P(A + B) = 4/52 + 4/52 - (4/52)^2 = 25/169$. For another case, we consider the probability of drawing two hearts when two cards are drawn from the same deck:

$$P(\text{2 hearts}) = P(\text{1 heart}) \times P(\text{1 heart} \mid \text{1 heart}) = \frac{13}{52} \times \frac{12}{51} = \frac{1}{17}.$$

This may serve as a warning to those who would draw two cards to a flush.

Jacques Bernoulli brought the theory of the calculus of variations to full maturity. In this area the problems date back to antiquity: the legend of Carthage, the isoperimetric problem of Archimedes, etc., are but a few of them. The questions are directed toward calculations of the extreme (maximum or minimum) value of an integral or function (we recall that Fermat's "Least Time Principle" was just this). Typical of the solutions of Jacques Bernoulli was that of finding the curve connecting two points, A and B, in space along which a particle starting from rest and sliding under the influence of gravity would accomplish the transition in the least time. The solution is called the "brachistochrone" and is a cycloid.* The calculus of variations served as a field of study for many later mathematicians.

Jacques also introduced polar coordinates, which have been discussed earlier in connection with the conservation of angular momentum, and extended the study of infinite series. He was one of the first to focus on questions of divergence. As an example of clever manipulation, he summed the finite sequence

$$\frac{1}{1\cdot 2} + \frac{1}{2\cdot 3} + \frac{1}{3\cdot 4} + \cdots + \frac{1}{n(n+1)}$$

by considering the harmonic series

$$S_n = 1 + \frac{1}{2} + \frac{1}{3} + \frac{1}{4} + \cdots + \frac{1}{n+1}$$

and subtracting the same sum displaced one position to the right:

$$S = 1 + \frac{1}{2} + \frac{1}{3} + \frac{1}{4} + \cdots + \frac{1}{n+1}$$

$$-S = \quad -1 - \frac{1}{2} - \frac{1}{3} - \cdots - \frac{1}{n} - \frac{1}{n+1}.$$

Then,

$$0 = 1 - \frac{1}{1\cdot 2} - \frac{1}{2\cdot 3} - \frac{1}{3\cdot 4} - \cdots - \frac{1}{n(n+1)} - \frac{1}{n+1},$$

or

$$\left(1 - \frac{1}{n+1}\right) = \frac{n}{n+1} = \frac{1}{1\cdot 2} + \frac{1}{2\cdot 3} + \frac{1}{3\cdot 4} + \cdots + \frac{1}{n(n+1)}.$$

More important than this, he showed that the infinite series

$$\frac{1}{\sqrt{1}} + \frac{1}{\sqrt{2}} + \frac{1}{\sqrt{3}} + \frac{1}{\sqrt{4}} + \cdots + \frac{1}{\sqrt{n}} + \cdots$$

diverged because every term was greater than the corresponding term in

$$\frac{1}{1} + \frac{1}{2} + \frac{1}{3} + \frac{1}{4} + \cdots + \frac{1}{n} + \cdots,$$

* The cycloid is the curve traced out by a point on a circle which is rolling along a straight line.

which had been shown to be infinite by his brother Jean.* The famous Reimann zeta function $\zeta(2)$ for an argument 2,

$$\zeta(2) = 1 + \frac{1}{2^2} + \frac{1}{3^2} + \frac{1}{4^2} + \cdots + \frac{1}{n^2} + \cdots,$$

baffled Jacques. This series was summed by Euler and later generalized for any integral argument by Riemann.

Jean Bernoulli (1667–1748), the younger brother of Jacques, became professor and held the same chair of mathematics after Jacques' death. He too was a vigorous champion of Leibniz. The calculus of two and three dimensions was fully developed by him, and his efforts are characterized by interesting manipulations or tricks for solving very complex problems. With his brother he shares credit for the calculus of variations, and to him we owe the discovery of the exponential function, which can be generated in many different ways and which was brought to full development by Euler. In order to achieve compactness here, we shall consider two approaches to the exponential function, first the original approach, and second the function as a series solution to a differential equation. It should be remarked that the importance of the exponential function and the associated trigonometric functions lies in the fact that, as solutions to a particularly important differential equation, they are the most fundamental functions of mathematical physics. The necessity of the exponential function becomes apparent when we consider the natural logarithm.

The natural logarithm of a quantity y can be labeled x where $x = \log_e y$. Although we know that this implies $y = e^x$, let us assume that the inversion gives an unknown function $y = E(x)$. The differential equation for $E(x)$ may be readily obtained from the primitive function of $1/y$ (if $\log_e y = \int dy/y = x$, then $dy/dx = y$). This provides two basic equations: $y = E(x)$, and $dE(x)/dx = E(x)$, where in modern notation $E(x) = e^x$. The natural exponential function is equal to its derivative. One can observe from the properties of the logarithm that $E(x)$ is a number "e" to the power x; for example, $E(a) \cdot E(b) = E(a + b)$; from $\log E(a) \cdot E(b) = \log E(a) + \log E(b)$, etc.

The earliest forms of $E(x)$ were obtained by considering

$$\frac{1}{y} = \frac{dx}{dy} = \underset{\Delta y \to 0}{\text{limit}} \frac{\log (y + \Delta y) - \log y}{\Delta y}.$$

By taking $\Delta y = 1/n$ and allowing n to become arbitrarily large, we find

$$\frac{1}{y} = \underset{n \to \infty}{\text{limit}} \frac{\log (y + 1/n) - \log y}{1/n} = \underset{n \to \infty}{\text{limit}} \log \left(1 + \frac{1}{ny}\right)^n.$$

Now allow $z = 1/y$ to give

$$z = \underset{n \to \infty}{\text{limit}} \log_e \left(1 + \frac{z}{n}\right)^n.$$

* The sum of $1/n$, where n is an integer, represents the sum of a set of rectangles of unit width whose heights vary as $1/n$. This sum of rectangles to $n = N$ is greater than the integral of $1/x$ between 1 and $N + 1$. From the integral one can then show that the sum of $1/n$ is greater than $\log (N + 1)$, which is undefined as N approaches infinity.

From the definition of the exponential, then, $e^z = \text{limit}_{n\to\infty}(1 + z/n)^n$. The term $(1 + z/n)^n$ can be expanded in terms of the binomial coefficients:

$$\left(1 + \frac{z}{n}\right)^n = 1 + \frac{1}{1!}z + \frac{1}{2!}\left(1 - \frac{1}{n}\right)z^2 + \frac{1}{3!}\left(1 - \frac{1}{n}\right)\left(1 - \frac{2}{n}\right)z^3 + \cdots;$$

and as n becomes arbitrarily large,

$$e^z = 1 + \frac{1}{1!}z + \frac{1}{2!}z^2 + \frac{1}{3!}z^3 + \frac{1}{4!}z^4 + \cdots.$$

This same result may be obtained in another fashion by assuming a power series expansion of e^x, where the coefficients are to be adjusted to satisfy the differential equation. Assume that

$$E(x) = e^x = a_0 + a_1x + a_2x^2 + a_3x^3 + \cdots = \sum_{m=0}^{\infty} a_m x^m.$$

The values of the coefficients a_m are determined by the differential equation developed in the previous discussion: $dE(x)/dx = E(x)$. Differentiating the series above term by term (recall that $dx^n/dx = nx^{n-1}$), one obtains:

$$\frac{dE(x)}{dx} = a_1 + 2a_2x + 3a_3x^2 + 4a_4x^3 + \cdots = \sum_{m=0}^{\infty} (m + 1)a_{m+1}x^m.$$

Equating the two series for $E(x)$ and for dE/dx,

$$\sum_{m=0}^{\infty} (m + 1)a_{m+1}x^m = a_1 + 2a_2x + 3a_3x^2 + \cdots$$

$$= a_0 + a_1x + a_2x^2 + \cdots = \sum_{m=0}^{\infty} a_m x^m.$$

Because the value of x is arbitrary, these two series can only be equal if the coefficients of like powers of x are equal:

$$a_1 = a_0,$$
$$2a_2 = a_1,$$
$$3a_3 = a_2,$$
$$4a_4 = a_3,$$

or in general,

$$a_{m+1} = \frac{a_m}{(m + 1)}.$$

Starting with $a_0 = 1$, the value of the a_n is obtained by successive substitution:

$$a_0 = 1,$$

$$a_1 = \frac{a_0}{1} = \frac{1}{1},$$

$$a_2 = \frac{a_1}{2} = \frac{1}{1\cdot 2},$$

$$a_3 = \frac{a_2}{3} = \frac{1}{1\cdot 2\cdot 3} = \frac{1}{3!},$$

and

$$a_m = \frac{1}{m!}.$$

Substituting these values for the a_m in the assumed series,

$$E(x) = e^x = 1 + \frac{1}{1!}x + \frac{1}{2!}x^2 + \frac{1}{3!}x^3 + \frac{1}{4!}x^4 + \cdots + \frac{1}{m!}x^m + \cdots.$$

This result introduces for the first time the powerful method employed in solving differential equations by assuming a power series expansion in the independent variable.

The Bernoullis extended the calculus to include two and three independent variables or dimensions. Such an extension requires that the definition of differentiation be modified to provide a consistent methodology in the presence of the additional degrees of freedom. To appreciate the problem, consider a function $f(x, y)$ of two independent variables. The surface $f(x, y)$ is shown in the accompanying figure.

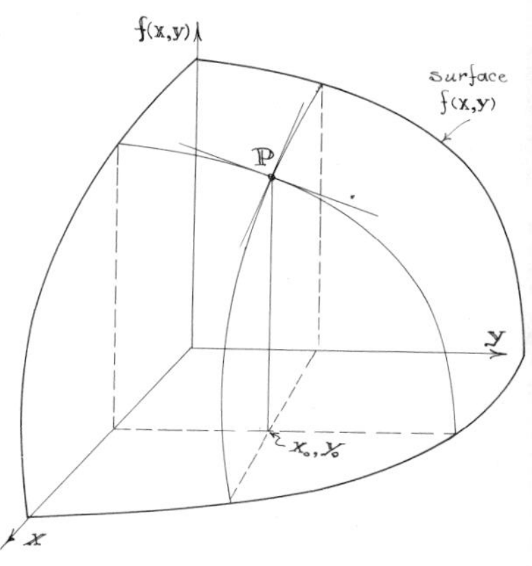

For any pair of coordinates (x_0, y_0) a point P is specified on the surface at $f(x_0, y_0)$. It is quite clear that any number of lines can be passed through P tangent to the surface, $f(x, y)$; thus the slope at P must be given a more precise meaning. If planes are passed through P parallel to the (f, x) plane and parallel to the (f, y) plane, two curves of intersection between the planes and the surface are formed. Partial derivatives at P are the respective derivatives of the two curves of intersection at P. Thus the partial derivative (written $\partial f/\partial x$) of $f(x, y)$ with respect to x is obtained by holding y constant in the operation:

$$\frac{\partial f(x, y)}{\partial x} = \lim_{\substack{\Delta x \to 0 \\ y \text{ held constant}}} \frac{f(x + \Delta x, y) - f(x, y)}{\Delta x}.$$

In the same manner, the partial derivative of f with respect to y (i.e., $\partial f/\partial y$) is written,

$$\frac{\partial f(x, y)}{\partial y} = \lim_{\substack{\Delta y \to 0 \\ x \text{ held constant}}} \frac{f(x, y + \Delta y) - f(x, y)}{\Delta y}.$$

Higher derivatives are handled in the same manner. To give an example of this operation, regard $f(x, y) = ax^2 + by^2 + cxy$; then $\partial f/\partial x = 2ax + cy$, and $\partial f/\partial y = 2by + cx$.

Both Jacques and Jean Bernoulli in their studies of probabilities arrived at the law of large numbers. This law focuses upon the difference between actual occurrences and ideal probability. For instance, in coin-tossing problems it is often claimed that the probability of tossing "heads" (with a perfectly balanced coin) is 1/2. This does not mean that in n tosses heads will occur exactly with the frequency $n/2$; rather, $h(n)$ heads will occur in n tosses. One cannot even show that $h(n)/n$ approaches 1/2 when n becomes arbitrarily large. Therefore, a new tack is taken, and the probability that $|h(n)/n - 1/2|$ is greater than some number ϵ may be calculated. The law of large numbers states that the probability that $|h(n)/n - 1/2|$ is greater than some ϵ, for n approaching infinity, is equal to zero.* This result led the Bernoullis to propose the law of determinism, which became the basis of the philosophy of classical physics: "If all events from now to eternity were constantly observed, it would be found that everything occurs for a definite reason."

* In these expressions the vertical lines bracketing the quantity $h(n)/n - 1/2$, i.e., $|h(n)/n - 1/2|$, mean that only the *magnitude* of the difference is to be considered.

In a famous controversy with Colin Maclaurin, Jean Bernoulli introduced the concept of the modern delta function, a mathematical device which becomes arbitrarily large in an infinitesimally small interval of the independent variable in such a manner that the total area under the curve remains constant and finite. Although Bernoulli lost the official contest in his own lifetime, his analysis was the correct one, while the winner, Maclaurin, has been judged wrong today. The question posed by the French Academy concerned the collision of two elastic spheres in the limit in which the spheres become infinitely hard or rigid. Maclaurin concluded that the collision became completely inelastic in the limit of absolute hardness, reasoning that the forces acting upon contact became arbitrarily large unless a discontinuity occurred in the elastic constant. Bernoulli, on the other hand, realized that, although the force tended to infinity as the hardness became infinite, the time of contact and the deformation of the spheres during contact approached zero. Therefore, he postulated that the total impulse, force integrated over time, remained the same, independent of the condition of hardness. Because the momentum change in the scattering is equal to the impulse (by Newton's second law), the net scattering remained the same and was unaffected by the hardness. The concept of a function which becomes undefined in an arbitrarily short interval such that (the area under the curve) the integral remains finite is today an accepted and highly useful mathematical object. It is known as a generalized distribution, or a delta function.

Classical mechanics was still beset by differences between the Cartesian and Leibnizian schools. The adherents of Leibniz had analyzed the conservation of energy for bodies moving in the earth's gravitational field of force in terms of the relative behavior of objects starting from different initial positions. For instance, it was recognized that a body falling vertically from rest through a distance h_1 would achieve a velocity downward of v_1. The velocity v_2, corresponding to a fall from a height h_2, was then obtained from the ratio $v_2^2/v_1^2 = h_2/h_1$, which suggests that the square of the velocity is proportional to the distance through which a body falls. Jean Bernoulli arrived at the equivalent result without the necessity of the ratio, by showing that $v^2 = 2gh$, where g is the acceleration of gravity. The symbol g was assigned to this quantity by him. He further proposed the existence of conservative forces which gave rise to the scalar potential term in the law of conservation of energy.

Jean Bernoulli had two sons, Nicolaus (1695–1726) and Daniel (1700–1782), both of whom were mathematicians. Devoted mainly to astronomy, physics, and hydrodynamics, Daniel became the more famous. In his *Hydrodynamica* (1738) appeared the conservation relation between the pressure and velocity of an uncompressible liquid and also the first hint of a mathematical formulation of a kinetic theory of gases. His interest in partial differential equations led to a number of advances in mechanics, exemplified by such fundamental questions as the motion of the vibrating string, the oscillating chain, and the theories of hydrodynamics and of gases. The vibrating string is an interesting application of Newton's second law to a segment of a string to provide a general differential equation determining the motion. Functions of more than one variable must be considered in such a problem: this is accomplished in much the same fashion as it is with one variable, except that we must consider independent variations of the variables. Consider a function $y(x, t)$ where x and t are independent variables. The derivatives with respect to either x or t are *partial derivatives* because of the presence of the second variable. Partials are written with a sign ∂ in place of the standard d. Functions of the type $y(x, t)$ are surfaces in the three-dimensional space formed by x, t, and y.

As shown, the partial derivative of $y(x, t)$ with respect to x is

$$\frac{\partial y(x, t)}{\partial x} = \underset{\substack{\Delta x \to 0 \\ t \text{ held constant}}}{\text{limit}} \frac{y(x + \Delta x, t) - y(x, t)}{\Delta x}.$$

Here we notice again that when x is varied by Δx, t is held constant. Analogously, the partial derivative of $y(x, t)$ with respect to t is performed holding x constant:

$$\frac{\partial y(x, t)}{\partial t} = \underset{\substack{\Delta t \to 0 \\ x \text{ held constant}}}{\text{limit}} \frac{y(x, t + \Delta t) - y(x, t)}{\Delta t}.$$

From the diagram one observes that the first derivative with respect to x, $\partial y/\partial x$, corresponds to a variation y when x is changed by an amount Δx

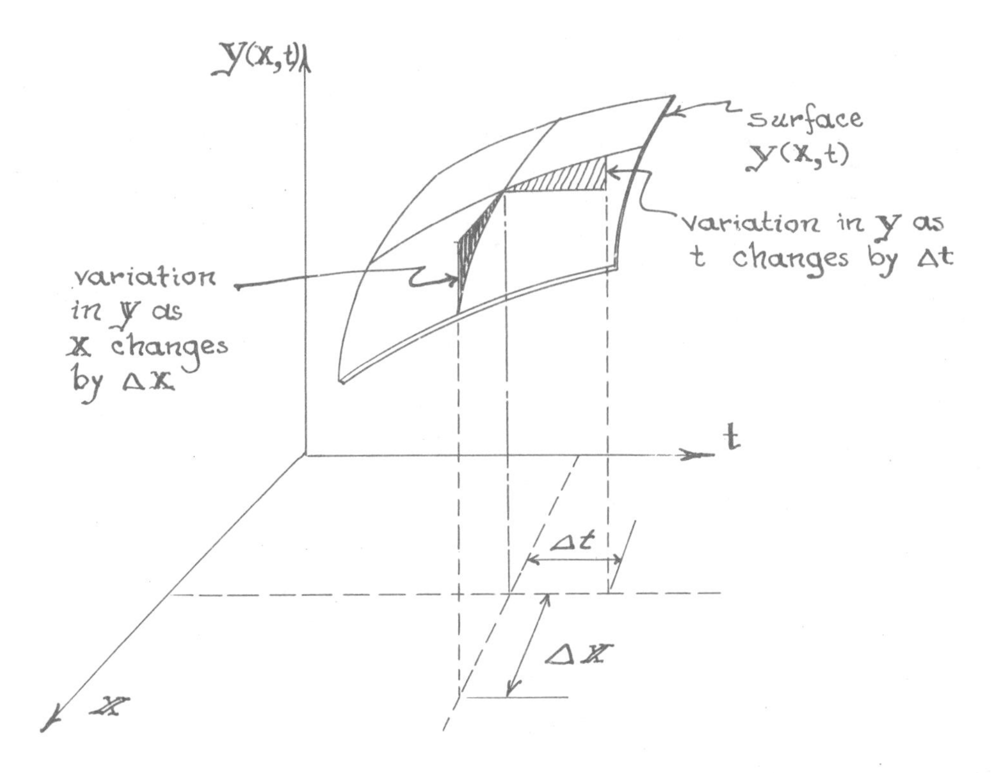

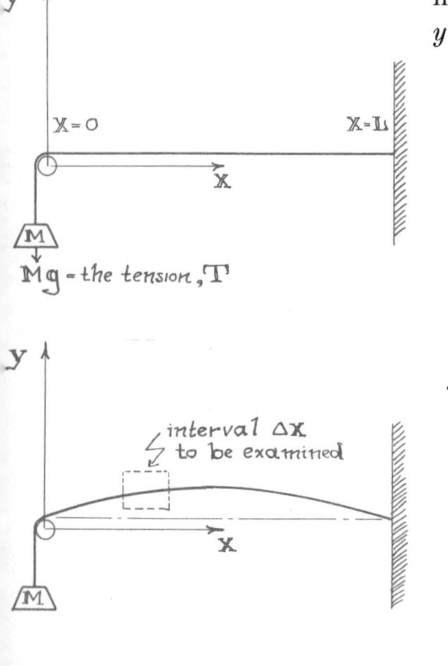

in a direction perpendicular to the t axis: $\partial y/\partial t$ corresponds to a change in y when t is varied by Δt along a direction perpendicular to the x axis.

Regard a string having a mass per unit length μ stretched between the points $x = 0$ and $x = L$, at a tension T, where T is the force directed along the string. If the string is stretched by means of a mass M hanging over a pulley, the tension equals Mg.

Assume that the string is plucked; at any point x and at a time t, the vertical displacement of the string is $y(x, t)$. To first order we assume that the tension is constant and therefore the amplitude of the vibration $y(x, t)$ is small (i.e., much less than L). We examine a small portion of the displaced string at x.

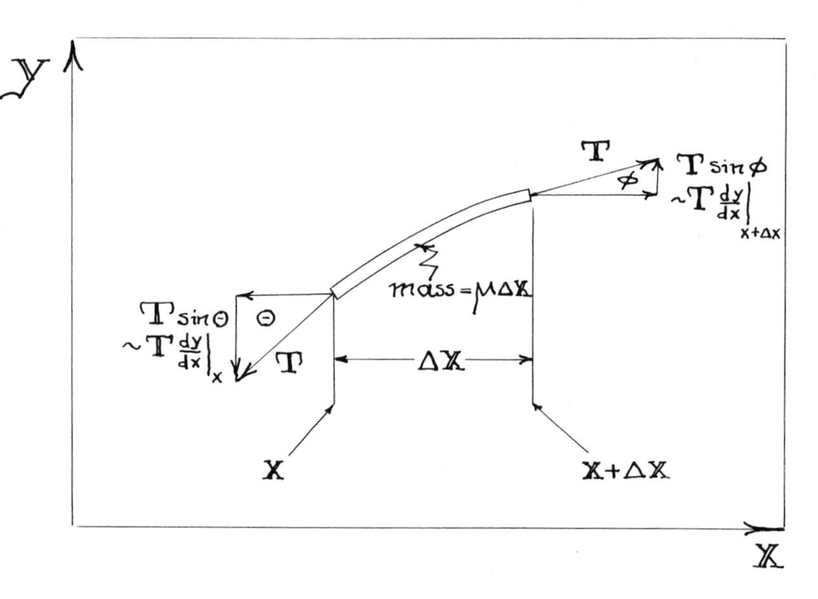

The length of the element is approximately Δx; therefore, the mass of the element is $\mu\,\Delta x$, while the total vertical force on the element is $T \sin \phi - T \sin \Theta$. Because the angles are small we can use $\partial y/\partial x$ evaluated at x for $\sin \Theta$ and $\partial y/\partial x$ evaluated at $(x + \Delta x)$ for $\sin \phi$. Equating the force on the element to the mass of the element times its vertical acceleration, one obtains

$$T \sin \phi - T \sin \Theta \simeq T\left[\frac{\partial y}{\partial x}\right]_{x+\Delta x} - T\left[\frac{\partial y}{\partial x}\right]_{x} \simeq (\mu\,\Delta x)\,\frac{\partial^2 y}{\partial t^2}\,.$$

Dividing the expression by Δx and taking the limit as Δx approaches zero, we find that the expression on the left becomes a second partial derivative with respect to x:

$$T \lim_{\Delta x \to 0} \frac{[\partial y/\partial x]_{x+\Delta x} - [\partial y/dx]_x}{\Delta x} = T\,\frac{\partial^2 y}{\partial x^2};$$

giving

$$T\,\frac{\partial^2 y}{\partial x^2} = \mu\,\frac{\partial^2 y}{\partial t^2},$$

or

$$\frac{\partial^2 y}{\partial x^2} - \frac{\mu}{T}\frac{\partial^2 y}{\partial t^2} = 0.$$

This last expression is the partial differential equation for the vibrating string and represents a form of Newton's second law. The constant μ/T has the dimensions of 1 over a velocity squared and is in fact the inverse of the square of the velocity of propagation, c (i.e., $c^2 = T/\mu$). This derivation may appear overly sophisticated to the uninitiated because the equation is thought of as an abstract device. Viewing this equation as a form of Newton's second law may make it seem somewhat more reasonable; the solutions to it are sinusoidal waves which are constrained to have nulls or zeros at the end points of the string, that is, at $x = 0$ and $x = L$. A typical motion is illustrated in x, t, y space.

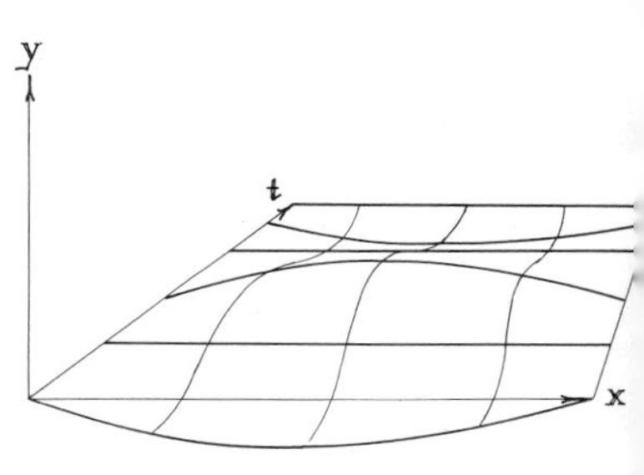

The velocity of propagation in the wave equation came to have a major role in the expansion of physical theory at the turn of this century. After Maxwell, the electromagnetic theory of light contains, as a basic equation, the wave equation. In this theory the term c^2 of course is not tied to the mechanical quantities tension and mass per unit length. However, electromagnetic theory gives the same type of equation with the term c appearing as the velocity of light.

Some one hundred years before the postulate of the mechanical theories of heat by Helmholtz and William Thomson, Daniel Bernoulli suggested the equivalence of heat and energy. This concept was developed out of his father's and his own views that a gas was constituted of microscopic particles moving at random and colliding elastically.

VII

Argonautica

While English mathematics languished in the reflected glory of Newton, on the Continent the influence of Leibniz was carried forward by the professors Bernoulli and their students. Leonhard Euler (1707–1783), who studied under Daniel Bernoulli, perhaps best exemplifies the European school of classical mathematics at its height. Euler, a Swiss by birth, spent most of his adult life in Russia, where the newly founded Academy of Sciences invited a small group of men who were destined to become the leading mathematicians of the age. In 1727 Euler joined Daniel and Nicholas Bernoulli at the Russian Academy, and in 1733 he was given the chair of mathematics there. In 1741 he left Russia for Berlin, at the invitation of Frederick the Great, but returned to St. Petersburg and the Academy in 1766. He remained there until his death in 1783. Toward the end of his life Euler went blind, but despite the enormity of this handicap and its attendant distresses, his productivity remained unaltered. Of his many important manuscripts, five volumes are the most prominent. Broadly speaking, these deal with functions and numbers, the calculus in two volumes, maxima and minima, and mechanics and hydrodynamics.

In his first major work Euler extended the concept of functions, classifying them as algebraic or transcendental and as rational or irrational, and ultimately subdividing the rational functions as integral or fractional.

We say that $f(x, y \ldots) = u$ is an algebraic function,* if u can be defined explicitly by an equation $F(x, y, \ldots, u) = 0$, where F is a polynomial in $x, y, \ldots$, and u. In other words, if u satisfies an algebraic equation it is an algebraic function; functions which do not are known as transcendental. When the function u satisfies an algebraic equation with rational roots, it is a rational function; division into integral and fractional functions follows from this. In the same manner, if u satisfies an algebraic equation with irrational roots, then it is an irrational function. Such an approach to functions stems from the division of the real numbers into algebraic and transcendental numbers. Algebraic numbers are roots of algebraic equations with integer coefficients; the remaining numbers are transcendental. Some algebraic numbers are rational, and the others are irrational. All transcendental numbers are irrational. The real numbers can be classed in two equivalent diagrams.

Some simple examples may prove useful. The positive integers are called the natural numbers, i.e., 1, 2, 3, 4, 5, ..., n.... These numbers are closed under the operations of addition and multiplication; that is to say, when either of these two operations is performed upon the set of natural numbers, the resultant is a member of the original set. The integers formed of the positive and negative whole numbers and zero are closed under the operations of addition, subtraction, and multiplication. These sets are not closed under division, which may produce fractions.†

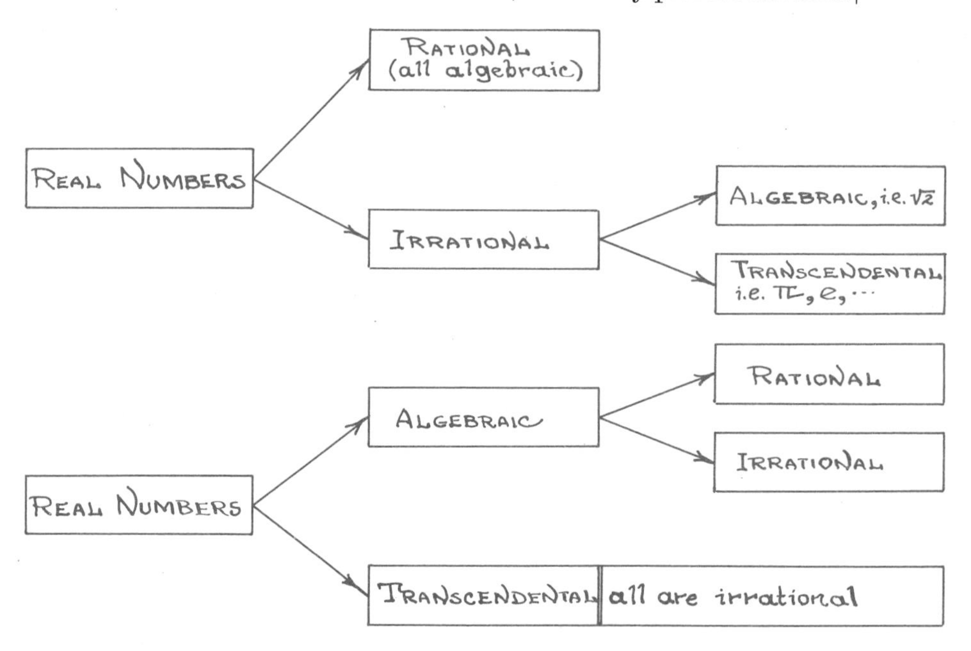

* For an expanded discussion, see Richard Courant, *Differential and Integral Calculus* (New York: Interscience, 1937), 1: 485.

† Notice that a somewhat modern phraseology has been introduced. This is done in the spirit of brevity in order that ideas need not be revised time and again. The concepts of fields, sets, and groups are appropriate to the nineteenth century and will occur naturally when we cover that period.

A rational number is a number which can be put in the form n/d, where n and d are integers and d is not zero. Rational fractions, then, are rational numbers, and when expressed in the decimal representation, these numbers are either terminating or periodic decimals, which repeat after a finite number of terms.* All numbers which are not rational are irrational. In the decimal representation, the irrationals are infinite decimals and are nonperiodic. Rational numbers are all algebraic in that they can be obtained as roots of algebraic equations. As simple examples, $3x - 2 = 0$, and $151x + 3 = 0$ have rational roots. Irrational numbers are either algebraic or transcendental. The roots of $x^2 + x - 1 = 0$ are irrational and algebraic; all irrationals which are not algebraic are transcendental. Common examples of transcendental numbers are π, the exponential number e, the natural logarithms of the natural numbers, and most of the values of the trigonometric functions.

Euler's first volume also made a significant contribution to infinite series. Despite his rather uneven approach, his results were usually correct. Recognizing that many of his series had to converge to be meaningful, he nonetheless paid little attention to the criteria for convergence. In the expansion in an infinite series of the logarithm of $(1 + x)$, it was held that this expansion was valid only for values of x less than 1 and greater than -1. Euler found that a combination of $\log(1 + x)$ and $\log(1 - x)$, however, will provide a converging series for all the logarithms of positive numbers greater than zero. Regard the difference of the two indefinite integrals:

$$\int \frac{dx}{1+x} - \int \frac{dx}{1-x} = \log_e \left(\frac{1+x}{1-x}\right)$$

$$= 2\left(x + \frac{1}{3}x^3 + \frac{1}{5}x^5 + \cdots + \frac{1}{2n+1}x^{2n+1} + \cdots\right);$$

[for $-1 < x < 1$, i.e., for x greater than -1 and less than $+1$]. Not only does the righthand side converge faster than that for $\log(1 + x)$, but every positive number $y = (1 + x)/(1 - x)$ greater than zero can be

* As an example of a periodic decimal, consider the number 1.245245245..., which has a period of three intervals after 1.000. This number may be factored and written as:

$$1 + 0.245 + 0.000245 + \cdots = 1 + 0.245(1 + 10^{-3} - 10^{-6} + \cdots)$$

$$= 1 + 0.245\left(\frac{1}{1 - 10^{-3}}\right) = 1 + .245\left(\frac{1000}{999}\right).$$

In general, if a cycle of N intervals, which we represent as $a_1a_2a_3\cdots a_N$, appears after a noncyclic part, M, the periodic number can be written as:

$$M + 10^{-m}(0.a_1a_2a_3\cdots a_N) \times \sum_{n=0}^{\infty} (10^{-N})^n = M + 10^{-m}(0.a_1a_2a_3\cdots a_N)\left(\frac{10^N}{10^N - 1}\right),$$

where m provides the appropriate displacement to the right or the left of the decimal point.

manufactured with the values of x in the interval greater than -1 and less than $+1$. For example:

$$\log_e 5 = \log_e \left(\frac{1 + 2/3}{1 - 2/3}\right) = 2\left[\frac{2}{3} + \frac{1}{3}\left(\frac{2}{3}\right)^3 + \frac{1}{5}\left(\frac{2}{3}\right)^5 + \cdots\right].$$

Euler is credited with the incorporation of the trigonometric functions into analysis. As demonstrated earlier by the expansion theorem for trigonometric functions, the derivative of $\sin x$ is $\cos x$ (i.e., $d \sin x/dx = \cos x$), and the derivative of $\cos x$ is $-\sin x$ (i.e., $d \cos x/dx = -\sin x$). By differentiating each of these equations again with respect to x, one obtains:

$$\frac{d}{dx}\left(\frac{d \sin x}{dx}\right) = \frac{d}{dx}(\cos x) = -\sin x = \frac{d^2 \sin x}{dx^2},$$

and

$$\frac{d}{dx}\left(\frac{d \cos x}{dx}\right) = \frac{d}{dx}(-\sin x) = -\cos x = \frac{d^2 \cos x}{dx^2}.$$

This result is unique in its simplicity, stating that for both $\sin x$ and $\cos x$ the second derivative of these functions is equal to minus the same function, a result similar to that for the exponential, where the first derivative of the function equals the function. From these results one might guess correctly two properties of $\sin x$ and $\cos x$: they have a series expansion in powers of x, and they are related to the exponential function.

The series solution, although tedious, is fundamental as an approach to differential equations. Therefore we consider the general trigonometric differential equation for a function $f(x)$,* $d^2f(x)/dx^2 = -f(x)$, and attempt a power series solution of the form

$$f(x) = a_0 + a_1x + a_2x^2 + \cdots + a_mx^m + \cdots = \sum_{n=0}^{\infty} a_nx^n,$$

where the coefficients a_n must be adjusted to satisfy the differential equation. By differentiating the series twice,

$$\frac{d^2f(x)}{dx^2} = 1\cdot2\cdot a_2 + 2\cdot3\cdot a_3x + 3\cdot4\cdot a_4x^2 + \cdots + (n+1)(n+2)a_{n+2}x^n + \cdots$$

$$= \sum_{n=0}^{\infty} (n+1)(n+2)a_{n+2}x^n.$$

Substitution of these series into the differential equation gives

$$\sum_{n=0}^{\infty} \{(n+1)(n+2)a_{n+2} + a_n\}x^n = 0.$$

* This differential equation is equivalent to that discussed with respect to the simple pendulum. If $d^2f/dx^2 + f = 0$ is multiplied by df/dx, it can be written as

$$\frac{d}{dx}\left[\frac{1}{2}\left(\frac{df}{dx}\right)^2 + \frac{1}{2}f^2\right] = 0,$$

or as $(df/dx)^2 + f^2 = 1$ with suitable normalization. The last equation has the Pythagorean form with solutions $f = \sin x$ and/or $\cos x$.

For this series to vanish for arbitrary x, each coefficient of x^n must vanish; therefore: $a_{n+2} = -a_n/(n+1)(n+2)$. This relation for the a_j defines two series, one starting with a_0 giving a series consisting of all of the even powers of x, and one beginning with a_1 providing a series in all of the odd powers of x. The even series is $a_0 \cos x$. The succession of coefficients can be determined by evaluating the first few:

$$a_2 = \frac{-a_0}{1 \cdot 2},$$

$$a_4 = \frac{-a_2}{3 \cdot 4} = \frac{a_0}{1 \cdot 2 \cdot 3 \cdot 4},$$

and

$$a_{2n} = (-1)^n \frac{a_0}{(2n!)}.$$

The constant a_0 which occurs in each term is the constant of integration and may be set equal to unity, giving

$$f(x) = a_0 \cos x = a_0\left(1 - \frac{x^2}{1 \cdot 2} + \frac{x^4}{1 \cdot 2 \cdot 3 \cdot 4} - \cdots\right),$$

$$\cos x = 1 - \frac{x^2}{2!} + \frac{x^4}{4!} - \frac{x^6}{6!} + \frac{x^8}{8!} - \cdots = \sum_{n=0}^{\infty} \frac{(-1)^n x^{2n}}{(2n)!}.$$

The coefficients of the odd series are determined in the same way:

$$a_3 = \frac{-a_1}{2 \cdot 3},$$

$$a_5 = \frac{-a_3}{4 \cdot 5} = \frac{a_1}{1 \cdot 2 \cdot 3 \cdot 4 \cdot 5},$$

and

$$a_{2n+1} = (-1)^n \frac{a_1}{(2n+1)!};$$

a_1 is the second constant of integration; there are two for a second-order differential equation.* To exhibit $\sin x$, we set a_1 to unity:

$$\sin x = x - \frac{x^3}{3!} + \frac{x^5}{5!} - \frac{x^7}{7!} + \cdots = \sum_{n=0}^{\infty} \frac{(-1)^n x^{2n+1}}{(2n+1)!}.$$

Recall now that the power series expansion of e^x is $\sum_{n=0}^{\infty} x^n/n!$, and notice that the even terms are similar to the cosine series, while the odd coefficients are equal in magnitude to the sine series.

If we now write out the expansion of e^{ix}, where $i = \sqrt{-1}$, we find that

$$e^{ix} = \sum_{n=0}^{\infty} \frac{(ix)^n}{n!} = 1 + i\frac{x}{1!} - \frac{x^2}{2!} - i\frac{x^3}{3!} + \frac{x^4}{4!} + i\frac{x^5}{5!} - \cdots.$$

Separating this series into an even series and an odd series,

$$e^{ix} = \left\{1 - \frac{x^2}{2!} + \frac{x^4}{4!} - \cdots\right\} + i\left\{x - \frac{x^3}{3!} + \frac{x^5}{5!} - \cdots\right\},$$

or $e^{ix} = \cos x + i \sin x$. This basic relationship between the exponential function and the trigonometric functions forms the basis of the trigonometric calculus. By consideration of e^{-ix} the exponential forms of $\cos x$ and $\sin x$ can be obtained algebraically as $e^{-ix} = \cos x - i \sin x$. Adding this to e^{ix}, one obtains $\cos x = (e^{ix} + e^{-ix})/2$; subtracting $\sin x = (e^{ix} - e^{-ix})/2i$. With these results, De Moivre's theorem becomes obvious, for

$$e^{inx} = \{e^{ix}\}^n = \cos nx + i \sin nx = \{\cos x + i \sin x\}^n.$$

* In general, a second-order differential equation must have two arbitrary constants of integration. Thus in general for this case, $f(x) = a_0 \cos x + a_1 \sin x$.

Included in Euler's first volume, *Introductio in Analysin Infinitorum* (1748), are sections on number theory and an analysis of curves of second degree, including the first discussion of curvature. Volumes 2 and 3 are a systematic exposition of the calculus. This is followed by a thorough investigation of differential equations of second order. The gamma and beta functions which play a fundamental role in the theory of many higher functions, were discovered by Euler. The factorials for integers occurred previously in the coefficients of the exponential and the binomial coefficients. Factorial n, written $n!$ was defined as $n! = 1 \cdot 2 \cdot 3 \cdot 4 \cdot \cdots (n-1)n$. The gamma function is a generalized factorial for real positive numbers, integral and nonintegral. By definition, this function is specified by

$$\Gamma(r) = \int_0^\infty x^{r-1}e^{-x}\,dx$$

for the real part of r greater than zero. When r is an integer, this function corresponds to a factorial.* One may integrate by parts to obtain the relation $\Gamma(r) = (r-1)\Gamma(r-1)$ when r is a positive integer, $\Gamma(r) = (r-1)!$.

The beta function, written as $B(r, s)$, is obtained from a similar integral and is related to the gamma function:

$$B(r, s) = \int_0^1 x^{r-1}(1-x)^{s-1}\,dx = \frac{\Gamma(r)\Gamma(s)}{\Gamma(r+s)},$$

for the real part of $r > 0$, and for the real part of $s > 0$. Again, this is modern notation; it does, however, show the form of these two important functions.

Generalizations of the properties of topological surfaces became primary to the creation of topology in the nineteenth century. Euler's studies of the properties of the polyhedra had a profound influence upon some of the basic theorems of topology. In fact, Euler's formula was later generalized to general topological surfaces. By polyhedra one means a solid whose closed surface consists of a number, F, of polygonal faces. When a solid is regular, all of the polygons are congruent. The term "simple surface" implies that there are no holes in the surface; therefore a simple polyhedron can be transformed continuously into a spherical surface. Both Descartes and Euler discovered independently that in

* The integration of a definite integral by parts is achieved in the following manner:

$$\int_a^b u(x)\,dv(x) = u(b)v(b) - u(a)v(a) - \int_a^b v(x)\,du(x).$$

In the instance of the gamma function, let $u(x) = x^{r-1}$, $du(x) = (r-1)x^{r-2}\,dx$, and $dv(x) = e^{-x}\,dx$ with $v(x) = -e^{-x}$. Then,

$$\Gamma(r) = \int_0^\infty x^{r-1}e^{-x}\,dx = -(r-1)[x^{r-2}e^{-x}]_0^\infty + (r-1)\int_0^\infty x^{r-2}e^{-x}\,dx.$$

By evaluating the first term on the right at the end points (it vanishes), and

$$\Gamma(r) = (r-1)\int_0^\infty x^{r-2}e^{-x}\,dx = (r-1)\Gamma(r-1) = (r-1)(r-2)\Gamma(r-2), \text{ etc.}$$

simple polyhedra the number of vertices, V, minus the number of edges, E, plus the number of faces, F, is equal to 2: $V - E + F = 2$.

Like his teacher Daniel Bernoulli, Euler expanded the field of differential equations, in particular the partial differential equations. His fourth volume was concerned with maxima, minima, isoperimetric curves, and the calculus of variation. In *Mechanica, sive motus scientia analyticè exposita*, his fifth great work, Euler abandoned the geometric methods of Newton and provided a complete and original work on dynamics. His analytic treatment was the foundation upon which Lagrange and Laplace based their great structures. The basic equations for rigid body motion still bear his name. Rigid body motion in the laboratory coordinates may be described by a complex set of differential equations, the basics of which involve the torques on the body and the rotations of the body about some point, usually the center of mass. Complications arise because each equation involves all of the three different components of the angular velocity and all of their derivatives. By transforming the problem to a set of coordinates fixed in the rigid body, the Euler equations of rigid body motion introduce a fundamental simplification in providing a set of three equations, each containing only the derivative of one angular velocity component.

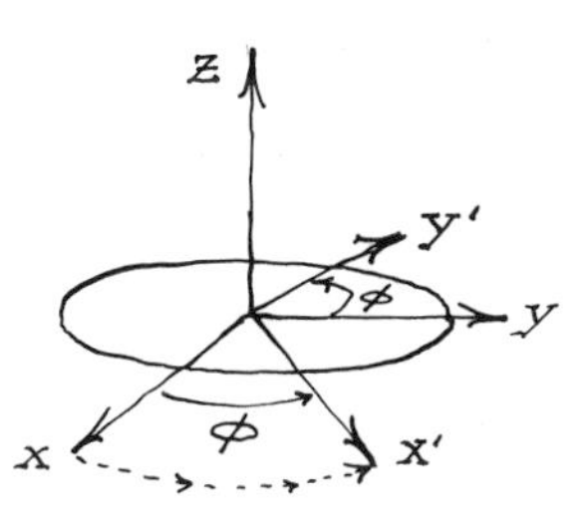

ROTATION ABOUT Z THROUGH φ

A basic advance in the problem of the rotation of a rigid body was achieved by the introduction of three independent rotations, which enables one to express a given orientation of the body in terms of three independent variables. The three angles* are specified by: (1) a rotation through an angle ϕ in the x, y plane about the z axis of the laboratory coordinates, taking $[x, y]$ to $[x', y']$, leaving z unchanged, (2) a rotation through an angle Θ about the new x' axis, changing $[y', z]$ to $[y'', z']$, with x' left unchanged, (3) finally, a rotation through an angle Ψ in the $[x', y'']$ plane about the new z' axis, which takes x' to x'' and y'' to y'''. These three rotations are illustrated in the diagram.

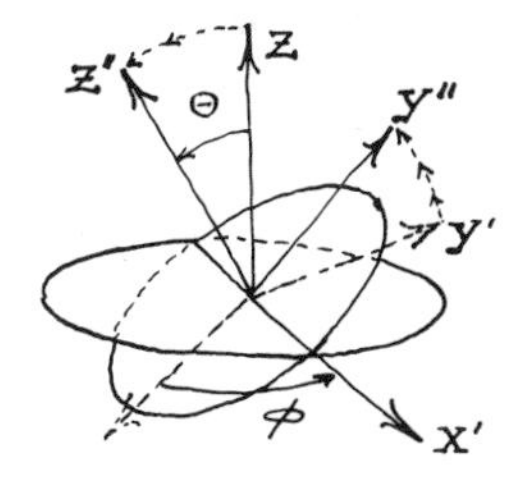

ROTATION ABOUT X' THROUGH Θ

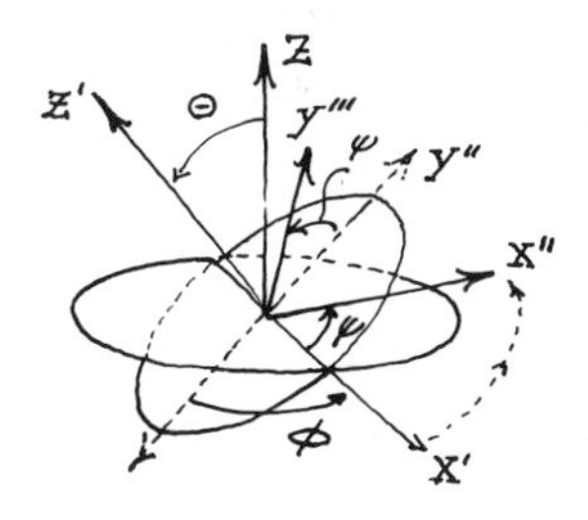

ROTATION ABOUT Z' THROUGH Ψ

Euler also assembled a treatise on algebra, *Vollstandige Anlietung Zur Algebra* (1770), in which he enlarged on the theory of equations. In his failure to solve the quintic (fifth-order) equation, and in his study of approximate solutions, he emphasized the paradox of the higher degree equations, which was finally settled by Abel and Galois. Optics was another interest of Euler's, and he suggested methods for correcting both spherical and chromatic aberrations of lenses. Concurring with Newton, he believed that color depended upon wave length; however, unlike Newton, he was a wave theorist. Anticipating Maxwell by a century, Leonhard Euler assumed that both light and electricity were transported through the same elastic medium (named the "aether"), and in this viewpoint light and electricity were interrelated.

The year 1730 represents a turning point in the history of electromagnetism. Almost all of the prominent men of science born between 1730

* These are not the original set of angles but are related to them and represent those used at this time.

and 1770 were in some degree instrumental in the discovery, development, and refinement of the theories of electromagnetism and light. In 1729 Stephen Gray showed that electrical charges could be transferred from one body to another—a glass tube was used to electrify other bodies—and electrical activity was found to depend upon the materials of which bodies are composed. By placing conductors on cakes of resin, Gray found that bodies could be insulated; further experiments with solid and hollow oak blocks demonstrated that the different blocks, when electrified, exhibited the same electrical properties. These experiments gave birth to the concept of an electric fluid which flowed from one body to another, and then to the more important idea that when charged the electric fluid resided only on the surface of a conductor.

The gold leaf electroscope was created by Charles François du Fay (1698–1739), and for the first time crude quantitative measurements of electrification were possible. Du Fay mistakenly believed that there were two types of electricities, that of transparent bodies and that of resinous bodies. To explain attraction and repulsion he postulated the two-fluid theory of electricity, in which the fluids were supposed to be separated by friction and to neutralize each other when combined.

The storage of electric charge (as we know it) was difficult in these early experiments because of discharge in a humid atmosphere. Pieter van Musschenbroek (1692–1761) invented the first practical condenser, the Leyden jar, in 1746. About this time Benjamin Franklin (1706–1790) postulated the one-fluid principle along with the conservation of charge: that the amount of positive and negative charge in the universe is constant. Franklin observed that electricity is present in all material, and that charging is a matter of transferring a single fluid from one body to another. He noted the insulating property of glass and theorized that the two surfaces of a dielectric charged with charges of equal magnitude and opposite signs. In this manner Franklin arrived at the concept of electrical action at a distance.

Joseph Priestley (1733–1804), the discoverer of oxygen, found that cork spheres suspended inside a charged metal cup were wholly unaffected by the electricity on the surface of the cup. To appreciate Priestley's ingenious conclusion, we must take into account the fact that Newton had shown that the force field inside a spherical shell of mass m was zero because of the inverse square law. This is a unique result applying only to fields which vary as the inverse square of the distance. To make this plausible in an elementary discussion, we will introduce the concept of angular measure in three dimensions. Recall that two-dimensional angular measure in radians is defined in terms of the length of arc subtended on a circle of unit radius. Analogously, solid angle measure, i.e., measure of three-dimensional solid angles, is defined in terms of the area subtended upon the surface of a sphere of unit radius.

Thus if an area A is specified at a distance r from a point P at the center of a unit sphere by the similarity of the solid cone-like figures, the solid angle Ω is specified in terms of A and r as the ratios $\Omega/1^2 = \Omega = A/r^2$.

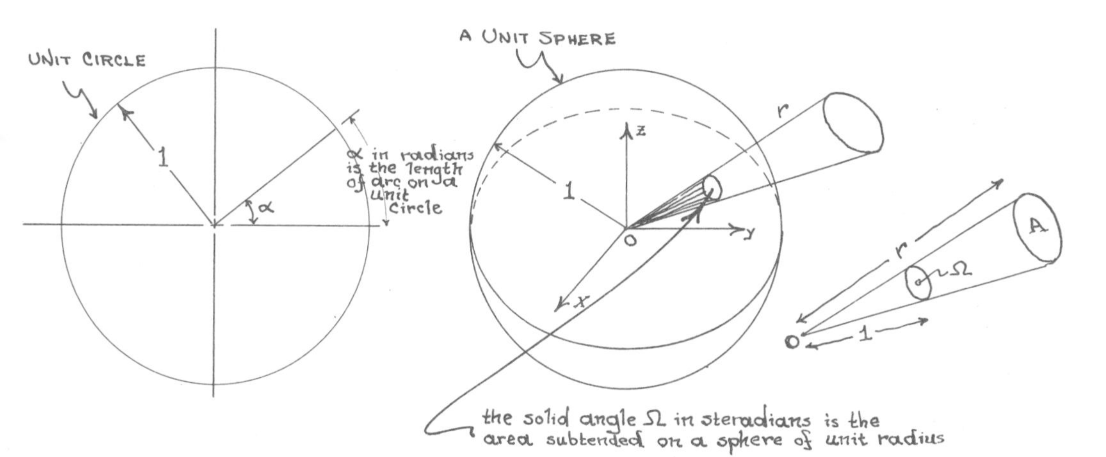

As a simple example of the inverse square law, consider a point P inside a hollow spherical shell of mass per unit area σ. The thickness t is considered to be uniform.

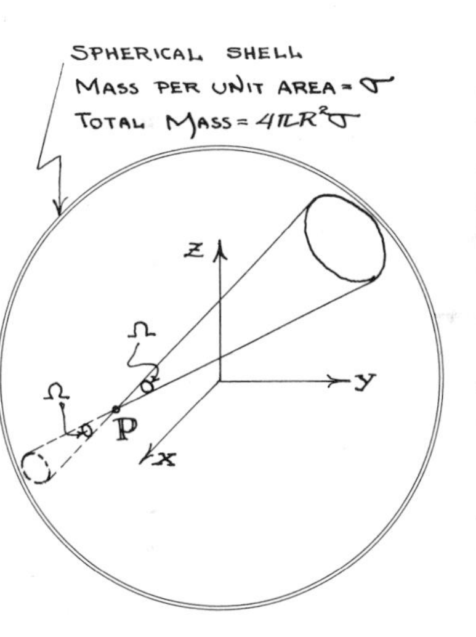

If we take any very small element of solid angle $\Delta\Omega$ at P subtending on the sphere an area A_1 on one side and an area A_2 on the other, then

$$\Delta\Omega = \frac{A_1 \cos \Theta}{r_1^2} = \frac{A_2 \cos \Theta}{r_2^2},$$

giving $A_1/r_1^2 = A_2/r_2^2$. The magnitude of the total force per unit mass at P caused by the mass elements $A_1\sigma$ and $A_2\sigma$ is zero and is given by the inverse square law:

$$\text{force per unit test mass at } P = \frac{GA_1\sigma}{r_1^2} - \frac{GA_2\sigma}{r_2^2} = G\sigma\left\{\frac{A_1}{r_1^2} - \frac{A_2}{r_2^2}\right\} = 0,$$

where G is the gravitational constant.

By taking all of the elements of solid angle about P (they total 4π on the unit sphere) we observe that the segments spanning the sphere surface appear in opposing and canceling pairs. Thus the total force at P is zero. From this knowledge Priestley postulated that the force law between charged particles would vary as the inverse square of the distance. One must exercise some care in extrapolating from the spherical shell with gravitational attraction, where the sources or masses are rigidly fixed in position, to the electrical case, where the charge resides in an infinitesimal layer on the surface of a charged hollow conductor and is free to arrange itself to minimize the energy of the charged system. The rule concerning the electric field inside the charged hollow conductor is much more general, in that the field inside any charged conducting body is zero. In other words, this rule for electrical charge is independent of shape because the charge always resides on the surface in a shell of infinitesimal thickness.

Priestley's postulate initiated an entire school of tests of the inverse square law. The English investigators essentially sought to determine the limits of accuracy to which one could ascertain a null measure of the electric field inside a charged conductor. Henry Cavendish (1731–1810) computed the manner in which charge is distributed between a circular disc electrically connected to a conducting sphere of the same radius. In 1771 he also measured the accuracy of the inverse square law by investigating the electric forces on one conducting sphere inside another. Unfortunately, Cavendish's work was not published and therefore had little impact on the general scientific thought of his time.

Charles Augustin de Coulomb (1736–1806) made a direct measurement of the force law between charged bodies in 1785. Using an elastic suspension, he created torsion balance, which for small rotations provides a torque proportional to the angle through which it is rotated. By charging two pith balls equally, one being fixed and the other attached to the suspension, he was able to measure the force of repulsion between them, and determined that it varied as the inverse of the square of the distance between centers, and was proportional to the product of the charges on the two balls. This law also assumes that the force is directed along the line of centers of the two charged particles.

A torsion balance consists of a vertical elastic fiber or suspension which carries a long moment arm or rod in the plane of rotation. The interacting bodies, usually spheres, are placed at the ends of the moment arm. When the fiber is twisted through an angle, say ϕ, the elasticity exerts a restoring torque which is proportional to the angle of rotation. Because the forces involved in the experiment are exerted at the ends of the moment arm and perpendicular to it, the angle of rotation of the elastic suspension is proportional to the force exerted.

By establishing the fractional sharing of charge that occurs when one charged sphere is brought into contact with an equal but uncharged sphere, the relative amount of charge can be varied at will. Thus in a series of charging and dividing charge experiments it can be shown with the aid of the torsional balance that the force between a charged sphere fixed in the laboratory and one fixed at the end of the moment arm of the balance varies as the product of the two charges. Next, by varying the distance between these same two spheres with corresponding changes in the angle of rotation of the arm, it can be shown that the force between the fixed charge and that carried at the end of the moment arm varies as the inverse square of the distance between them. The resultant law for the magnitude of the force between two charges q_1 and q_2 is $F_{12} = k(q_1q_2/d^2)$, where k is a constant and d is the distance of separation between the charges. F_{12} is actually a vector quantity and acts along the line of centers, either parallel or antiparallel, depending upon whether the force is attractive or repulsive.

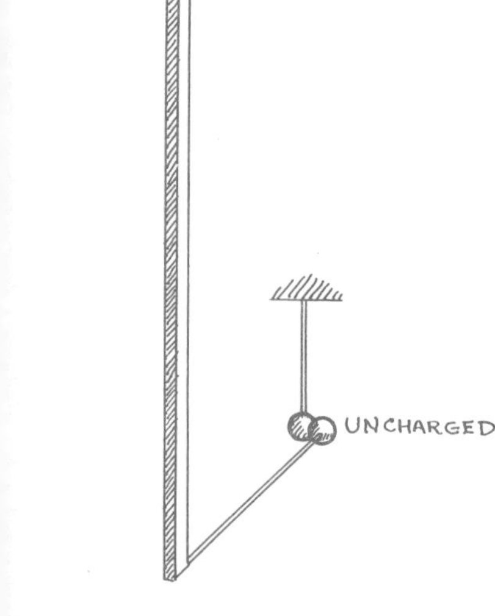

Coulomb was an adherent of the two-fluid theory. He also established the inverse square law of force for magnetic monopoles. These demonstrations, however, do not have a fundamental physical significance and now serve only to provide a convenient representation for a few experiments involving bar magnets.

The inverse square law of force between point charges was the important step in the theory of electrostatics. In principle, once this law is stated, mathematical analysis provides concepts of energy storage and capacity. The concept of electrical conduction and the conductivity of materials were those of Cavendish, and he actually proposed the condenser and measured the dielectric capacity of several insulators, but, as these researches were not published in his lifetime, the credit of discovery must be assigned to Coulomb.

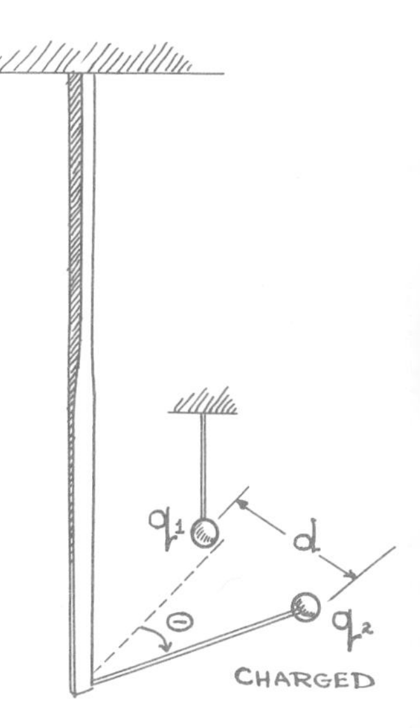

Dynamic electricity or current electricity was first systematically exploited as a new science by Aloisio Galvani (1737–1798). Galvani's current indicator was the reaction of biological muscle to electric charge. He showed that a muscular contraction could be produced in the leg of a frog by touching a charged scalpel to an exposed nerve. His experiments were suggestive but of little or no fundamental value; however, they drew the attention of Alessandro Volta (1745–1827). Volta was particularly struck by Galvani's discovery that two dissimilar metals in contact produced a contraction in frog muscle if their extremities were each touched to the muscle at different points. Volta discovered that a silver and a gold coin, when connected with a metal wire, produced a bitter taste on the tongue, and from this theorized that contact between two dissimilar metals produced a new kind of electricity. To generate larger amounts of this electricity, he created a pile of alternating layers of copper and zinc separated by pasteboard. When one of the layers was removed, he discovered that it was charged. On immersing the pile in an acid solution, Volta found much larger electrical effects at its terminals. In March of 1800 he communicated his results to the President of the Royal Society, one Joseph Banks. Six weeks later Nicholson and Carlisle published a description of the pile and obtained a priority over the original inventor, Volta. This unhappy episode marred the otherwise significant contributions of William Nicholson and Sir Anthony Carlisle, who systematically studied the decomposition of water. This had been done earlier by Ash at Oxford, Fabbroni in Florence, and Creve in Mainz, but it was left for Nicholson and Carlisle to demonstrate that the liberated gases were oxygen and hydrogen. The creation of the pile, Volta's primitive battery, was a fundamental step in the understanding of electrical charge and had a decisive impact on science in the years which followed, although twenty years elapsed before current electricity ushered in electromagnetism.

As in the example of the definition of mass, the introduction of the words "charge" and "charged body" presented another difficulty. In what manner can charge and the property of being charged be defined? An elementary approach may be established in terms of an operational recipe: a body may be defined as charged when forces characteristic of the charging process are observed. The earliest charging process involved frictional electricity, but with the invention of the pile, or battery, another source of charge came into being. Regardless of the sophistication of the field of physics, the basic definitions of the elementary sources of force fields such as mass or charge remain rather rudimentary. Another method of definition is the axiomatic one which postulates the existence of certain quantities and their interactions; the consequences may be worked out in detail, with the test of validity resting on experience. If all observations are consistent with the predictions of the axiomatic model, the initial postulates are assumed to be valid.

During this period the foundations of mathematics were becoming quite formal. Traditionally, analytic mechanics kept in step with the advances in mathematics. Perhaps the foremost mathematician of the eighteenth century was Joseph Louis Lagrange (1736–1813). He was born of French ancestry in Turin and became a professor of mathematics there at the age of nineteen. At the age of twenty, he produced a general solution to some isoperimetric problems that had baffled the leading mathematicians of the day. In the course of these solutions, Lagrange developed the calculus of variations, which established his reputation as one of the foremost mathematicians of his time. He made fundamental contributions to the theory of differential equations, number theory, and algebra, and part of his early fame rests upon his work in celestial mechanics. In 1766, at the invitation of Frederick the Great, he moved to Berlin. His greatest work, *Mécanique analytique*, laid the foundations of modern analytic dynamics: his mechanics was based upon Leibnizian concepts involving the kinetic and potential energies. Using a variational approach and the principle of least action, Lagrange derived the equivalent of Newton's laws in a generalized form. His result was essentially that of D'Alembert, but the method was much more general. Lagrange created the concept of the scalar potential function from which certain conservative vector forces could be derived—in many problems of dynamics and electrostatics, this function made the solutions much easier. While discussing D'Alembert's formulation of the equations of motion, he introduced the concept of the work done by a force acting through a distance l as the projection of the force on each element of length $d\vec{\mathbf{l}}$ summed over all the elements. This is illustrated in the diagram.

In this diagram the force is assumed to vary from point to point along the curve. The variation is shown graphically by indicating the direction and magnitude of $\vec{\mathbf{F}}_j$ at the beginning of each line segment, $d\vec{\mathbf{l}}_j$. The projection of $\vec{\mathbf{F}}_j$ on a given line segment is defined as the magnitude of $\vec{\mathbf{F}}_j$ at that point times the cosine of the angle between $\vec{\mathbf{F}}_j$ and the line

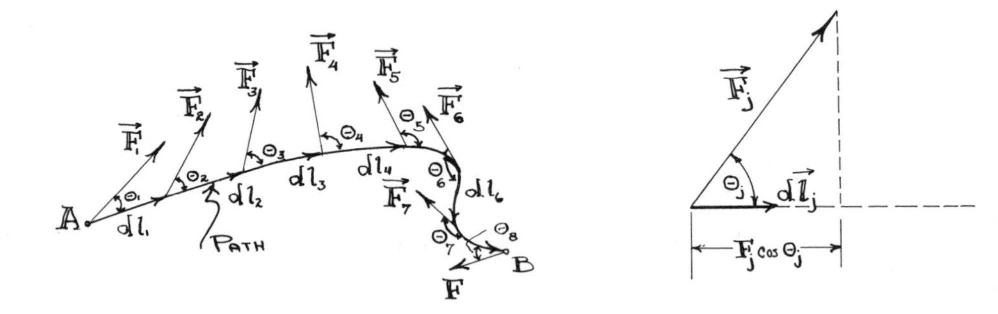

segment $\vec{dl}_j$. The work to move from one end of the line to the other is approximately the sum of the terms $F_j\,dl_j \cos \Theta_j$.

$$W_{AB} = \text{the work to move from } A \text{ to } B = \sum_{\text{all } j} F_j\,dl_j \cos \Theta_j$$
$$= F_1\,dl_1 \cos \Theta_1 + F_2\,dl_2 \cos \Theta_2 + \cdots + dl_n \cos \Theta_n.$$

In the limit, as the lengths dl_j become arbitrarily small, this sum becomes an integral:

$$W_{AB} = \int_A^B F(x, y, z) \cos \Theta\, dl.$$

As a simple example, consider a force $F = -kx$, where k is a constant, and $\vec{\mathbf{F}}$ is directed along the negative x axis when x is positive, and vice versa; then the work to move from $x = a$ to $x = b$ is:

$$W_{ab} = -k \int_a^b x\,dx = -k\left(\frac{x^2}{2}\right)_a^b = -k\left(\frac{b^2}{2} - \frac{a^2}{2}\right).$$

Certain vector forces can be generated from scalar functions of the type $V(x, y, z) = $ a constant. These are known as conservative forces, and the scalar functions V are called potential functions. Consider a force $F_x(x)$ directed only parallel or antiparallel to the x axis and depending only upon x. If a function $V(x)$ exists such that $F_x(x) = -dV(x)/dx$, the function $V(x)$ is the scalar potential for $F_x(x)$. Now consider the former problem of computing the work of $F_x = -kx$ between $x = a$ and $x = b$. In this special case, it is immediately discernible that

$$F_x(x) = -kx = -\frac{d}{dx}\left(\frac{1}{2}kx^2\right);$$

therefore, the potential function corresponding to $-kx$ is $V(x) = (1/2)kx^2$.

From the result of the previous example it is clear that the work done is equal to the difference between $V(x)$ evaluated at $x = a$ and at $x = b$:

$$W_{ab} = \int_a^b F(x)\,dx = -\int_a^b \frac{dV(x)}{dx}\,dx = -\int_a^b dV = -V(b) + V(a).$$

Thus the work done between two points a and b is the negative of the difference in the potential between a and b. When the potential difference is negative, the work is positive, and the agent exerting the force does work constituting a loss of energy to the agent. Where the potential difference is positive, the work done is negative, and the agent exerting F has work done upon it, constituting a gain of energy by the agent. If a ball rolls down a hill under the action of the gravitational field of the earth, it moves from higher to lower points of gravitational potential, gaining kinetic energy; therefore, the gravitational force field of the earth does positive work, with a consequent loss of potential energy by the ball and with a corresponding gain in kinetic energy.

In three dimensions V is a scalar function of the three coordinate variables x, y, and z. Then the three components of the vector force, which is derivable from V, are given by

$$F_x = -\frac{\partial V(x, y, z)}{\partial x},$$

$$F_y = -\frac{\partial V(x, y, z)}{\partial y},$$

and

$$F_z = -\frac{\partial V(x, y, z)}{\partial z}.$$

All the conclusions drawn from the one-dimensional case still hold. The function $V(x, y, z) =$ a constant represents a surface in three dimensions, and the consequences of our definitions of F_x, F_y, and F_z in terms of V lead to a vector, $\vec{\mathbf{F}}$, which is perpendicular to the surface V at every point on the surface. As different values of the constant are taken in the expression $V =$ a constant, new surfaces are generated in space.

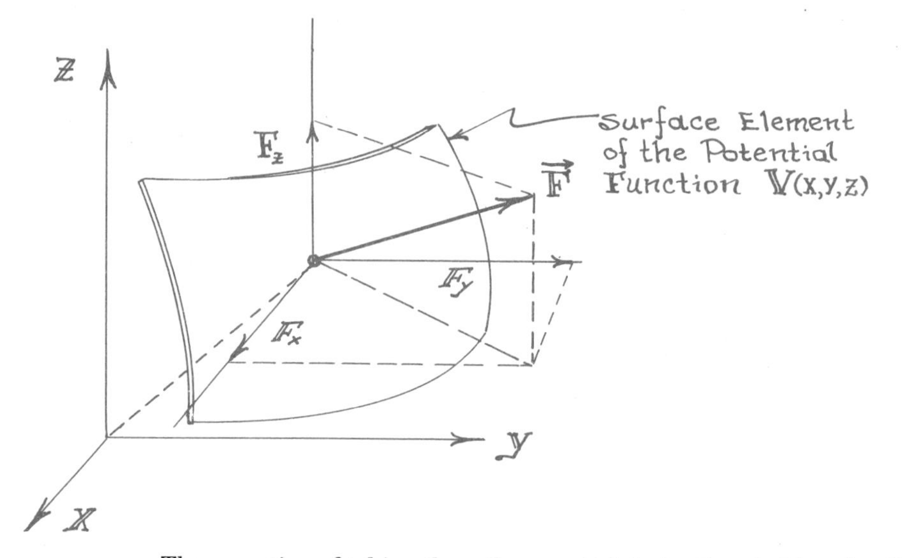

The operation of taking these three partial derivatives (with an implication of an associated direction) is known as the gradient operation. A two-dimensional analogue of the potential function is produced by observing the x, y projection of a ball starting from rest and rolling down a hill. It always moves perpendicular to the contour lines (curves of the hill mapped out on the hill at a fixed altitude, i.e., the curves produced by taking plane horizontal slices through the hill). Several paths of rolling are shown on the page opposite.

The forces are not only perpendicular to the equipotential surfaces (curves in two dimensions) at every point, but, as can be seen from the definition, the magnitude is given by the rate at which the potential function changes value along the direction perpendicular to the surface.

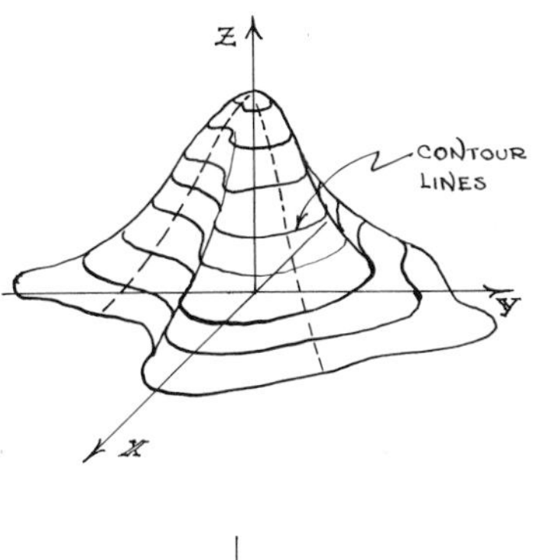

In our two-dimensional analogue this implies that the force is greater when the contours are closer together, suggesting that the hill is steeper in these regions, i.e., the gradient is larger.

When a force is conservative the total energy—kinetic plus potential—is conserved; this is to say that the total energy is constant, and this is most easily seen from a one-dimensional motion. Let

$$F(x) = \frac{d}{dt}\left(m\,\frac{dx}{dt}\right);$$

then, multiplying by dx/dt and using the conservative condition that $F = -dV/dx$, one finds that

$$F(x)\,\frac{dx}{dt} = -\frac{dV(x)}{dx}\,\frac{dx}{dt} = -\frac{dV}{dt},$$

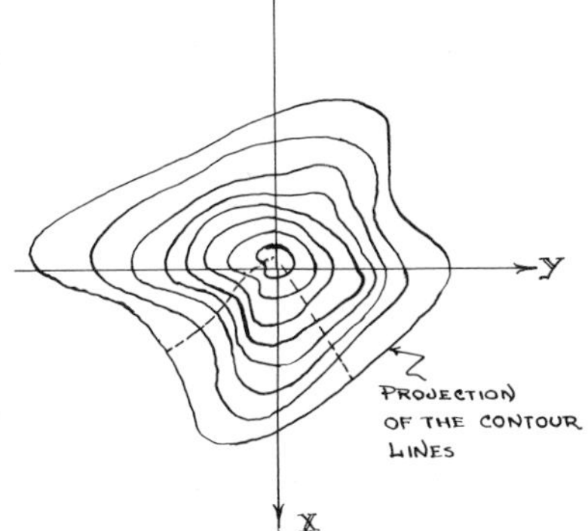

which in turn equals

$$\frac{dx}{dt}\,\frac{d}{dt}\left(m\,\frac{dx}{dt}\right) = \frac{d}{dt}\left(\frac{1}{2}\,m\left[\frac{dx}{dt}\right]^2\right) = \frac{d}{dt}\left(\frac{1}{2}\,mv^2\right).$$

Bringing both time derivatives to the lefthand side gives

$$\frac{d}{dt}\left\{\frac{1}{2}\,mv^2 + V(x)\right\} = 0.$$

When a derivative is zero the function differentiated is a constant; thus $(1/2)mv^2 + V(x) = \text{a constant} = E$. This equation represents the conservation of energy for a particle moving in one dimension under the action of a conservative force. Closer examination suggests that another constraint must be imposed upon $\vec{\mathbf{F}}$ and V if energy is to be conserved: the force must not be an explicit function of time; otherwise, the total energy will vary in time as the potential function varies in time.

Lagrange's formulation of dynamics begins with a function L for the system called Lagrangian. L is defined as the *difference* between the kinetic and the potential energies (this is not the total energy E, which is the sum). In many examples, the function L can be generated more readily than Newton's second law. By assuming that the variation of the integral of (Ldt) between two fixed points A and B is a minimum, one obtains a set of equations which correspond to Newton's second law. The advantages of this method lie in the fact that L may be obtained relatively easily for very complex ensembles and for quite elaborate coordinate systems. If one of the generalized coordinate variables is q_j, with a time rate of change $\dot{q}_j$, the equivalent of Newton's second law for the jth coordinate is:

$$\frac{d}{dt}\left(\frac{\partial L}{\partial \dot{q}_j}\right) - \left(\frac{\partial L}{\partial q_j}\right) = 0.$$

The quantity $\partial L/\partial \dot{q}_j$, is known as the jth component of the generalized momentum. Many problems involve more than three coordinates. Three-point masses interconnected by springs have 3×3 or 9 coordinates. Equations of motion for such systems are difficult to write down directly.

The Lagrangian for such a system is particularly simple: beginning with the function L, the complicated equations of motion can be developed straightforwardly. Indeed, one of the major goals of quantum field theorists in the twentieth century was to discover the appropriate Lagrangian which would adequately describe a system of particles.

Long before the theory of relativity, Lagrange expressed the belief that mechanics must consist of a geometry of four dimensions, three spatial coordinates plus a time coordinate. The theory of numbers held his lifelong interest, and he solved many of the problems which had been put forward by Fermat. For instance, he found all of the integral solutions of the expression $x^2 - ay^2 = 1$, where a is a nonsquare number. Solution of the ancient riddle of π was carried another step forward by Lagrange's proof that π was irrational; that it was transcendental as well was not proved until 1882.

In addition to his work on the calculus of probabilities, Lagrange stimulated inquiry into the theory of groups. Since this theory has had such a decided impact upon physical theory in the middle of our own century, it is most useful to examine the beauty of Lagrange's solution to the quartic equation. It is not the problem itself which is so striking—this equation had been solved by Cardano and others—but rather the approach to it, or methodology, which set the stage for the algebraic revolution of Galois and Abel. For two centuries no one doubted that equations of degree higher than four would be solved in terms of rational radicals of the coefficients. In other words, it had been assumed that solutions would be obtained which involved only addition, subtraction, multiplication, division, and radicals (with rational positive exponents) of the integral coefficients of a polynomial. Lagrange did not believe that polynomials of degree higher than four could be solved in this manner; he further recognized that the theory of permutations was of fundamental importance in the solution of algebraic equations. Regard the polynomial

$$f(x) = x^N + a_{N-1}x^{N-1} + \cdots + a_1x + a_0 = \sum_{n=0}^{n=N} a_nx^n = 0$$

of degree N, with roots $x_1, x_2, x_3, \ldots, x_{N-1}, x_N$. Then $f(x)$ can be factored to the form

$$f(x) = (x - x_1)(x - x_2)(x - x_3)\cdots(x - x_N) = \prod_{j=1}^{N} (x - x_j) = 0,$$

where the symbol $\prod_{j=1}^{N}$ is an instruction to take the repeated product from $j = 1$ to $j = N$.

In particular Lagrange viewed the quartic

$$x^4 + mx^3 + nx^2 + px + q = (x - x_1)(x - x_2)(x - x_3)(x - x_4) = 0,$$

with roots x_1, x_2, x_3, and x_4. Beginning with an expression $x_1 + x_2 - x_3 - x_4$, which presents the four roots in an additive array with two positive and two negative signs, Lagrange wrote down all permutations of these numbers with respect to the sign. There are $4! = 24$ different arrangements taken two at a time; therefore there are $4!/(4-2)!\,2!$ or six arrays of the numbers:

$$\begin{aligned} X_1 &= x_1 + x_2 - x_3 - x_4, \\ X_2 &= x_1 + x_3 - x_2 - x_4, \\ X_3 &= x_1 + x_4 - x_3 - x_2, \\ X_4 &= x_3 + x_4 - x_1 - x_2 = -X_1, \\ X_5 &= x_2 + x_4 - x_1 - x_3 = -X_2, \\ X_6 &= x_3 + x_2 - x_1 - x_4 = -X_3. \end{aligned}$$

With these six permutations, X_k, a sixth-order polynomial with six roots, X_k, can be formed:

$$(y - X_1)(y - X_2)(y - X_3)(y - X_4)(y - X_5)(y - X_6) = 0.$$

This polynomial has roots X_j and will have coefficients (according to Viete's formulas) which do not vary with all twenty-four permutations of the original roots x_k. Permutations of the x_k show that the coefficients of this sixth-degree polynomial cannot depend upon the order in which the roots are taken. Because the X_k appear in pairs, i.e., X_1 and $-X_1$, etc., the sixth-order equation can be written as a product of terms of second degree: $(y^2 - X_1^2)(y^2 - X_2^2)(y^2 - X_3^2) = 0$. Letting $y^2 = t$, we obtain a cubic in t whose roots are $t_j = X_j^2$ or $X_j = \sqrt{t_j}$. Using the coefficient m of the x^3 term in the original polynomial, we can invert the algebraic expressions for the X_j to obtain the proper roots x_k. Remember that (by Viete's formula) $m = -x_1 - x_2 - x_3 - x_4$; therefore,

$$x_1 = \frac{1}{4}(-m + X_1 + X_2 + X_3) = \frac{1}{4}(-m + \sqrt{t_1} + \sqrt{t_2} + \sqrt{t_3}),$$

$$x_2 = \frac{1}{4}(-m - X_1 - X_2 + X_3) = \frac{1}{4}(-m - \sqrt{t_1} - \sqrt{t_2} + \sqrt{t_3}),$$

$$x_3 = \frac{1}{4}(m + X_1 - X_2 + X_3) = \frac{1}{4}(m + \sqrt{t_1} - \sqrt{t_2} + \sqrt{t_3}),$$

$$x_4 = \frac{1}{4}(m - X_1 + X_2 + X_3) = \frac{1}{4}(m - \sqrt{t_1} + \sqrt{t_2} + \sqrt{t_3}).$$

In this manner the solutions of the fourth-degree polynomial have been reduced to expressions involving the roots of a cubic. This demonstration is equivalent to Cardano's manipulation; however, this approach reveals the formal nature of the solution in a clear and elegant manner, and the conditions of solubility are inherent in it, as they are not in that of Cardano.

When Lagrange applied this method to an equation of degree five, he achieved an auxiliary equation of degree six, which was irreducible to a lower order. Notice that in the application of the method to the equation of degree four, the auxiliary equation of degree six was reducible to degree three. Degree four seemed to be a transition point. For all equations of degree four or lower, the auxiliary equations reduced to degrees lower than that of the initial equation. For all equations of degree higher than four, Lagrange found that the auxiliaries were of a higher degree than the initial equation, and all of these appeared irreducible. This suggested to him that it was impossible to effect rational radical solutions of polynomials having a degree higher than four, but he was not able to prove this conclusively.

In this period the French established themselves as the masters of analysis and *mécanique rationale.* Following Lagrange, Pierre Simon, Marquis de Laplace (1749–1827), climaxed this era of excellence in dynamics by achieving spectacular results in the analysis of planetary motions. Newton and his followers had taken the point of view that the inherent complexity of the forces involved in the solar system indicated that it was in all probability unstable. Laplace established the fundamental stability of planetary motions. His great work on celestial mechanics was published in five volumes from 1799 to 1825. His studies of the gravitational attractions of ellipsoids led him to the definition of the potential function and to defining differential equations for potentials in free space: the equation (known as the Laplacian in Cartesian coordinates) appears in the particularly simple form $\partial^2 V/\partial x^2 + \partial^2 V/\partial y^2 + \partial^2 V/\partial z^2 = 0$, where V is a function of x, y, and z. This form stresses a highly important mathematical concept. Earlier we discussed the gradient operator which produced a vector function directed perpendicular to the surface of V and of magnitude equal to the rate of change of V along a direction perpendicular to the surface. Although the strict definition of the "divergence operator" is a more modern development, it serves our purpose to introduce the concept here. The divergence operator must operate upon a vector field, that is, upon an array of vectors defined in direction and magnitude at each point of space. By "the divergence of a vector field at a point," one implies that when a small closed surface is drawn about the point and the vectors on this surface are counted in terms of their projections perpendicular to the surface, the divergence is positive if there are more vectors coming out of the surface than going into it. If there are more vectors directed into the surface than coming out, the divergence is negative: similarly, if the net count of those extending outward minus those extending inward is zero, the divergence is zero. In line with this idea, when the divergence is positive at a point, the point is said to act as a "source" of the vector field, and, conversely, as a "sink" when the divergence is negative. Finally, when the divergence is zero at a point, the point is called "source-free."* The three possible divergences at a point P are shown in the illustration.

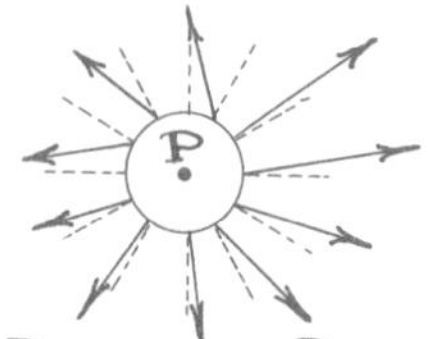

DIVERGENCE NEGATIVE
P IS A SINK

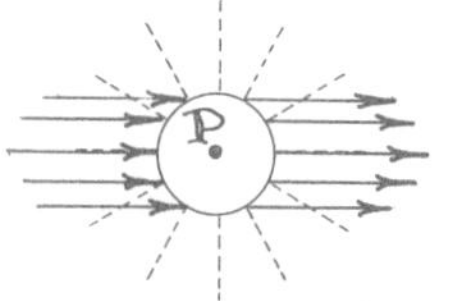

DIVERGENCE POSITIVE
P IS A SOURCE

DIVERGENCE ZERO
P IS SOURCE-FREE

* Actually, the earliest reference to the divergence operator was in terms of a "convergence." The possible confusion of this word with the idea of the convergence of a series led to the adoption of the word "divergence" to represent the "source" or "sink" properties at a point.

Returning to Laplace's equation, it may be written symbolically as

$$\frac{\partial}{\partial x}\left(\frac{\partial V}{\partial x}\right) + \frac{\partial}{\partial y}\left(\frac{\partial V}{\partial y}\right) + \frac{\partial}{\partial x}\left(\frac{\partial V}{\partial z}\right) = 0,$$

or as

$$\text{divergence }\{\text{gradient of } V\} = 0.$$

Grad V produces a vector field, whereas, according to Laplace's equation, the divergence of this field is equal to zero, and there is an implication that this field is source-free at the points where div (grad V) $= 0$. This, of course, is the original condition for Laplace's equation. These concepts of the field and the potential surface are easily pictured for the gravitational force of a point mass or for the electrostatic force of a point charge: both forces vary as the inverse of the square of the distance r from the source point. Because the forces are central, the force fields must be isotropically distributed about the source point. This is to say that the fields are radial and everywhere perpendicular to spherical surfaces concentric with the source point. As a result the potential surfaces must be concentric spheres.

Notice in the figure that the lines can be viewed as continuous; then the area density of the fixed number of lines falls off as 1 over the surface area of the sphere, or as $1/4\pi r^2$. The concept of line density as associated with the field strength was put forward by Faraday, and it is introduced here as a useful graphic concept.

The theory of probability was another of Laplace's major interests. His *Théorie analytique des probabilities*, although ponderous, contains a

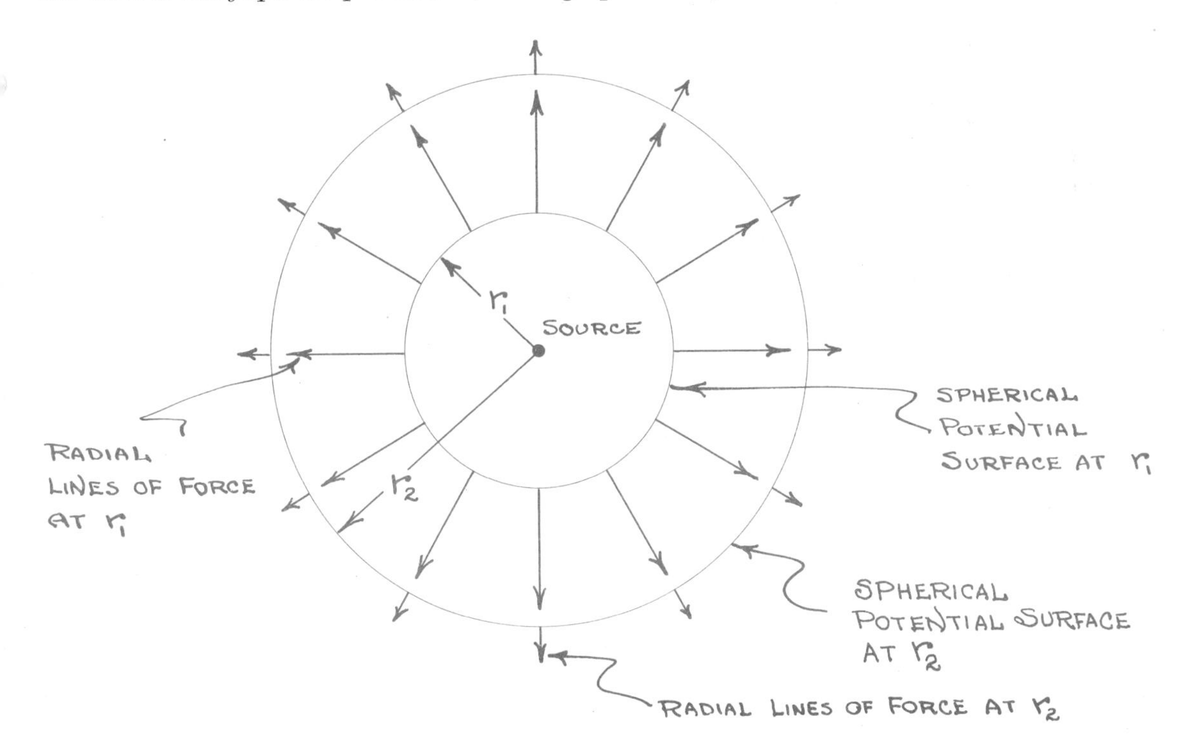

general survey of the methods for solving problems and includes many of the contributions of earlier authors. Differential equations and their solutions were first introduced by Leibniz, and, as we have seen, the methods for solving the partial differential equations were expanded by the Bernoullis and Euler, who uncovered some of the earliest special functions associated with each differential equation. While investigating the oscillation of heavy chains, Daniel Bernoulli encountered a quantity which was later to be named the Bessel coefficient of zero order. The vibration of stretched membranes led Euler to more general Bessel coefficients in 1764, and seven years later Lagrange encountered the same functions in elliptic motion.

The earliest method for solving differential equations, and still the most useful, is by separation of variables. This technique was actually stumbled upon by Leibniz. Jacques Bernoulli later solved the wave equation by the method of separation of variables. By 1723 it was recognized that ordinary differential equations of first order do not necessarily have solutions which can be expressed in terms of elementary functions; and in 1762 Euler solved second-order differential equations by an infinite series and invented the method of variation of parameters. The solutions to Laplace's equation provided the scientists and mathematicians with a valuable array of special functions; the versatility of this equation comes in part from its form and its solution in various coordinate systems. Laplace's equation in Cartesian coordinates x, y, and z provides exponential or sinusoidal solutions; in cylindrical coordinates one obtains Bessel functions, trigonometric solutions, and exponentials; and in spherical coordinates the solutions evolved are power series in the distance from the origin, polynomials of trigonometric functions (called Legendre polynomials), and exponential functions.

To illustrate the basic method of separation of variables in Laplace's equation, consider the formulation in Cartesian coordinates and assume that the potential function $V(x, y, z)$ can be written as the product of three functions $X(x)Y(y)Z(z)$, each depending upon only one of the independent variables: $V(x, y, z) = X(x)Y(y)Z(z)$. Substituting this trial form into Laplace's equation, one obtains

$$Y(y)Z(z)\frac{d^2X(x)}{dx^2} + Z(z)X(x)\frac{d^2Y(y)}{dy^2} + X(x)Y(y)\frac{d^2Z(z)}{dz^2} = 0$$

(notice that because two of the functions do not contain the derivative variable they are brought outside). Dividing by V and bringing one term to the right, we find that

$$\frac{1}{X(x)}\frac{d^2X(x)}{dx^2} + \frac{1}{Y(y)}\frac{d^2Y(y)}{dy^2} = \frac{-1}{Z(z)}\frac{d^2Z(z)}{dz^2}.$$

The term on the righthand side depends only upon z. If the equality is to hold for arbitrary values of all of the independent variables, both the right and left sides of the equation above must be equal to the same constant, $-m^2$:

$$\frac{1}{Z(z)}\frac{d^2Z(z)}{dz^2} = -m^2;$$

or

$$\frac{d^2Z}{dz^2} + m^2Z = 0,$$

and

$$\frac{1}{X}\frac{d^2X}{dx^2} + \frac{1}{Y}\frac{d^2Y}{dy^2} = m^2.$$

We recognize at once that the solution for $Z(z)$ is either a trigonometric function or a real exponential, depending on whether m^2 is a positive or negative constant. The previous process can be repeated, separating the equation in X and Y again, giving the same types of solutions with different separation constants.

Initially, one is struck by the fact that the solutions are not uniquely determined; in other words, they can be either exponentials or sines and cosines. The specific form is determined when account is taken of the constraints placed upon the problem. Usually these constraints amount to stating the values of $V(x, y, z)$ on the boundaries. If, for example, the problem is confined inside a cube, the values of $V(x, y, z)$ must be specified on the surface of the cube. The separation of variables gives rise to three functions, and in Cartesian coordinates these are all of the same general type. For other systems this is not the case; therefore different special functions are generated whenever new coordinates are employed.

Adrien Marie Legendre (1752–1833), a contemporary of Laplace, was first appointed to the École Militaire in Paris, through the influence of D'Alembert, and later to the École Normale. His interests ranged from number theory and geometry to analysis. For his early studies on the motion of projectiles he was awarded a prize by the Berlin Academy. In his discussion of the theory of numbers he evolved the law of quadratic reciprocity, which was later given a rigorous proof by Gauss. This law states that when the squares of the natural numbers 1, 2, 3, etc., are divided by a number n, there is a unique finite set of integers which are the remainders; these are called the quadratic residues of n. The law of quadratic reciprocity states that in congruences $x^2 \equiv q \pmod{p}$ and $x^2 \equiv p \pmod{q}$, p and q both being prime, both congruences are solvable if one of p and q is of the form $(4m + 1)$. If p and q are both of the form $(4m + 3)$, then one of the congruences is solvable and the other is not. The symmetry of this result led Legendre to apply the term "reciprocity" to the theorem. Since the notation employed above will be encountered again, it is necessary to outline clearly what is meant by $x^2 \equiv p \pmod{q}$, etc. Two integers are said to be congruent (written $\equiv$) modulo q if they leave the same remainder on division by q. As an example, 2, 7, 12, 22, $\ldots, -3, -8, -13, -18, \ldots$ are all congruent modulo 5 (note that $-1 \times 5 + 2 = -3$).*

Legendre extended the theory of extrema. A basic difficulty of this problem was to establish the necessary conditions. Euler had found these for one variable and Lagrange for two, but neither had been able to establish criteria to distinguish between maxima and minima.† This was done by Legendre.

* R. Courant and H. Robbins, *What Is Mathematics?* (Oxford: Oxford University Press, 1941), p. 32.

† Scott, *A History of Mathematics*, p. 198.

The study of the length of an arc and of the large amplitude vibrations of the simple pendulum had led Bernoulli and Euler to the elliptic integral. In 1694 Jacques Bernoulli, working with a problem in elasticity, expressed the opinion that these integrations were impossible in terms of elementary functions. Euler proposed that the elliptic integrals be recognized as new transients. Legendre's contributions to this subject were invaluable: he showed that a rational function of x and the square root of x or a quadratic function of x can be integrated in terms of the elementary functions. The class of rational functions of x and the square root of a cubic or quadratic in x cannot be handled in the customary manner. Elliptic integrals are integrals of the latter type of rational function. A characteristic elliptic integral known as an integral of the first kind is:

$$u(k, x) = \int_0^x \frac{dt}{\sqrt{(1 - t^2)(1 - k^2t^2)}}\cdot$$

In all there are three general types of elliptic integrals and Legendre devoted himself at one time or another throughout his life to these functions; among other things, he computed elaborate tables of these integrals. Of foremost concern was the algebra of the integrals and the differentiation theorems, but in spite of his thoroughness and concentration, he missed the crucial simplification of these integrals, which was to be supplied in his lifetime by Abel.

A significant characteristic of these integrals was that they were doubly periodic. Single periodicity was formalized for the exponential functions when Euler found in 1748 the real and imaginary periods of the trigonometric and hyperbolic functions. Nothing in these studies suggested a more general function with two distinct periods, from which both the circular and hyperbolic functions could be derived as special cases. To illustrate the relation between $u(k, x)$ above and the singly periodic functions, regard the case for $k = 0$:

$$u(0, x) = \int_0^x \frac{dt}{\sqrt{1 - t^2}} = \sin^{-1} x \quad \text{(or arc sine } x\text{)};$$

or $x = \sin u(0, x)$. This clearly suggests the connection. Again, the double periodicity was discovered by Abel only in 1825. Nonetheless, Legendre's interest in both the elliptic integrals and the gamma and beta functions provided a rich field for testing the methods of complex integration which were later invented by Cauchy.

To physicists the name of Legendre is immediately associated with the special functions which are generated from the angular part of Laplace's equation in spherical coordinates. The fact that one differential equation with a single set of boundary conditions has a large number of solutions leads to the dilemma of deciding which of the many special functions is to be used as a solution. After examining the problem of the vibrating string, Daniel Bernoulli gave a solution formed of a series of sinusoidal functions of the kind now named after Fourier. A controversy raged for years over the existence of seemingly multiple solutions of one natural problem. The paradox was almost resolved by Lagrange, who regarded the continuous string of mass per unit length μ as the limiting case of a finite set of equal mass particles equally spaced on a weightless string. The example of mass points on a weightless string illustrates the existence of independent solutions, as well as the need to add these independent solutions in order to represent arbitrary configurations of the system. Each mass vibrates parallel to the vertical axis in these demonstrations.

As an example, consider three mass points, m, equally spaced upon a string of length L. When the problem is solved subject to the boundary condition that the amplitude of the string at the end points $x = 0$ and $x = L$ must be 0, three independent solutions are obtained. By independent, one means that it is impossible to obtain any one of the three configurations by a linear combination of the other two. This is analogous to the statement in vector algebra that the projection of a vector upon any one of three mutually perpendicular axes cannot be represented in any manner by projections upon the other two axes. Thus the statement concerning the independence of solutions is analogous to the concept of dimensionality of a space; in fact, at the end of the nineteenth century these independent solutions in the case of differential equations will be regarded explicitly in terms of the dimensionality of an abstract function, space. The independent configurations, or modes, of vibration of three mass points, together with the corresponding modes of a continuous string of the same length, are illustrated in the figure.

Notice that the boundary condition, that the end points be fixed, is satisfied in each case. The validity of using the continuous solutions in series is clearly demonstrated by the fact that any arbitrary configuration of the oscillating mass points can be represented by a linear superposition of the

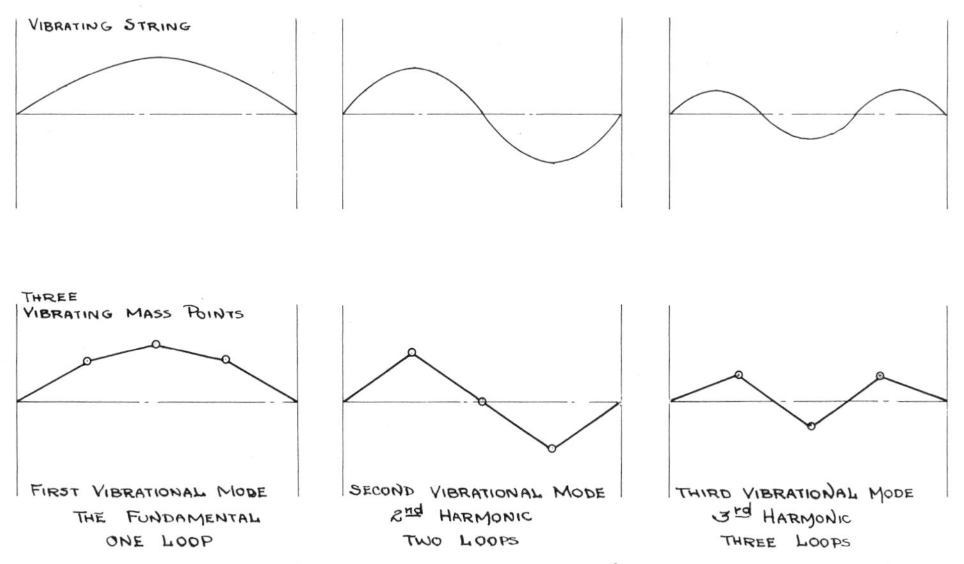

(for vibrating mass points the total number of modes equals the number of points) three independent modes. Suppose one asks for a configuration of one mass up and all other mass points at zero. This can be done by adding diagrams 1, 2, and 3 with different weights assigned. Diagram 1 must be multiplied by a number less than 1 in order to bring the center mass of diagram 3 to 0 when the two are added. This leaves an excess in the difference between the sum of 1 and 3 and diagram 2. By a relative scaling of the sum and diagram 2, the difference of the two will produce a configuration with only mass point 1 above 0.

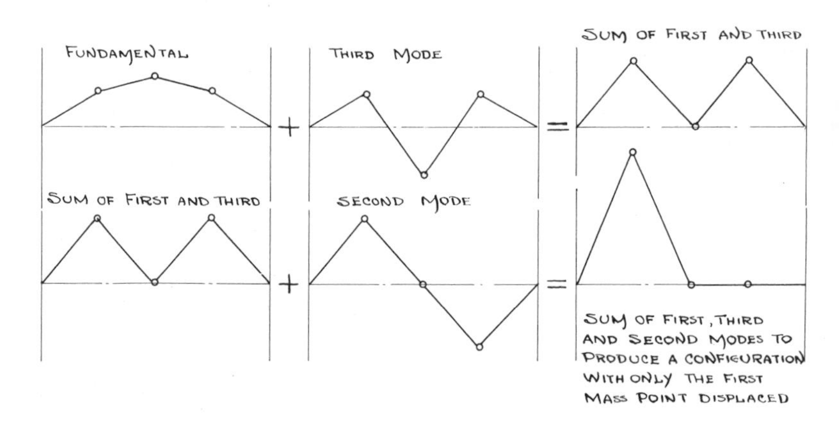

The fact that different configurations can be produced by linear combinations of the infinite set of solutions to the continuous string is a completely different question than that of generality, i.e., can any arbitrary configuration be represented by a linear superposition of all the independent solutions of a given problem? This question was not settled until Hilbert attacked the general problem at the end of the nineteenth century.

By creating the continuous string problem as a limiting case of N point masses on a weightless string, Lagrange came very close to the discovery of the general theorem for determining the expansion, or Fourier, coefficients, that is, the relative weights with which the independent solutions are added. Lagrange reached a step in the analysis at which it was necessary to interchange the order of a sum and an integral; his mathematical conscience restrained him because he was unable to show rigorously that this interchange was valid.

Jean Baptiste Joseph Fourier (1768–1830) submitted a memoir on heat conduction in 1807 to the Paris Academy in which he confronted the problem and postulated the Fourier series. The panel considering the memoir, made up of Lagrange, Laplace, and Legendre, rejected it because it lacked rigorous proof. Although much criticized for this decision, these men demonstrated an understanding of the difficulties associated with this question. What could not be easily realized and posed a formidable problem was that the proof of the infinite series of trigonometric functions was complete, in that it contained all of the functions necessary to provide a representation of an arbitrary configuration of the system! Fourier, like others before him, recognized the versatility of the series and considered this sufficient evidence of its validity and acceptance.

With the normal modes of the continuous vibrating string shown earlier, the manner in which nonsinusoidal waves are built up can be indicated by superposing harmonics. This expansion theorem was later generalized to include all complete sets of functions derived from the solution of linear differential equations and became the cornerstone of electromagnetic theory, elasticity, hydrodynamics, and, finally, quantum mechanics.

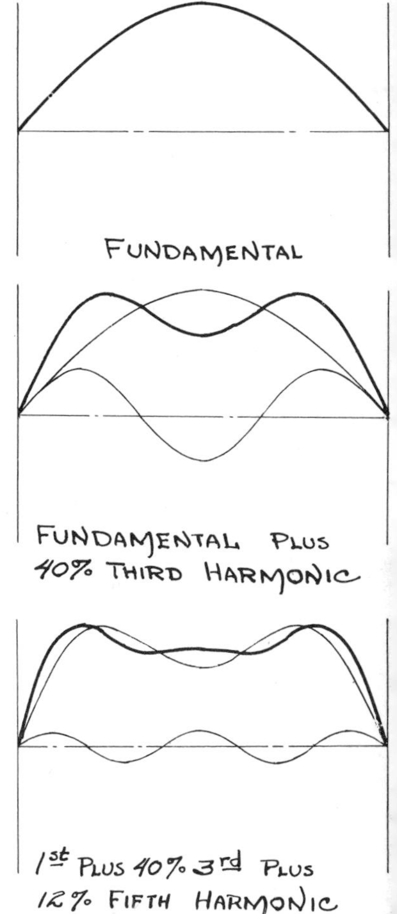

Implicit in Cardano's solution for the cubic equation was the imaginary number $\sqrt{-1}$, and the first hint at complex numbers occurred when John Wallis in 1797 represented $a + \sqrt{-1}\,b$ as a doublet of numbers (a, b). In 1710 Roger Cotes explicitly employed $\sqrt{-1}$ in Cotes's formula, and this was followed by the equivalent application in De Moivre's theorem. By 1750 in the exponential and trigonometric calculus the complex function was an important aspect of analysis. Some approach to a graphic illustration of complex numbers had been made by Wallis, who remarked that negative areas were pictorially acceptable once the negative line was introduced. He further argued that whereas $\sqrt{ab}$ was the mean proportional between $+a$ and $+b$, by analogy $\sqrt{-ab}$ was the mean proportional between $-a$ and $+b$ (or $+a$ and $-b$). He then conceived of measuring one length perpendicular to the other, taking one axis as the imaginary.

A Norwegian surveyor, Caspar Wessel (1745–1818), gave the first geometric treatment of complex numbers in 1797. Unaware of Wessel's paper, Jean Robert Argand in 1806 rediscovered the geometry of $a + \sqrt{-1}\,b$, and the corresponding diagram was named for him. This geometry of complex numbers was a trivial step, however, coming after

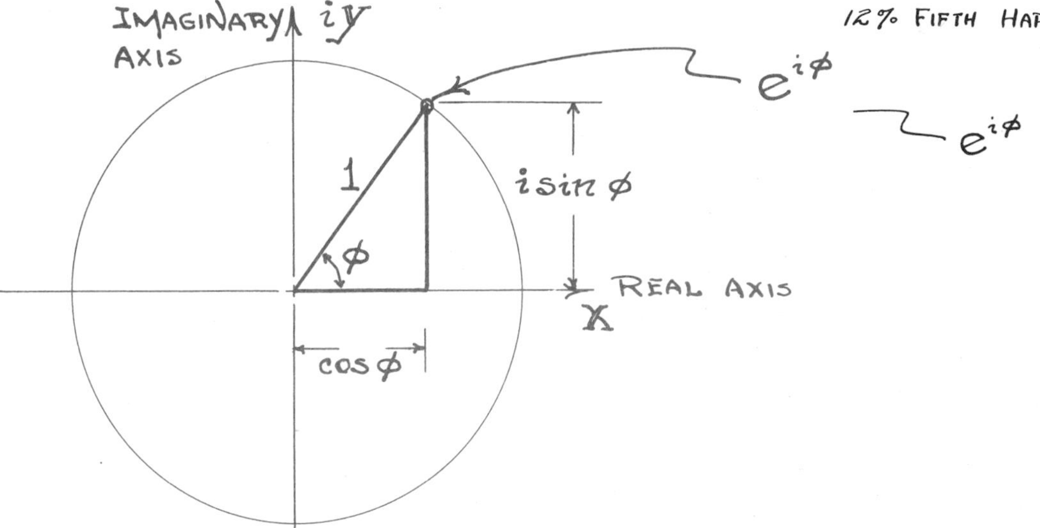

Euler's proof that $e^{i\phi} = \cos\phi + \sqrt{-1}\sin\phi$, where the projection of a unit radial line on two mutually perpendicular axes is implied.

Before examining the complex plane further, some of the properties of complex numbers, addition, multiplication, and division should be noted. These were referred to by John Wallis as early as 1673. It is customary to assign the symbol i for $\sqrt{-1}$. To add and multiply by $\sqrt{-1}$, symbols such as $2i$, $5i$, $-i$, $2 + 7i$ must be formed, or, more generally, $a + bi$, where a and b are any two real numbers. The symbol $a + bi$ (or $a + ib$—notice here that the number b and the symbol i can be commuted) is called a complex number with a real part, a, and an imaginary part, ib. Operations of addition and multiplication are performed as though i were an ordinary real number, except that i^2 should always be replaced by -1, and i^n is $(-1)^{n/2}$. Addition of two numbers $a + ib$ and $c + id$ entails addition of the real parts and addition of the imaginary: $(a + ib) + (c + id) = (a + c) + i(b + d)$.

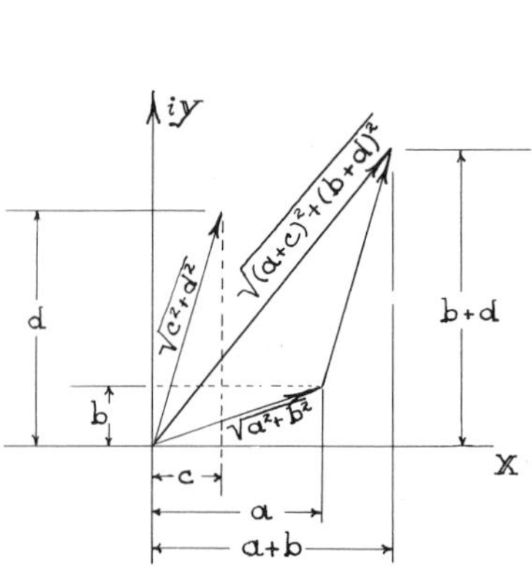

This operation has an appealing representation in the Argand diagram as the addition of two vectors of lengths $\sqrt{a^2 + b^2}$ and $\sqrt{c^2 + d^2}$. Multiplication gives four terms which can be grouped into a real product and an imaginary product: $(a + ib)(c + id) = (ac - bd) + i(ad + bc)$. In other words, $(a + ib)(c + id) = ac + iad + ibc + i^2bd$. Division is defined for nonzero complex numbers; that is to say, division by zero is excluded. A complex zero is a number for which both the real and imaginary parts are zero. Division is ordinarily handled by converting the denominator to a real number:

$$\frac{a + ib}{c + id}\cdot\frac{(c - id)}{(c - id)} = \frac{(ac + bd)}{(c^2 + d^2)} - i\frac{(ad - bc)}{(c^2 + d^2)}.$$

To exhibit the real and imaginary parts of the division illustrated above, we multiply the numerator and denominator by the complex conjugate of the denominator. The complex conjugate of a number $(a + ib)$ is determined by changing the sign of the imaginary component, giving $(a - ib)$. This is usually denoted by an asterisk: if z is complex, z^* is the complex conjugate. A unique property of the complex conjugate times the original is a real number (which is the square of the magnitude of the original number: $z\cdot z^* = (a + ib)(a - ib) = a^2 + b^2$, or $z\cdot z^* = |z|^2$. The field of complex numbers can be represented as the collection of points in a two-dimensional space. The projection of a point on the horizontal axis is the real part, and the projection of the vertical axis (labeled i) is the imaginary part.

In the diagram the point P represents the complex number $a + ib$, and the line OP measuring the distance of P from the origin O is designated as the modulus (or magnitude) of the number and is written as the square root of the product of the number and its conjugate:

$$\text{modulus } (a + ib) = |a + ib| = \sqrt{(a + ib)(a - ib)} = \sqrt{a^2 + b^2}.$$

Here the form $(a^2 + b^2)$ obtained from $(a + ib)(a - ib)$ suggests the

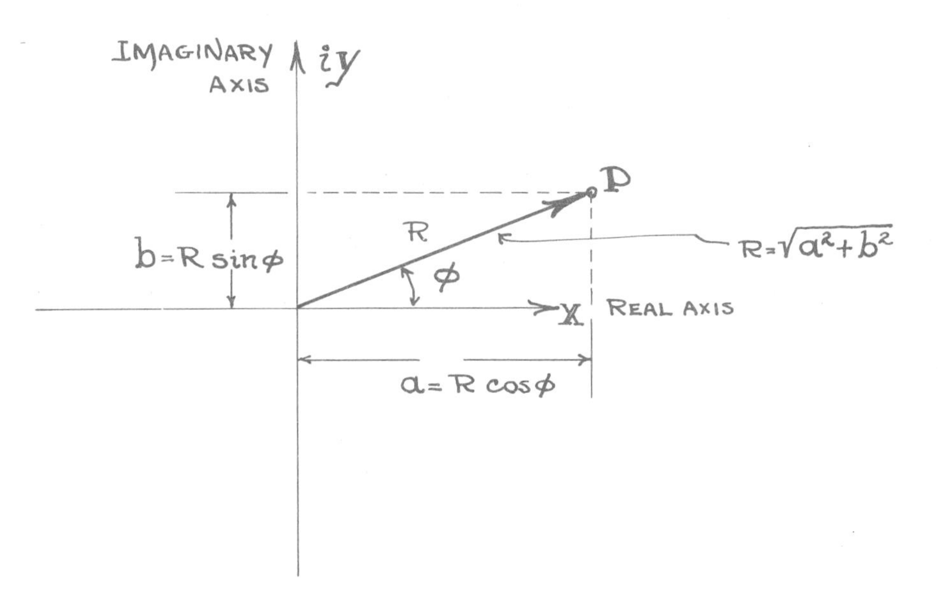

Pythagorean theorem, and a vector representation in the complex plane provides geometric meaning.

Euler's expansion of $e^{i\phi}$ can be utilized to provide a polar form of the complex number:

$$a + ib = \sqrt{a^2 + b^2}\left\{\frac{a}{\sqrt{a^2 + b^2}} + i\,\frac{b}{\sqrt{a^2 + b^2}}\right\},$$
$$= |a + ib|(\cos\phi + i\sin\phi) = |a + ib|e^{i\phi},$$

where $\tan\phi = b/a$.

This reduction illustrates the connection between the modulus, $\sqrt{a^2 + b^2}$, and the argument (or angle), ϕ. Just as the Cartesian form has two parts, the real and imaginary, the polar form is composed of two terms, the modulus and the argument. The polar form provides a convenient multiplication for two numbers $(a + ib)$ and $(c + id)$:

$$(a + ib)(c + id) = \sqrt{a^2 + b^2}\, e^{i\phi} \times \sqrt{c^2 + d^2}\, e^{i\theta}$$
$$= \sqrt{(a^2 + b^2)(c^2 + d^2)}\, e^{i(\phi + \theta)},$$

where $\tan\phi = b/a$ and $\tan\theta = d/c$. From this result one observes that the magnitude of a product is the product of the magnitudes

$$|(a + ib)(c + id)| = |(a + ib)|\,|(c + id)|.$$

This description of the properties of complex numbers anticipated the work of Gauss, who in 1831 expanded the geometric representation and deduced the properties of complex numbers from accepted postulates of common arithmetic. Graphically, the addition of two complex numbers is completely analogous to the addition of two vectors. Complex multiplication has sufficient freedom that several vector multiplications are implied. The diagram illustrates two specific numerical problems. The vector

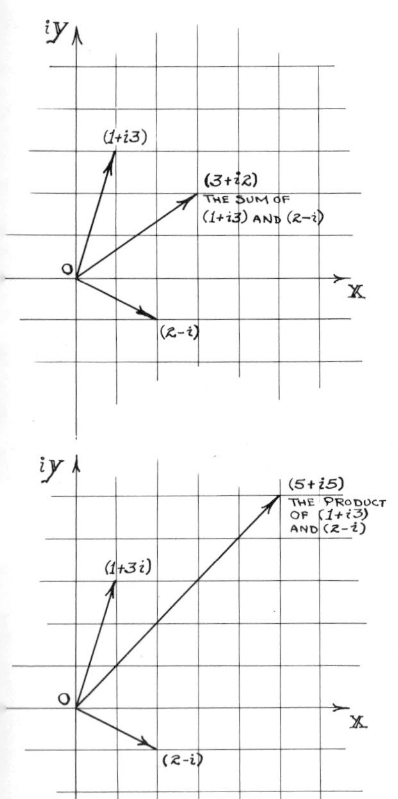

analogue is pictorially obvious in the addition. Vector multiplication has an analogue when one complex number, say $a + ib = \rho_1 e^{i\phi_1}$, is multiplied by the complex conjugate of a second number, $(c + id)^* = c - id = \rho_2 e^{-i\phi_2}$. Then

$$(a + ib)(c + id)^* = (a + ib)(c - id) = \rho_1\rho_2 e^{i(\phi_1 - \phi_2)}$$
$$= \rho_1\rho_2 \cos(\phi_1 - \phi_2) + i\rho_1\rho_2 \sin(\phi_1 - \phi_2).$$

Here the real part of the product is the product of the magnitude of one vector and the projection of the second vector on the first, i.e., $|a + ib|\ |c + id|$ times the cosine of the angle between them. The imaginary part of this product is the product of the magnitudes times the sine of the angle between the two vectors. Such vector products arise only in multiplication between one complex number and the complex conjugate of the second; this is apparent from the fact that only in this form is the angle between the two vectors generated, i.e., $(\phi_1 - \phi_2)$. Straight multiplication without conjugating one of the complex numbers gives the sum of the angles ϕ_1 and ϕ_2, which does not express the angle between the two vectors in any manner.

By the end of the eighteenth century the controversy between the theorists supporting the ray-corpuscular theory of light and those supporting the wave theory began to resolve in favor of the wave theory. A new and stimulating champion of the wave theory arose, Thomas Young (1773–1829). Although trained in the practice of medicine, he is best remembered for his analysis of the interference of light. The experiment named after him shows that when monochromatic light from a single source is split into two narrow beams by an opaque mask with two slits in it, the distribution of light falling on a screen behind the slits shows alternating light and dark hues. Young explained that this resulted from waves; his original analysis employed the analogy of water waves progressing along two narrow channels and finally combining into one channel. He pointed out that if the waves combine in phase (with the maxima coinciding) the resultant wave would exhibit enhanced vibration. Conversely, if the waves combine out of phase, with the maxima of one falling on the minima of the other, no vibration results. Mathematically, this can be illustrated by adding two sinusoidal waves.

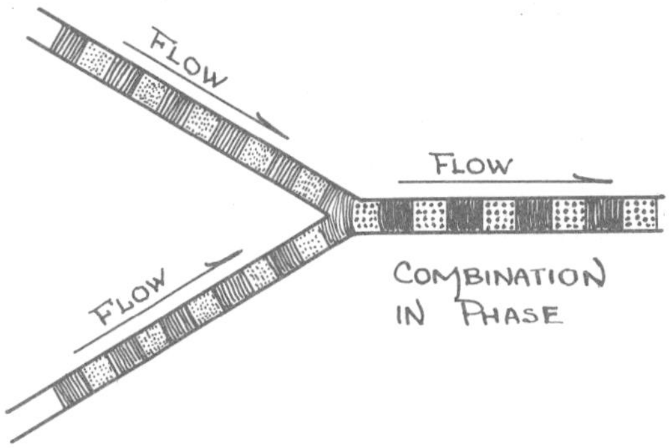

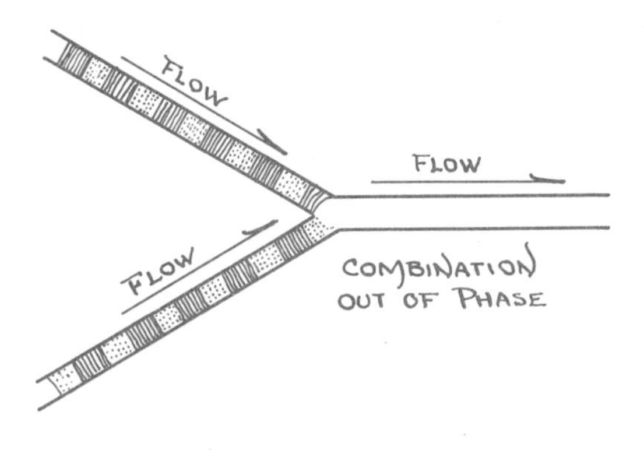

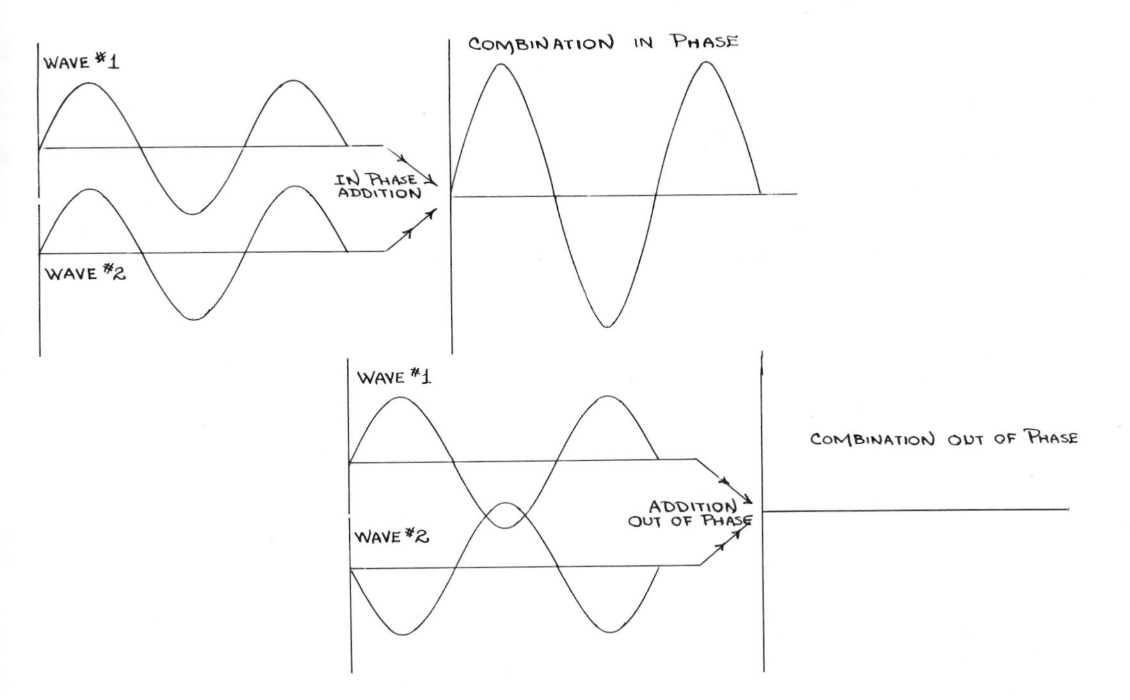

Young's experiment with light can then be illustrated, as shown here. With the light source S at the left, the crests and valleys are shown at any instant t. Since these are traveling waves, they must be conceived of as moving to the right with the velocity of light c. If they are traveling waves of a single frequency, mathematically the sinusoidal function is of the form $\sin [(2\pi/\lambda)x - 2\pi ft]$, where f is the frequency and λ is the wave length of the light. The total phase Φ of such a wave is $[(2\pi/\lambda)x - 2\pi ft]$. By following a point of constant phase (Φ = const), the velocity of motion of the wave can be determined from $d\Phi/dt$. If Φ = a constant, then the time rate of change is 0:

$$\frac{d\Phi}{dt} = 0 = \frac{2\pi}{\lambda}\frac{dx}{dt} - 2\pi f,$$

and the associated velocity of the point of constant phase is $dx/dt = \lambda f = c$ = the velocity of light. This demonstration provides us with the fundamental relation between frequency f and wave length λ. The wave length is the distance between adjacent maxima (or minima) at any instant.

Location of the dark hues relative to the center line can be found by observing that to reach the screen on either side of the center line the paths from the two slits are different in length by $d \sin \Theta$, where d is the distance between the slips and Θ is the angle between the line connecting the point of examination to the point halfway between the slits and the center line. When this path difference is equal to an odd number of half wave lengths, the two contributions arrive at the screen out of phase by π (or 180°) and destructive interference takes place, producing a null or dark point. The dark lines then are located according to $d \sin \Theta_{\text{null}} = (2n + 1)\lambda/2$;

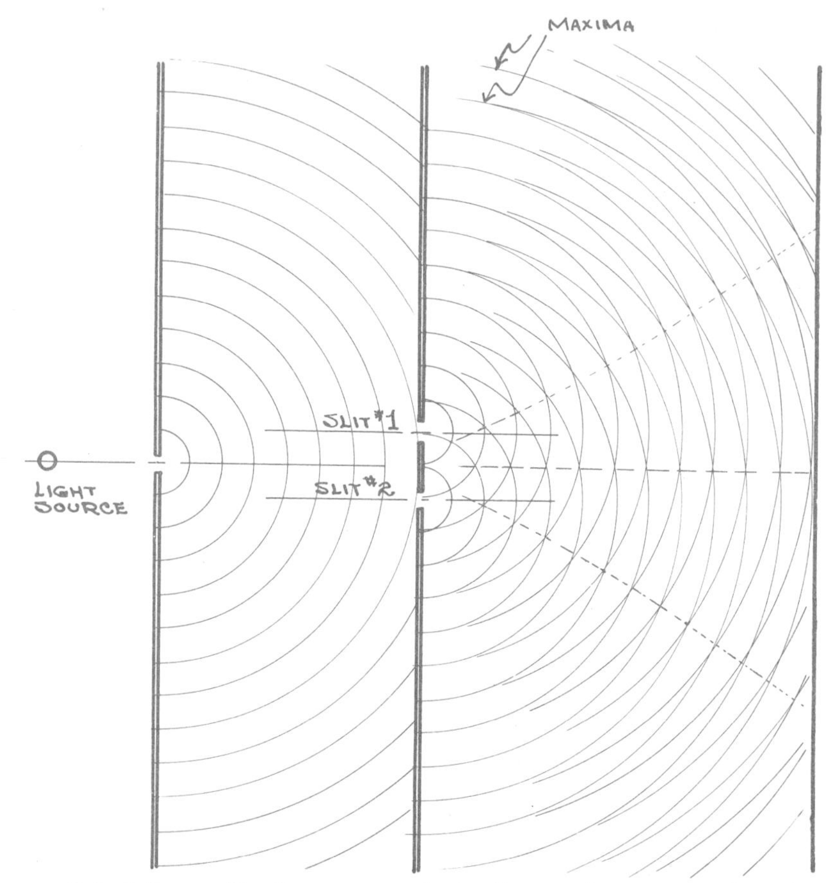

$n = 0, 1, 2, 3, \ldots$. By the same analysis, the maxima occur when the phase difference is a multiple of 2π (or 360°): $d \sin \Theta_{\max} = n\lambda$; $n = 0, 1, 2, 3, \ldots$.

Examination of the diagram reveals that Huygens' construction may be applied in that the light wave front, after being limited by a slit, is mapped out by assuming that each point on the emerging front radiates in spherical wavelets. Therefore, the light passed through the slit does not retain the geometric form imposed on it by the slit. Claiming that light would bend into the shadow if composed of wave motion, Newton had dismissed the wave theory, but by examining the shadow of a card interposed in a cone of light emerging from a single pinhole, Young found faint fringes of color at the edges of the shadow and, within the shadow itself, a sequence of faint dark and light bands parallel to the edges of the card.

In 1801 Young proposed his law of interference and explained the colors of thin transparent plates. The light reflected from the top surface of a plate will have a phase difference relative to light that has traveled to the back surface and is then reflected to combine with the first. This phase difference is related to the time of transit down and back through the plate. If the thickness of the plate is δ, and the velocity in the transparent plate is v (the index of refraction, $n = c/v$), the time of transit is $2\delta/v =$

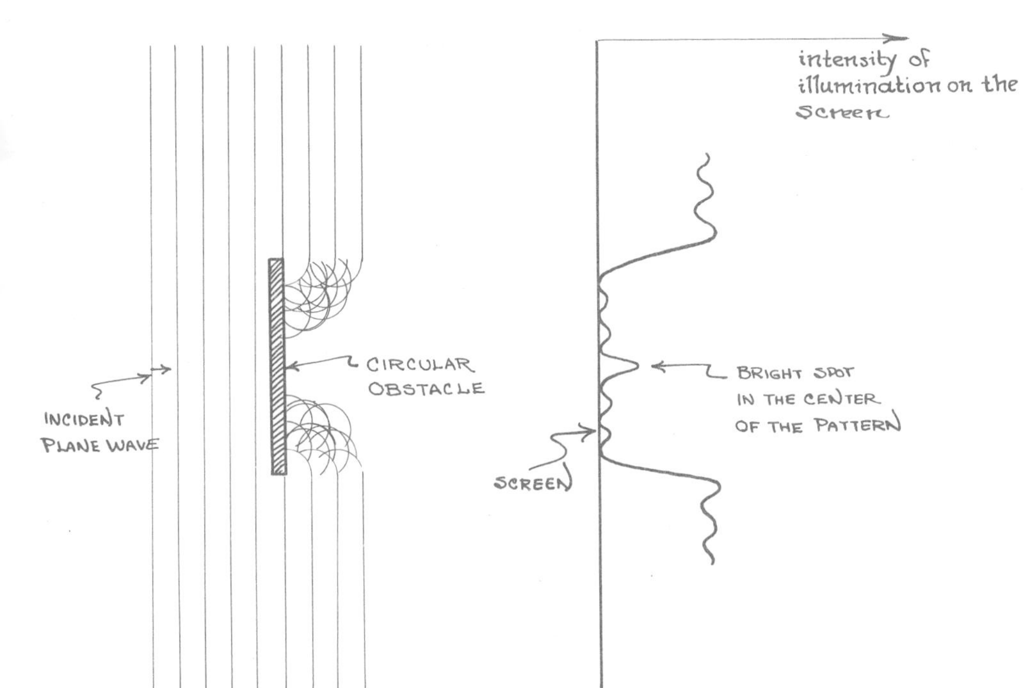

$2\delta n/c$. One would expect that the total phase shift ϕ is $2\pi f$ times the transit time, or $\phi = 2\pi f \times 2n\delta/c$. Thus if ϕ equals an integral multiple of 2π, one should expect constructive interference or reinforcement. Young found that this was not the case and that the crucial test of this formula arises for plates of much, much less thickness than a wave length. If upon reflection from each of the surfaces the phases are both zero, then the combination of the two reflected rays should exhibit constructive reinforcement. Experimentally this was not observed: very thin plates appeared dark or black; thus it was found that reflection taking place at an intersurface going from a less dense to a dense medium has a phase shift of π (or 180°), while reflection taking place going from dense to less dense has a zero phase shift. Consequently, the total phase shift difference between the two light components from a thin plate is $4\pi n\delta f/c - \pi$, or $\pi(4\pi\delta/\lambda - 1)$. Constructive interference occurs for phase differences of 0, 2π, 4π, etc., or $\delta = \lambda/4n$, $3\lambda/4n$, etc., and destructive interference occurs for $\delta = \lambda/2n$, λ/n, etc. White light consists of all colors, and when reflected from a thin plate reinforcement takes place for one color, while the others are canceled to some extent by destructive interference.

Further results upon the wave nature of light were offered by Étienne Louis Malus (1775–1812) and David Brewster (1781–1868). Malus found that light reflected from glass was similar to one ray transmitted by a calcite crystal, and this he attributed to "polarization" of the light—an anomalous term at the time because it was not known whether the vibration was transverse or longitudinal to the direction of propagation. Longitudinal waves occur in sound waves where the direction of vibration is along the direction of propagation. Brewster extended the researches of

Malus and found that the reflected light was completely polarized when the angle between the reflected and the refracted rays is 90° (or $\pi/2$ radians). At this time the expression "completely polarized" implied that the light was identical in behavior to one of the rays transmitted by a calcite crystal.

As the theory of light began to mature, the theory of electromagnetism was born. In 1820 Hans Christian Oersted (1777–1851) discovered the magnetic interaction between a magnetized needle and a current-carrying wire. Oersted had previously attempted to discover a connection between electrostatic charge and magnetism, without success. After observing the deflection of a compass needle in a thunderstorm, he began a series of experiments to study the behavior of the magnetized needle near a current-carrying wire. When the needle was at right angles to the wire, it aligned itself perpendicular to the wire; when it was parallel to the wire, the needle was deflected. Moreover, Oersted observed that when the wire was below the needle the direction of deflection was the reverse of the direction when it was above the needle.

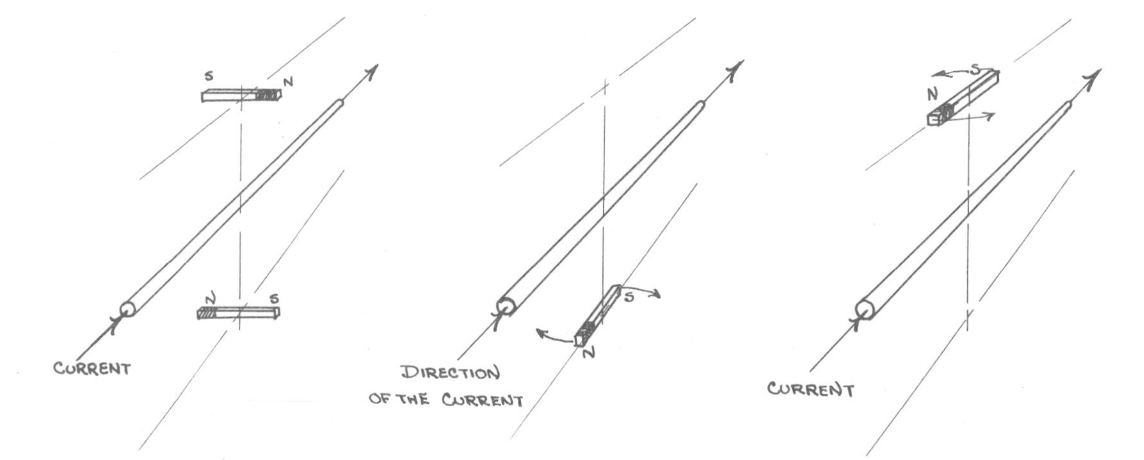

Following Oersted, Jean Baptiste Biot (1774–1862) and Félix Savart (1791–1841) gave an expression for the torque exerted on a bar magnet by a current-carrying wire. At the same time François Arago demonstrated that a current-carrying coil attracted iron filings, hitherto unmagnetized, which adhered to the coil as long as current passed through it. Since the filings dropped off the coil as soon as the current was cut off, he correctly deduced that the coil acted as a magnet when a current was passed through it.

The use of a magnetized needle as an indicator of magnetic fields was fortunate, in that the classical theory of magnetism stated that such a needle lines up along the magnetic field with the North Pole, indicating the direction of the field when the needle is in stable equilibrium. Further, if the needle is suspended in the magnetic field by an elastic suspension having a linear restoring torque, the frequency of the small oscillations of the needle about the equilibrium alignment is proportional to the strength of the magnetic field. The magnetic field vector is now denoted as $\vec{\mathbf{B}}$ and is called the magnetic induction vector, a somewhat inappropriate name. A magnetized needle has a magnetic moment, $\vec{\mathbf{M}}$, which will be defined more specifically later. The moment of a magnetized needle was traditionally described as the product of the magnitude of the magnetic pole strength at the ends times the distance between the poles. The direction of the vector $\vec{\mathbf{M}}$ is in the direction of the North Pole of the needle.

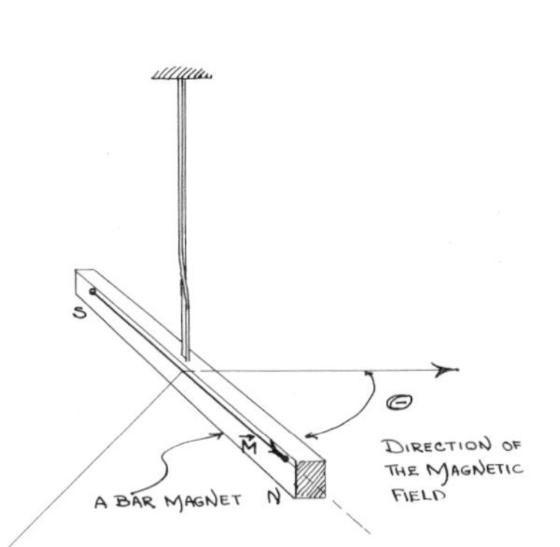

The restoring torque on $\vec{\mathbf{M}}$ when it is rotated through an angle away from alignment with $\vec{\mathbf{B}}$ is $-BM \sin \Theta$ (the minus sign means that the torque is directed opposite to the angular displacement, Θ). Newton's second law equates the time rate of change of the angular momentum of the needle, $I(d\Theta/dt)$ (I is the moment of inertia of the needle), to the torque. Thus with I constant,

$$\frac{d}{dt}\left(I\,\frac{d\Theta}{dt}\right) = I\,\frac{d^2\Theta}{dt^2} = -BM \sin \Theta,$$

and for small angles, where $\sin \Theta$ can be replaced by Θ,

$$\frac{d^2\Theta}{dt^2} + \left\{\frac{BM}{I}\right\}\Theta = \frac{d^2\Theta}{dt^2} + \omega^2\Theta = 0,$$

where the constant ω has been defined above as $\sqrt{BM/I}$. This expression is similar to that for the simple pendulum, and the reader will recognize the differential equation discussed earlier which has trigonometric solutions, $\sin \omega t$ and $\cos \omega t$. Here ω is 2π times the frequency of oscillation. By measuring the number of oscillations per unit time, we obtain $f = \omega/2\pi = (1/2\pi) \times [\sqrt{(BM/I)}]$, and $B = (I/M)(2\pi f)^2 = \text{constant} \times f^2$, where f is the frequency of oscillation.

With this background the method by which early experimenters could map out the direction and relative strengths of magnetic fields may be visualized. Biot and Savart discovered that the magnetic field of a long straight wire fell off as the inverse of the perpendicular distance from the long wire, and that the magnetic fields of a long straight wire were circles concentric with the wire. $\vec{\mathbf{B}}$ was proportional to I/r, where r is measured perpendicular to the axis of the wire, and the direction of $\vec{\mathbf{B}}$ was perpendicular to the plane formed by the axis of the wire and r. Because there are two directions perpendicular to a plane (positive and negative), a rule called the righthand rule was created to specify $\vec{\mathbf{B}}$. In the twentieth century this rule has been generalized mathematically into what is known as a "vector" or "cross" product. If $\vec{\mathbf{l}}$ and $\vec{\mathbf{r}}$ are to define a vector perpendicular to the plane of $\vec{\mathbf{l}}$ and $\vec{\mathbf{r}}$, we assign a positive direction by the following rule: The direction of $\vec{\mathbf{l}} \times \vec{\mathbf{r}}$ (pronounced $\vec{\mathbf{l}}$ cross $\vec{\mathbf{r}}$) is found by aligning the fingers of the right hand along the vector appearing first—in this example, $\vec{\mathbf{l}}$. The fingers are then curled or rotated toward the palm to form $\vec{\mathbf{r}}$ (this done through the smallest angle between $\vec{\mathbf{l}}$ and $\vec{\mathbf{r}}$). The thumb then points in the direction of the cross product, according to the right-hand rule (or the more modern vector product).

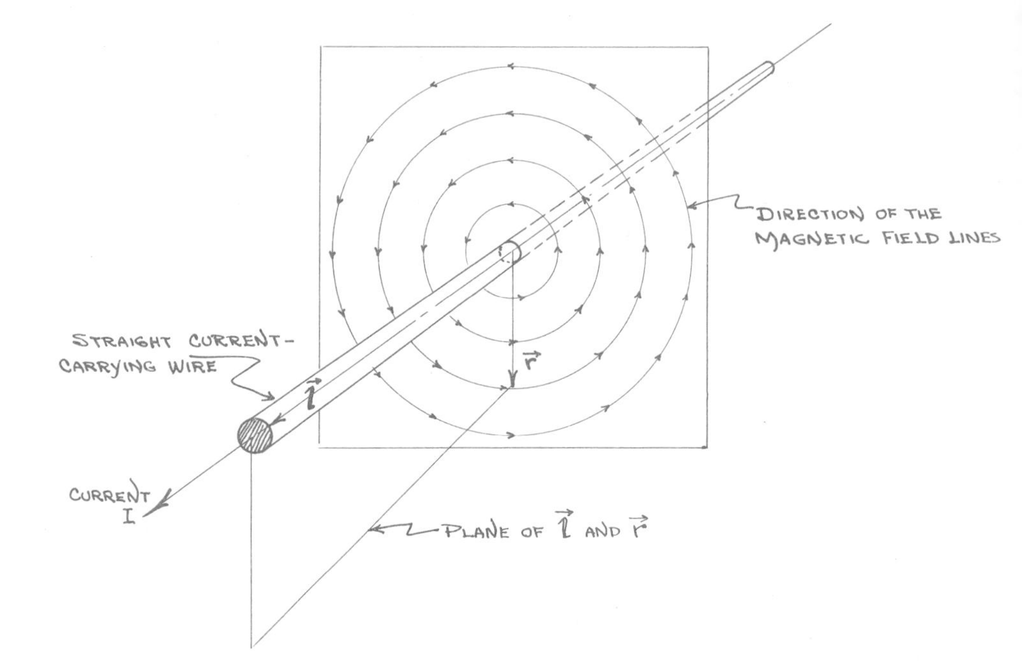

In modern vector terminology, then, $\vec{\mathbf{B}}$ is proportional to

$$\frac{I}{r}\left(\frac{\vec{\mathbf{l}}}{l} \times \frac{\vec{\mathbf{r}}}{r}\right).$$

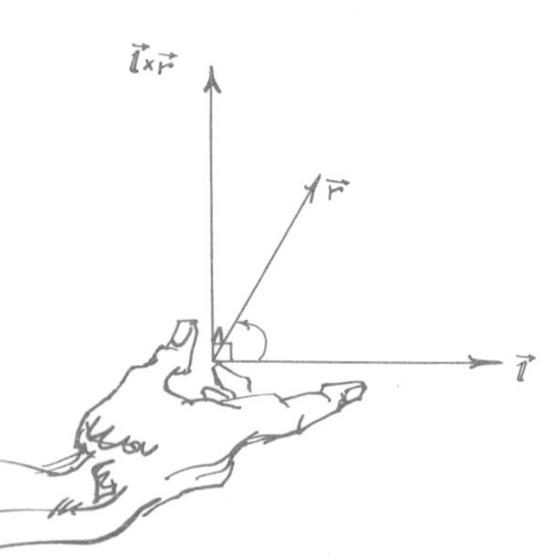

The vector terms in the parentheses have been divided by their magnitudes to provide vectors of unit magnitude. A vector or cross product not only has direction but also a magnitude equal to the product of the magnitudes of the two vectors times the sine of the angle between them. In the Biot and Savart law, for a long straight wire $\vec{\mathbf{r}}$ and $\vec{\mathbf{l}}$ were taken perpendicular to each other, therefore the sine of the angle between them is 1.

The first theory of electromagnetism was formulated by André Marie Ampère (1775–1836). Ampère analyzed the force between two parallel current-carrying wires and found that when the currents were parallel the wires were attracted, and when the current in one of the wires was reversed the force was repulsive. He further discovered that the force was proportional to the product of the magnitudes of the two currents and acted at right angles to the wires. The general formulation of Ampère's law for the force between two current elements is somewhat more complex in its geometric representation than the laws which had been formulated up to that time. The law of universal gravitation and the law of force between electrically charged bodies involve inverse square central forces. When two current elements interact, the magnetic fields are not central; rather, they are directed according to the righthand rule (or vector product) just discussed in connection with the Biot and Savart law. The force between two current elements, $I_1\,\Delta\vec{\mathbf{l}}_1$ and $I_2\,\Delta\vec{\mathbf{l}}_2$ (where $\Delta\vec{\mathbf{l}}_j$ is directed in the sense of positive current flow), is proportional to the product of the currents in the two wires $I_1 I_2$.

If the lengths $\Delta\vec{\mathbf{l}}_1$ and $\Delta\vec{\mathbf{l}}_2$ are small compared to the separation of the centroids, $\vec{\mathbf{r}}$, then the force is proportional to the product of the lengths times the sine of the angle Θ_1 between $\Delta\vec{\mathbf{l}}_1$ and $\vec{\mathbf{r}}$ and the sine of the angle Θ_2 between $\Delta\vec{\mathbf{l}}_2$ and a line perpendicular to the plane of $[\Delta\vec{\mathbf{l}}_1, \vec{\mathbf{r}}]$. Further, this force falls off as the inverse of r^2.

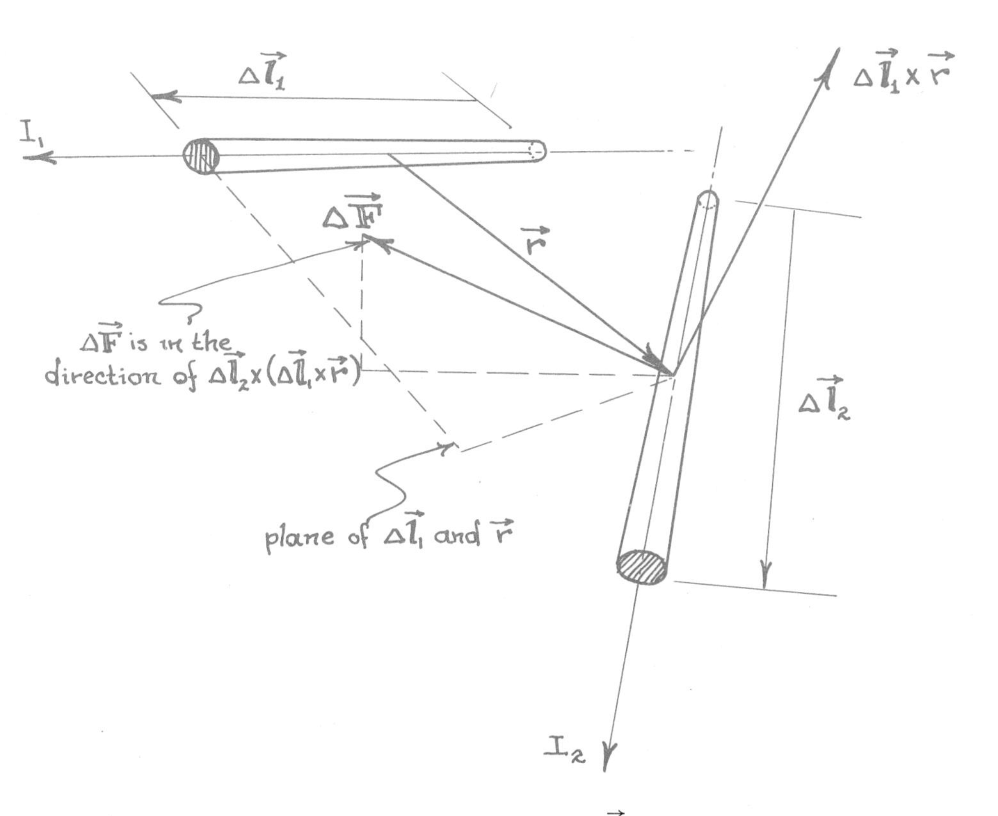

In terms of the more convenient cross product $\Delta\vec{\mathbf{F}}$, the force between $I_1\,\Delta\vec{\mathbf{l}}_1$ and $I_2\,\Delta\vec{\mathbf{l}}_2$ is proportional to $I_1 I_2\,\Delta\vec{\mathbf{l}}_2 \times (\Delta\vec{\mathbf{l}}_1 \times \vec{\mathbf{r}})/r^3$. When this

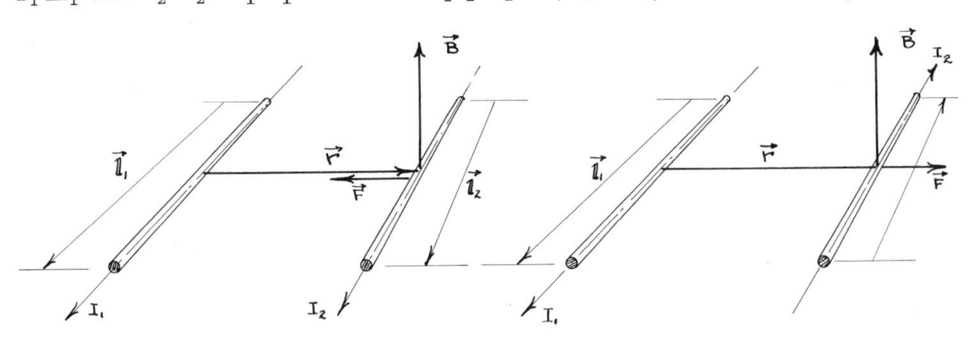

rule is applied to long straight wires, Ampère's observations appear in an elementary form. For parallel wires all vectors ($\vec{\mathbf{l}}_1$, $\vec{\mathbf{l}}_2$, and $\vec{\mathbf{r}}$) are either parallel or mutually perpendicular.

The constant of proportionality is a matter of units. Modern convention takes the constant of proportionality as 10^{-7} with the current I_j in "amperes" (coulombs per sec) and the lengths measured in meters. These units then give the force between the wires in "newtons." A newton is the force required to accelerate one kilogram of mass at one meter per second squared.

A more modern viewpoint permits Ampère's law to be factored into the field of one wire times the current and length of the other. The magnetic induction field of a current-carrying element $\Delta\vec{\mathbf{l}}$ is written as

$$\Delta\vec{\mathbf{B}} = \frac{\mu_0}{4\pi}\frac{I(\Delta\vec{\mathbf{l}} \times \vec{\mathbf{r}}/r)}{r^2},$$

where $\mu_0/4\pi = 10^{-7}$ is the constant proportionality for meter-kilogram-second (Mks) units, giving $\vec{\mathbf{B}}$ in webers per square meter. Examination shows that a weber per square meter corresponds to a newton-second per coulomb-meter. As stated, these are modern units and have been introduced here only to lend an air of reality to the proceedings.

In view of the form of $\Delta\vec{\mathbf{B}}$ above, the force $\Delta\vec{\mathbf{F}}$ exerted on the second current-carrying element $I_2\,\Delta\vec{\mathbf{l}}_2$ by the field $\Delta\vec{\mathbf{B}}_{21}$ of $I_1\,\Delta\vec{\mathbf{l}}_1$ is merely $\Delta\vec{\mathbf{F}} = I_2\,\Delta\vec{\mathbf{l}}_2 \times \Delta\vec{\mathbf{B}}_{21}$. The reader should be aware that successful as Ampère's law was, it was formulated historically in a much more complex form. Modern notation and some extrapolation of later forms have been used here only to make the formulation more concise and apparent. The magnetic induction field of a current element has been written as an element, $\Delta\vec{\mathbf{B}}$, because in the final result the total magnetic induction field at a point in the vicinity of a current-carrying wire must be computed by summing (i.e., integrating) the contributions from all of the current elements $\Delta\vec{\mathbf{l}}$. For instance, the magnetic field at the center of a current-carrying circular wire is perpendicular to the plane of the circle and directed according to the righthand rule. From the diagram we can see that the magnitude of the field $\vec{\mathbf{B}}_0$ at the center point is given by

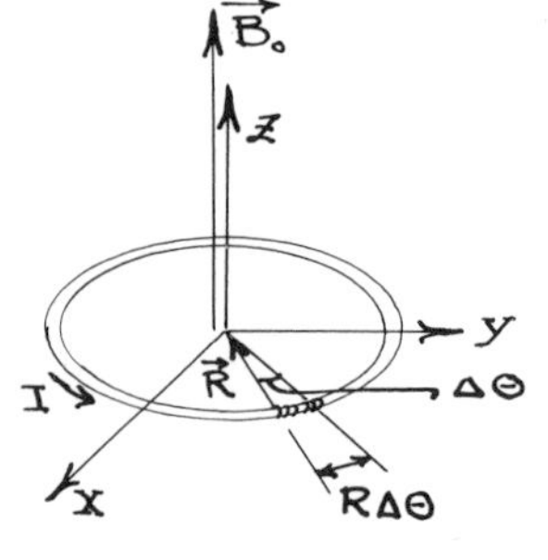

$$|\vec{\mathbf{B}}_0| \to \sum_{\text{all }\Delta l} \frac{\mu_0}{4\pi} I \frac{|\Delta l|}{R^2} \to \frac{\mu_0}{4\pi} I \int_0^{2\pi} \frac{R\,d\Theta}{R^2} = \frac{\mu_0 I}{2R}.$$

This example is particularly simple. Needless to say, the integrals for most current geometries are quite difficult. To realize this, one need only attempt to find a general expression for the field of a circular ring of current at a point off the axis of symmetry and not in the plane of the ring. The answer to this more general problem involves complete elliptic integrals of the first and second kind.

Electrostatic theory was advanced to the level of sophistication present in the theory of dynamics by Siméon Denis Poisson (1781–1840).

In 1813 he formulated the law of electrostatics as a differential equation, in much the same form as that of Laplace. Poisson based his formalism upon the potential function, and in contrast with Laplace's equation, which held for a source-free region, gave the general defining equation for regions which could contain charges or sources of the field. Obviously, his result would have to assume the form of Laplace's equation, where the charge densities were everywhere zero. The charge density ρ is defined as the charge per unit volume at a point; therefore the total charge in a small volume element $\Delta x\, \Delta y\, \Delta z$ would be $\rho(x, y, z)\, \Delta x\, \Delta y\, \Delta z$. The density was viewed as a function of the coordinates, a function which could vary from point to point in space. Poisson's equation defining the electrostatic potential V is:

$$\text{div (grad } V) = \frac{\partial^2 V}{\partial x^2} + \frac{\partial^2 V}{\partial y^2} + \frac{\partial^2 V}{\partial z^2} = -\rho,$$

where $V(x, y, z)$ is the potential function.

If we recall the definition of the divergence as a measure of the source or sink properties at a point, this differential equation can be read to say that the source function of the gradient of the potential V is the negative of the charge density ρ. Further, we might also recall that minus the gradient of V is a force field which has been defined as the electric field intensity.

The advantage of formulating Coulomb's law in terms of a differential equation involving a continuum of charge is that the solution can be expressed in a complete set of sinusoidal functions combined in linear combination to represent an arbitrary solution. Homogeneous differential equations provide complete sets of special functions, and the specific functions obtained depend upon the coordinate variable used to express the operator, "divergence of the gradient." The concept of completeness is actually tied in with the ability to represent arbitrary functions and was not thoroughly understood until the end of the nineteenth century; thus many problems involving complicated geometries and boundary conditions could be solved formally by expanding the solution as a linear sum of the complete set of these special functions. The useful solution arose when the solution was well approximated by a single function or the first few functions in the series, which allowed the remaining terms to be neglected.

VIII

Of Pride and Prejudice

When the curtain came down on the world of eighteenth-century mathematics, a comfortable aura of achievement had settled over its audience. Analysis still lacked a rigorous base, to be sure, but much had already been achieved in application, and the wide scope of its grasp left little doubt that the sound basis would soon be forthcoming. Mathematics appeared to be clearly delineated. A secure and fruitful future could be envisioned in classical physics, assured by the already breathtaking rise of electromagnetic formalism and the heights of sophistication to which the theory of dynamics had been carried. Unquestionably, the nineteenth century had earned the mantle it had inherited; as rigorous proofs of logic were assigned a leading role, small uncertainties which lay at the very basis of the classical mathematics of the eighteenth century were gradually uncovered.

Like Greek geometry, analytic geometry of two and three variables appeared to have reached the limits of its extension and had already yielded all that might be achieved with it. The limits upon its dimensionality, on the other hand, were felt to be still open to question, and more important, the turn of the century saw a reexamination of the fundamental postulates of geometry. Although the most important questions in algebra seemed to turn upon the existence theorems for roots of polynomials, the axioms of commutability were very early overthrown and broader avenues explored. Number theory was confronted with the existence of complex numbers, and investigations into these doublets caused great upheavals in algebra. These and many other perplexities were the natural inheritance of the nineteenth-century scientists and mathematicians, among the foremost of whom was Karl Friedrich Gauss (1777–1855).

Gauss received his early education in his native city of Brunswick; and if Fermat was the "prince of amateurs," Gauss may be acclaimed the "emperor of mathematics." His grasp of this discipline was as profound as it was productive and had its origins in his earliest days—before entering the university he not only had mastered the binomial theorem but was thoroughly aware of the subtle dangers involved in employing it uncritically. In the expansion of $(1 + x)^n$ the series is infinite if n is not a positive integer. From the geometric series we know that

$$(1 + x)^{-1} = 1 - x + x^2 - x^3 + x^4 - \cdots + (-1)^n x^n + \cdots.$$

This series is obviously divergent unless x is greater than -1; conversely, at $x = +1$ the function on the left is well defined, while the series expansion on the right is not; i.e., $1/2 = 1 - 1 + 1 - 1 + 1 - 1 \cdots$. Such paradoxes led Gauss to seek a rigorous method for testing and applying infinite series.

Confronted with the ancient Pythagorean sequence of the sum of the integers from 1 to N, while still a youth Gauss wrote the answer by displaying the sequence twice, once in ascending and once in descending order:

$$\begin{array}{cccccc} 1 + & 2 + & 3 + & 4 + \cdots + & (N-1) + N \\ N + (N-1) + & (N-2) + & (N-3) + \cdots + & & 2 + 1 \\ \hline (N+1) + (N+1) + & (N+1) + & (N+1) + \cdots + & & (N+1) + (N+1) \end{array}.$$

Obviously, in this presentation the two sequences result in $(N + 1)$ added N times to give the sum of one sequence (the triangular number of side N) as $(1/2)N(N + 1)$. Gauss, who had not yet entered the university, created the method of least squares, whereby the most probable value of a variable quantity could be obtained from a large number of observations. Let f be some unknown quantity that we wish to determine as accurately as possible. If a set of n readings yields values $x_1, x_2, x_3, \ldots, x_n$, it is customary to select the arithmetical mean m as the optimal value of f:

$$m = \frac{1}{n}\sum_{j=1}^{n} x_j = \frac{x_1 + x_2 + x_3 + \cdots + x_n}{n}.$$

This assumption can be shown to be justified by the method of least squares. If we let u be the value of f to be measured, then the differences $(u - x_1) \cdots (x - x_n)$ represent the deviation of the various readings. Following Gauss, we take the squares of these deviations as the appropriated measure of the inaccuracy. Thus the optimal value u can be obtained by minimizing the sum of the squares of the deviations; that is, we take u

in such a manner as to make the sum of the squares of the deviations as small as possible. Let the sum of the squares be S, and

$$S = (u - x_1)^2 + (u - x_2)^2 + \cdots + (u - x_n)^2.$$

Writing $(u - x_j)$ with the arithmetic mean included,

$$(u - x_j) = (m - x_j) + (u - m),$$

and

$$(u - x_j)^2 = (m - x_j)^2 + 2(m - x_j)(u - m) + (u - m)^2.$$

Substituting into the sum of the squares, one obtains

$$S = \sum_{j=1}^{n} (u - x_j)^2 = (m - x_1)^2 + (m - x_2)^2 + \cdots + (m - x_n)^2$$
$$+ n(u - m)^2$$
$$+ 2(u - m)(nm - x_1 - x_2 - x_3 - \cdots - x_n).$$

The last term is 0 because of the definition of m, and $n(u - m)^2$ is a positive number; therefore,

$$S \geqslant (m - x_1)^2 + (m - x_2)^2 + \cdots + (m - x_n)^2.$$

In this equation the equals sign holds only for $m = u$, which proves the theorem.

Gauss entered the University of Göttingen in 1795, seventeen years old, with a promise of brilliance which he was soon to fulfill. By the following year he had set himself to the ancient geometric task of identifying all of the regular polygons which could be constructed by ruler and compass alone. In addition to the known constructions of polygons of n sides with $n = 2^m$, 3, 6, etc., he showed that construction was possible when n is a Fermat number of the form $2^{2^p}+1$. This early effort formed part of the final and crowning section of his *Disquisitiones arithmeticae*.

In his doctoral thesis of 1799, Gauss provided a proof of the so-called fundamental theorem of algebra:* "Every algebraic equation in one variable has at least one root." This proof had been attempted by his predecessors, of course, Lagrange and D'Alembert among them, but none had been free of defect. D'Alembert, for instance, assumed that a continuous function given on a bounded and closed set of points has a minimum somewhere on the set. Although it was true, he was unable to offer proof of his assumption, which was not, in fact, achieved until after 1800. Gauss himself left the proof with one unsatisfactory gap when he assumed that a continuous function which is somewhere positive and somewhere negative must pass through zero.

In his thesis Gauss established the status of complex numbers on an equal footing with real numbers; this was a consequence of the general rule that all roots of algebraic equations are of the form $a + ib$, where $i = \sqrt{-1}$.

* Some criticize the name of this theorem in view of the fact that the proof may involve either the theory of complex functions or topology.

To suggest in brief form how a proof of the fundamental theorem of algebra can be attempted, consider the algebraic polynomial

$$f(z) = z^N - a_{N-1}z^{N-1} + \cdots + a_1z + a_0 = \sum_{n=0}^{N} a_nz^n,$$

where a_N has been normalized to 1. Here z is a complex variable, $z = x + iy$, and can take on all the values of x and y necessary to span the two-dimensional complex plane. Regard then a special plot of $f(z)$, namely, the modulus of $f(z)$ or magnitude of $f(z)$, written as $|f(z)|$. If z is a complex variable, $f(z)$ is complex and equals $u(x, y) + iv(x, y)$, where $u(x, y)$ and $v(x, y)$ are real functions in the two variables x and y. If the magnitude of $f(z)$ (called the modulus of f) is plotted on a third axis perpendicular to the complex z plane, this magnitude, $|f(z)|$, constitutes a surface. For larger values of z, $|f(z)|$ increases without bound as $|z^N|$, a surface of rotational symmetry. Now choose some large value of z, say z_L, for $|z_L| =$ a constant, a curve R is swept out on the surface $|f(z)|$.

For some value z_0 inside of R, $|f(z_0)|$ is less than the values of $|f(z_L)|$ on R and there is at least one minimum inside R. Notice that this has not been proved, only stated. The crux of the argument lies in the amplitude of this minimum. A lemma of D'Alembert states that if z_0 is a complex number such that $f(z_0) \neq 0$, then another complex number h can be found such that $|f(z_0 + h)| < |f(z_0)|$. In other words, if $f(z_0) \neq 0$, there is always a point in the vicinity of z_0 for which the modulus of f has a value smaller than at z_0. A minimum of this surface at z_0 has the property that $f(z_0)$ must be smaller than or equal to all values in the neighborhood of z_0. Taking D'Alembert's lemma without proof, we see that if there is a minimum in $f(z)$, then $f(z)$ at that point must be zero because any $f(z) \neq 0$ cannot be a minimum. When $f(z_0) = 0$, z_0 is a root of the polynomial. Summing up, there must be at least one minimum and it must occur at a point z_0, for which $f(z_0) = 0$; therefore, the point corresponding to the minimum must be a root of $f(z)$.

Once the existence of one root is established, the proof of the existence of N roots for a polynomial $f(z)$ of degree N is straightforward. Suppose the first root known to exist is α_1; then

$$f(z) = z^N + a_{N-1}z^{N-1} + \cdots + a_1z + a_0 = (z - \alpha_1)g_1(z).$$

Because α_1 is a root of $f(z)$, $g_1(z)$ is a polynomial of degree $N - 1$. By Gauss's theorem, then, there exists at least one root of $g_1(z)$, α_2. Therefore,

$$g_1(z) = b_{N-1}z^{N-1} + b_{N-2}z^{N-2} + \cdots + b_1z + b_0 = (z - \alpha_2)g_2(z).$$

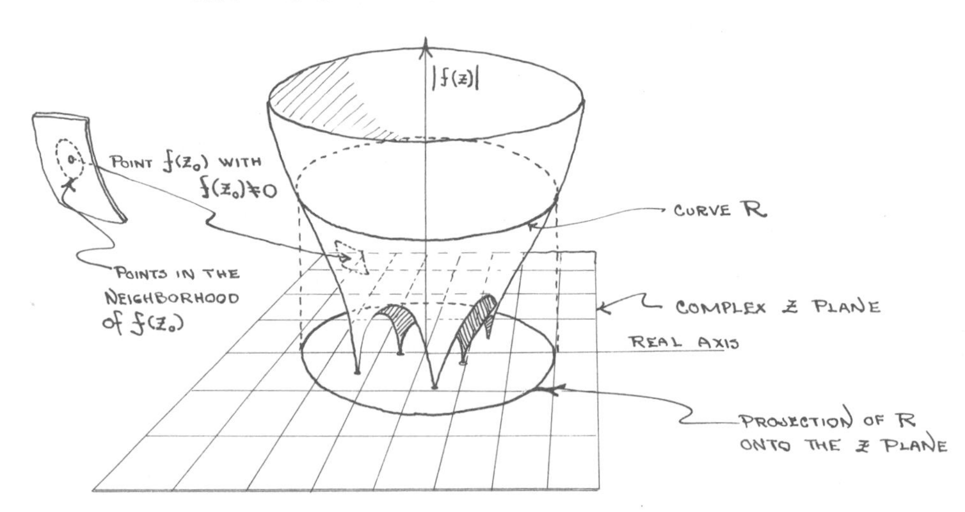

In repeated fashion g_2 is a polynomial of degree $N - 2$ and must have a root α_3. Repeating this reduction $N - 1$ times, one achieves the complete factorization:

$$f(z) = (z - \alpha_1)(z - \alpha_2)(z - \alpha_3)\cdots(z - \alpha_{N-1})(z - \alpha_N);$$

the repeated product is written as

$$f(z) = \prod_{j=1}^{N} (z - \alpha_j).$$

The initial assumption could have been made somewhat more general by writing $f(z) = (z - \alpha_1)g_1(z) + R$, where R is a remainder which must be evaluated. Because α_1 is a root we choose $z = \alpha_1$ to give $f(\alpha_1) = 0 = (\alpha_1 - \alpha_1)g_1(\alpha_1) + R$. Therefore the remainder R is equal to 0.

Gauss deduced the properties of the double algebra from the accepted postulates of common arithmetic. The notation is significant in that it suggests the generality of the hypernumber algebras which came later. Write (a, b) instead of $a + ib$, then $(a, b) = (c, d)$ implies $a = c$ and $b = d$. In the same manner, $(a, b) + (c, d) = (a + c, b + d)$. And by definition, $(a, d) \times (c, d) = (ac - bd, ad + bc)$. Generalizations of the number system as quaternions composed of number quadruplets grew out of this representation of the complex numbers as number doublets, as did Grassmann's number, n-tuples.

Gauss was led to a satisfactory theory of complex numbers in the process of devising a concise solution to the Diophantine problem: if p and q are prime, find the condition which both must satisfy in order that at least one of the two equations, $x^4 = qy + p$ and $z^4 = pw + q$, be solvable in integers x, y, z, and w.

In the theory of numbers Gauss founded the modern abstract arithmetic with the equivalence relation called "congruence." The basic technique in the theory of congruences is to map an infinite class of integers onto a finite subclass, a method known as homomorphism. Legendre had made important contributions to this theory, but it was Gauss who supplied the rigor. Implicit in the idea of congruence is the question of the divisibility of a set of integers by a fixed integer. If a number measures the difference between two numbers n and m, m and n are said to be congruent with respect to p. If p does not measure the difference, the numbers are incongruent modulo p. Examine* the remainders left when integers are divided by 5:

$0 = 0 \times 5 + 0$	$5 = 1 \times 5 + 0$	$10 = 2 \times 5 + 0$
$1 = 0 \times 5 + 1$	$6 = 1 \times 5 + 1$	$11 = 2 \times 5 + 1$
$2 = 0 \times 5 + 2$	$7 = 1 \times 5 + 2$	$12 = 2 \times 5 + 2$
$3 = 0 \times 5 + 3$	$8 = 1 \times 5 + 3$	$13 = 2 \times 5 + 3$
$4 = 0 \times 5 + 4$	$9 = 1 \times 5 + 4$	$14 = 2 \times 5 + 4$

* For further reading see Courant and Robbins, *What Is Mathematics?*, p. 31.

Three horizontal lines (the symbol $\equiv$) represent the congruence of two numbers: the form $a \equiv b \pmod d$ means that a and b are congruent modulo d. In the list of numbers modulo 5, we notice that 3, 8, 13, 18, . . ., -2, -7, -12, . . . are all congruent. For a fixed modulus many of the formal properties of ordinary arithmetic hold. As examples,

always $a \equiv a \pmod d$;
if $a \equiv b \pmod d$, then $b \equiv a \pmod d$;
if $a \equiv b \pmod d$ and $b \equiv c \pmod d$, then $a \equiv c \pmod d$;
if $a \equiv a' \pmod d$ and $b \equiv b' \pmod d$, then

$$a + b \equiv a' + b' \pmod d$$
$$a - b \equiv a' - b' \pmod d$$
$$ab \equiv a'b' \pmod d.$$

These expressions can be verified by using the equivalent algebraic forms for $a \equiv b \pmod d$, such as $a = b + nd$, where n is an integer, or $(a - b)/d$ = an integer n. Integers are often portrayed geometrically by their position on a line. Congruences have an equally appealing representation as points separated by equal arcs on the circumference of a circle. The hands on a clock, for example, indicate the hour modulo 12. Again, using modulo 5, the congruences on a circle may be exhibited as shown in the figure.

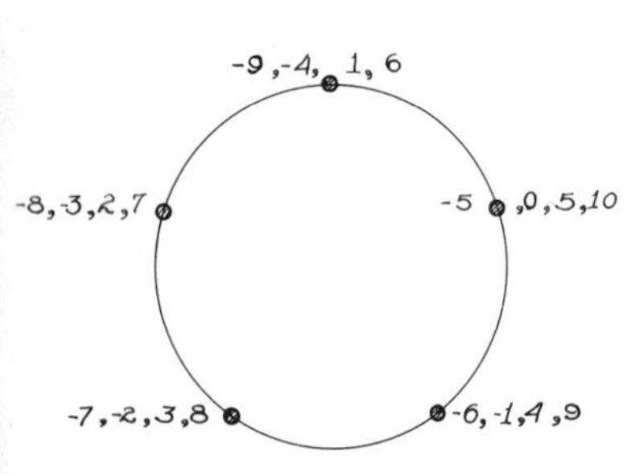

The multiplicative property provides some interesting insight into the base system of numbers. Consider the base 10 modulo 11: $10 \equiv -1 \pmod{11}$, $10^2 \equiv +1 \pmod{11}$, $10^3 \equiv -1 \pmod{11}$, $10^4 \equiv +1 \pmod{11}$.

Any number z expressed in the decimal system is of the form

$$z = a_0 + a_1(10) + a_2(10)^2 + \cdots + a_n(10)^n + \cdots,$$

where the a_n are integers in the range from 0 to 9.

Replacing the powers of 10 by their congruences, modulo 11 gives

$$t = a_0 - a_1 + a_2 - a_3 + a_4 \cdots.$$

If t and z are congruent modulo 11, the difference $z - t$ must be congruent to 0 modulo 11. Taking the difference, we find

$$z - t = a_1(11) + a_2(10^2 - 1) + a_3(10^3 + 1) + a_4(10^4 - 1) \cdots.$$

Each coefficient of a_n is congruent to 0 modulo 11; therefore it follows that a number is divisible by 11 with remainder zero if and only if the alternating sum of its digits (the a_n) is divisible by 11. As a further exercise, it can be demonstrated that a number is divisible by 9 or 3 if the sum of its digits is divisible by 9 or 3. For congruences modulo 7, the problem is more complex, since a number is divisible by 7 only if its digits (the a_n) are summed according to the rule that

$$a_0 + 3a_1 + 2a_2 - a_3 - 3a_4 - 2a_5 + a_6 + 3a_7 + 2a_8 - \cdots$$

is divisible by 7. Fermat's theorem—if n is any whole number and p is any prime, then $n^p - n$ is divisible by p—can be proven by rephrasing it in terms of congruences, where it then becomes $n^{p-1} \equiv 1 \pmod{p}$. The proof runs as follows. If n is divisible by p, the theorem is trivial. Assume that p does not divide n, and consider the first $(p - 1)$ multiples of n; i.e., $n, 2n, 3n, \ldots, (p - 1)n$. No two of these can be congruent modulo p, for then p would be a factor of $(s - r)$ where $(s - r)$ is less than p. In like manner the numbers $n, 2n, 3n, \ldots, (p - 1)n$ must be, respectively, congruent to $1, 2, 3, \ldots, (p - 1)$. From this it follows that

$$n\cdot(2n)\cdot(3n)\cdots(p-1)n \equiv 1\cdot2\cdot3\cdots(p-1) \pmod{p},$$

or

$$1\cdot2\cdot3\cdots(p-1)(n^{p-1}-1) \equiv 0 \pmod{p}.$$

Because $1, 2, 3, \ldots, (p - 1)$ is not divisible by p, $(n^{p-1} - 1)$ must be divisible by p, thus proving the theorem. As an example, if $p = 7$ and n is 3, then $3^{(7-1)} - 1 = 728$ must be divisible by 7. In the language of congruences, $3^6 \equiv 1 \pmod{7}$.

Gaussian congruences can be immediately extended to congruences between polynomials. If

$$F = \sum_{r=0}^{m} a_r x^{m-r}$$

and

$$M = \sum_{s=0}^{n} b_s x^{n-s}$$

are polynomials with $m \geqslant n$ and $a_0 b_0 \neq 0$, there is exactly one polynomial R of degree $\leqslant (n - 1)$ (i.e., less than or equal to $n - 1$) and one polynomial Q such that $b_0^{m-n+1}F = QM + R$. These relations were extended by Cauchy.

The generalization of arithmetic began in 1831, when Gauss initiated the study of biquadratic reciprocity, and reached its apogee with the development of mathematical logic in the latter part of the century. Legendre's law of quadratic reciprocity consisted of a study of the residues of x^2 modulo p. A number a, not a multiple of p, which is congruent modulo p to the square of some integer x is called a "quadratic residue" of p. A number b, not a multiple of p, which is not congruent to any square is known as a "quadratic nonresidue" of p. The Gauss-Legendre law of quadratic reciprocity concerns the behavior of two different primes p and q, and states that q is a quadratic residue of p if and only if p is a quadratic residue of q provided that the product

$$\left(\frac{p-1}{2}\right)\left(\frac{q-1}{2}\right)$$

is even. Whenever the product is odd, the result is changed, in that p is a residue of q if and only if q is a nonresidue of p. Gauss was the first to provide a rigorous proof of this theorem.

Biquadratic reciprocity involves the generalization that, when $a \equiv x^n \pmod p$, a is the $n$ic residue of p. A restatement of quadratic reciprocity occurs in the form

$$(p \mid q)(q \mid p) = (-1)^{[(p-1)/2][(q-1)/2]},$$

where the symbol $(p \mid q)$ denotes $+1$ or -1, according to whether $p \equiv x^2 \pmod q$ is or is not soluble for x. Gauss's investigations of $n = 4$ revealed that the tests for biquadratic reciprocity were simple only when he passed beyond the real numbers to the complex. For $n = 3$, he discovered that the reciprocity relation was based upon numbers of the form $a + b\rho$ where ρ is a root of $y^2 + y + 1 = 0$, with a and b integers.

For all n greater than 2, the $n$ic reciprocity laws depend upon algebraic number fields entering through polynomial equations of degree n. An algebraic number field of degree n consists of all of the rational functions of the roots of a given irreducible algebraic polynomial of degree n having rational integer coefficients. These results represent the importance of the concepts of number fields and algebraic polynomials, which in turn culminated in the group theory of Galois.

Gauss concluded his *Disquisitiones* with a thorough discussion on the nth root of 1—a problem equivalent to the division of the circumference of a circle into n equal parts. The equation $x^n = 1$ is equivalent to $x^n = e^{im2\pi}$, where m takes the values $0, 1, 2, \ldots, (n - 1)$. Thus for $n = 4$ the roots are

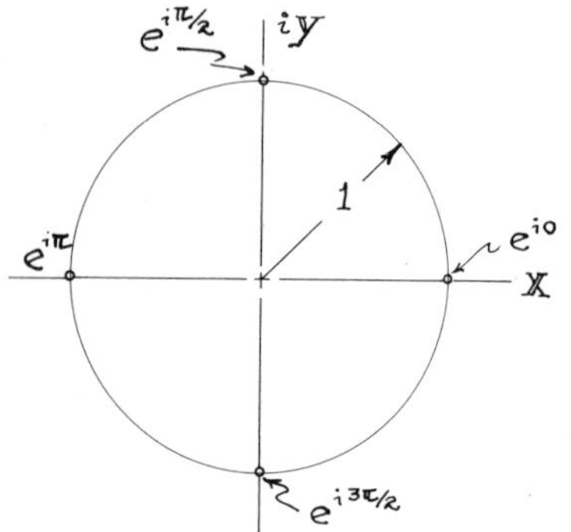

$$x_0 = e^{i0} = 1,$$
$$x_1 = e^{2\pi i/4} = +i = +\sqrt{-1},$$
$$x_2 = e^{4\pi i/4} = -1,$$
$$x_3 = e^{6\pi i/4} = -i = -\sqrt{-1}.$$

It was not until some years after Gauss died that his notebooks were found. These revealed the insight of the master who in so many ways anticipated the mathematical age to come. Gauss posed the fundamental question of mathematical structure: "If, in a given set of axioms, one or more of the axioms are suppressed, does one in fact obtain a more general theory?" It is apparent from his notes that long before the appearance of Lobachevsky's work he had planned a consistent system of geometry which excluded Euclid's fifth postulate.

Along with this massive contribution to mathematics came discoveries in astronomy as well as electricity and magnetism. In 1801 the planetoid Ceres was discovered, and Gauss, with the scant data available, computed its orbit quite accurately. One of our notable inheritances from Gauss is his law rephrasing the inverse square law of force in terms of the free sources of such a field. According to this law, the integral of the normal (perpendicular) component of the force field over a closed mathematical surface S is directly proportional to the sum of all of the sources enclosed inside of the surface S. A trivial demonstration of this general law can be made with a point electric charge q placed at the center of a spherical surface S. Because the force is central, the equipotential surfaces are spherical and the force is everywhere normal to S.

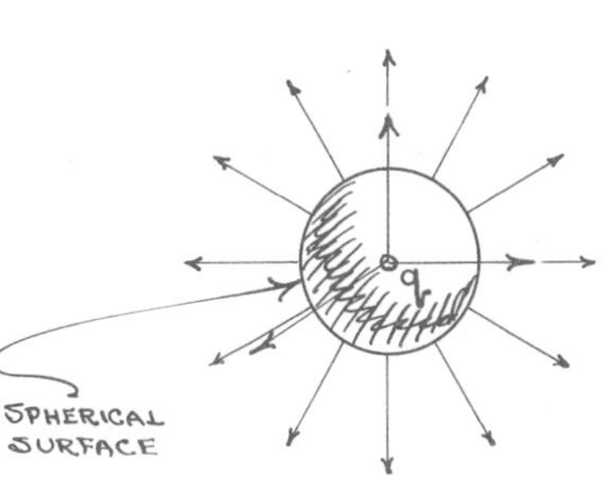

In this case the surface integral is just the product of the surface area times the force on the surface: $(q/R^2)4\pi R^2 = 4\pi q$. The general form of this law states that if the normal to the surface element dS is $\vec{\mathbf{n}}$ and the force at that point is $\vec{\mathbf{F}}$, the normal component is $\vec{\mathbf{F}} \cdot \vec{\mathbf{n}}$ (the projection of $\vec{\mathbf{F}}$ at that point onto $\vec{\mathbf{n}}$), with

$$\iint_{\substack{\text{closed} \\ \text{surface } S}} \vec{\mathbf{F}} \cdot \vec{\mathbf{n}}\, dS$$

equal to a constant times the sum of the sources of $\vec{\mathbf{F}}$ inside of S.

Gauss's form of the inverse square law was further generalized into the leading equation of the set now known as Maxwell's equations. Actually, for special cases this final form was discovered by Poisson.

The seventeenth, eighteenth, and nineteenth centuries mark an era in which mathematics and theoretical physics were pursued simultaneously by the mathematician-physicist. During the expansion of classical mathematical analysis, more often than not investigations of subtle advances in mathematics were stimulated by the pressing needs of physical models. Partial differential equations and the associated boundary value problems were traditionally tied to physical problems: electromagnetic theory and optics, which was to become a branch of the former, provided a vast area of application. At the turn of the nineteenth century the wave theory of light, although accepted, was but poorly understood, and its formulation was at best rather awkward. Two major questions lay unexplored, first, the direction of the light vibrations and, second, the medium through which light waves could be transported. It was realized that a given light wave had three possible orientations for vibration, of which only one was unique.

The two other degrees of freedom were perpendicular to the propagation vector $\vec{\mathbf{k}}$, which defined the direction of motion of the wave front. Vibration along either or both of these axes constituted a transverse vibration, as in a vibrating string. The third axis was parallel to the direction of propagation, and vibration along this axis is known as longitudinal vibration: for instance, it was recognized that the pressure waves in a gas constituting sound waves were of the longitudinal type in that the compression and rarification in the gas took place along the direction of motion.

As for the second major question, the material medium through which waves could be transported, the hypothesis for a supporting medium for the motion of light had been formalized by Newton. He postulated the aether, a weightless invisible medium having elastic properties similar to those of a transparent crystal. In 1817 the wave theory was still encumbered with great difficulties. "Polarization" was a word applied to describe an experimental phenomena which was not associated as yet with the orientation of the vibrations of light. Further, diffraction had not been explained satisfactorily. The successes of Laplace's corpuscular description of light encouraged its supporters to propose diffraction as the topic of the

French Academy's prize for 1818. The results of this competition were unforeseen: a manuscript by Augustin Fresnel (1788–1827) set the stage for the supremacy of the wave theory, and in the next seven years the corpuscular theory of light was completely discredited. Fresnel had been a soldier in a small army which attempted to prevent Napoleon's return from Elba. During a brief imprisonment he used his time to study interference and diffraction. In 1818 he presented an exhaustive memoir to the Academy in which he utilized the principles of Huygens and Thomas Young to account for the effects of diffraction. Interference effects occur when a wave front is split into at least two parts and later combined. The effects of diffraction arise when a wave front is limited, either by passing it through a slit or by blocking off part of it.

When he submitted his work on diffraction, Fresnel also published a paper discussing the aberration of light caused by the motion of the earth, and this interest led him to further questions concerning the motion of the aether relative to a moving, transparent body. Although the concept of the aether was to be of no consequence, Fresnel developed the equations which show that the velocity of light in a moving dielectric is different from the velocity in a stationary dielectric. These results were confirmed in 1851 by Armand Hippolyte Fizeau.

In 1816 François Arago (1786–1853) and Fresnel performed a critical experiment which seemed to resolve the question of the direction of polarization of light. Using a standard Young's interference setup, they introduced polarizers on the exit side of the two slits. When the orientations of the polarizers were mutually perpendicular, the two pencils of transmitted light no longer interfered (to achieve zero interference, the two polarizers were oriented with parallel axes by rotating in a plane perpendicular to the direction of motion of the light pencil). Arago communicated this result to Young, who immediately perceived in it a critical experiment proving that light was polarized in a direction perpendicular to the direction of propagation. From this time on, the property of transverse vibration assumed a primary role in the description of light. Attempts to incorporate some longitudinal vibration continued until the middle of the nineteenth century and were finally discarded in Maxwell's equations. Eighty years earlier, Bernoulli had considered the possibility of transverse vibrations and had rejected it.

When Fresnel saw the import of Young's hypothesis, he realized the direction which the theory of propagation of light in material bodies must take, and with this realization the theory of propagation of waves in elastic solids was born. Fresnel's remarkable insight into the mechanism by which light is propagated in elastic solids led him to propose a model capable of accounting for double refraction in crystals. He regarded a uniaxial crystal as an elastic medium in which the restoring forces on molecular matter resulting from displacements under the influence of the vibrating light are the same for all directions perpendicular to the characteristic axis of the crystal. Displacements parallel to the characteristic axis result in different restoring forces. According to Fresnel, the crystal is "repulsive" if the parallel restoring force is the stronger of the two, while the medium is "attractive" if the parallel force is the weaker. By connecting the restoring forces and the velocities of propagation of light through crystals, Fresnel concluded that the stronger restoring forces were associated with lower indices of refraction, n, or high velocities of propagation. Similarly, the weaker restoring forces were associated with higher indices of refraction, or lower velocities of propagation (remember that $n = c/v$ where c is the velocity of light in a vacuum). With this model Fresnel noted that the explanation of double refraction demanded that the polarization or oscillatory motion of light be executed at right angles to the direction of propagation, providing another indication of the transverse polarization of light.

The most obvious consequence of Fresnel's hypothesis on double refraction in crystals was that a wave propagated in any direction through

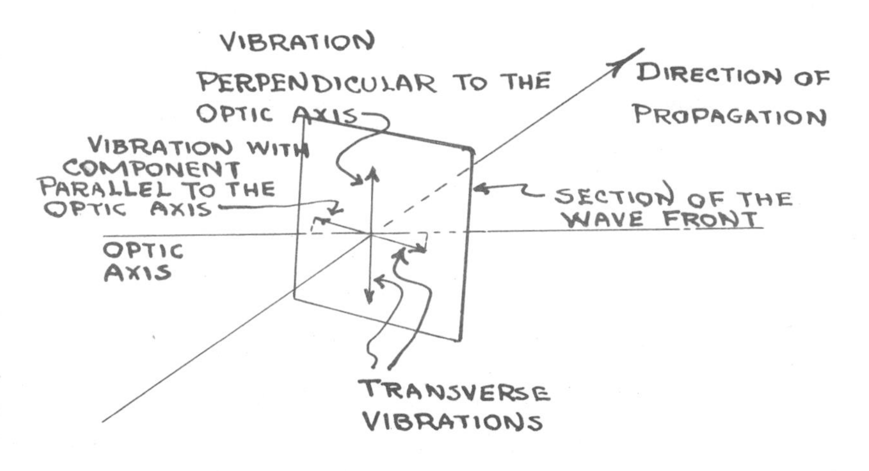

a uniaxial crystal could be resolved into two plane, polarized components, an ordinary ray polarized parallel to the principal axis and an extraordinary ray polarized in a plane perpendicular to the principal axis. These resolved components are then propagated through the crystal, with different velocities giving rise to double refraction. Since the electromagnetic properties of anisotropic dielectrics were not formulated for another half century, Fresnel's final analysis, based upon a dynamic substructure, was open to question. The general form of his final results, on the other hand, was correct if one reinterprets his parameters in the light of modern theories of matter. As Sir George Stokes observed, "If we reflect upon the state of the subject as Fresnel found it, and as he left it, the wonder is not that he failed to give a rigorous dynamical theory, but that a single mind was capable of effecting so much."*

Fresnel conceived of matter as being formed of molecular groups or particles separated by intervals which, though small, were not altogether insensible to the length of a wave. With this concept, he attempted in 1821 to formulate a description of dispersion. He then set forth the well-known Fresnel equations, which describe the reflection and refraction of plane polarized light at a dielectric interface. To achieve this derivation he employed the principle of the conservation of energy of the wave and a condition that the displacements of adjacent molecules at the interface be equal. Fresnel's equations were a generalization of a restricted form discovered earlier by Sir David Brewster. These results can scarcely be thought of as a theory, in view of the fact that the qualities of the media were undefined. To a large extent Fresnel worked backward from the known equations in an attempt to discover the mechanism to which these properties could be attributed. Although this illuminating career was abruptly ended by death at the early age of thirty-nine, Fresnel's work was recognized both in Europe and in England.

The firm position of the wave theory of light established by Young and Fresnel led to a formal representation of the aether as an elastic solid. This view of the aether was further expanded after Fresnel's death in a brilliant series of contributions by Claude L. M. H. Navier (1785–1836), Augustin Louis Cauchy (1789–1857), and Stokes (1819–1903). In 1821 Navier presented a memoir in which the correct equations of motion for an elastic solid were given for the first time. He showed that if $\vec{\epsilon}$ denotes the vector displacement of a particle whose equilibrium position is x, y, z, and if ρ denotes the density of the medium with η being a constant measuring the rigidity (or resistance to distortion), the equation of motion of the solid is

$$\rho \frac{\partial^2 \vec{\epsilon}}{\partial t^2} = -3\eta \text{ grad div } \vec{\epsilon} - \eta \text{ curl curl } \vec{\epsilon}.$$

This equation at first sight is a formidable thing. Earlier the concept of the

* E. Whittaker, *History of the Theories of the Aether and Electricity* (London: Thomas Nelson and Sons, 1951), p. 122; George Stokes, *British Association Reports* (1862), p. 254.

gradient and the divergence of a vector field were introduced: the divergence of a vector field was a measure of the source or sink properties of the field. In other words, if one counted the vectors crossing a closed surface, and if the net exits minus the entrances was other than zero, a point inside the closed surface was a source or a sink. Logically, one would then expect two types of fields, those with sources and sinks, such as the electrostatic and gravitational fields, and those with no sources or sinks. The first type of field (with a nonzero divergence somewhere) is called irrotational. The second type, with no sources or sinks, is known as a rotational field. Magnetic fields are rotational in that the field lines always close upon themselves, forming closed loops, neither starting nor stopping at any point in space. Remembering the diagrams of the magnetic field lines of a long, straight, current-carrying wire, one may recall that the magnetic field lines of this example are concentric circles falling in planes perpendicular to the wire with the center of the wire coincident with the center of the circle.

To define the curl or circulation of a vector field, the line integral of the projections of the field vectors upon a closed path in space must be considered. A simple example of such a line integral is shown here. The line integral of the projections of the $\vec{\mathbf{F}}_j$'s on the elements of the path $\Delta\vec{\mathbf{l}}_j$ can be written as a sum over all of the elements making up the closed loop called the path. If the path is broken up into N elements,

$$\oint \vec{\mathbf{F}} \cdot d\vec{\mathbf{l}} \Rightarrow \sum_{j=1}^{N} F_j \cos \alpha_j \, \Delta l_j$$

in the limit as Δl_j becomes arbitrarily small, and N becomes arbitrarily large in such a way that

$$\sum_{j=1}^{N} \Delta\vec{\mathbf{l}}_j = \text{the length of the path}.$$

The curl of a field $\vec{\mathbf{F}}$ at a point P is obtained by calculating the line integral which is shown above and the cap surface S which is enclosed by the loop.

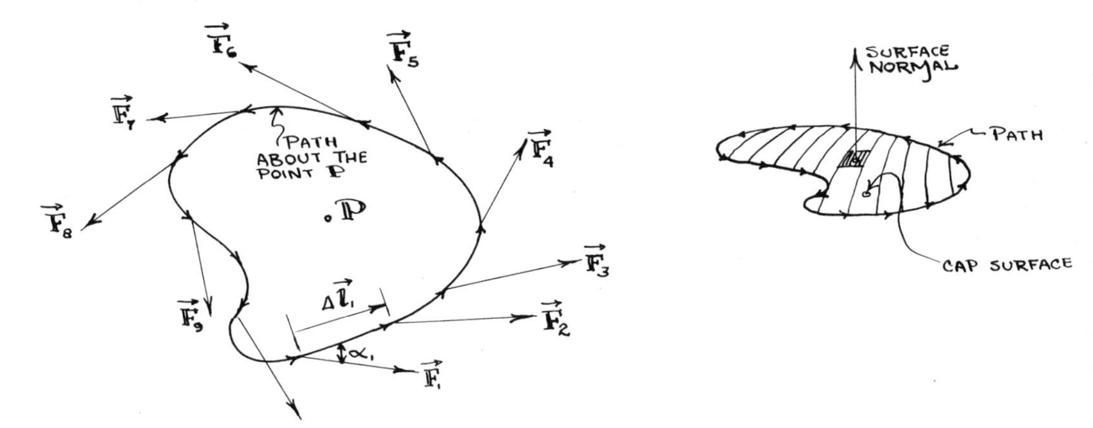

In order to define the curl at P one allows the loop to shrink about P. Then the curl of $\vec{\mathbf{F}}$ at P is the ratio of the line integral (measuring the circulation) to the surface S in the limit as S becomes arbitrarily small. The curl is a vector function, taking the direction of the vector normal (i.e., perpendicular) to S. Then

$$\text{curl}\,\vec{\mathbf{F}} = \lim_{S\to 0} \frac{\oint_{\text{path}} \vec{\mathbf{F}}\cdot d\vec{\mathbf{l}}}{S}$$

in a direction normal to the surface S. One further remark needs to be added: by convention the direction of $\vec{\mathbf{n}}$ is taken according to the right-hand rule. By allowing the fingers of the right hand to curl in the direction of the path of integration the direction of $\vec{\mathbf{n}}$ is along the thumb.

The reader is reminded once again that in the interest of clarity we have consistently introduced the modern forms of mathematical notation. This decision may occasionally create some anachronism. For instance, the vector concepts of divergence and curl were developed from the algebra of Hermann Günther Grassmann by Josiah Willard Gibbs at the end of the nineteenth century. On the other hand, many of the associated concepts such as irrotational and rotational vectors were introduced by Poisson in 1828, in connection with Navier's equation.

In 1828 Augustin Louis Cauchy published a modification of the Navier equation for the vibrations of an elastic solid wherein he viewed the elastic properties in bulk, rather than assuming that matter was an aggregate of point centers of force. An innovation arising from his work was the modulus of compression k, which is the ratio of the pressure to the cubical compression. With this new term incorporated into the derivation the coefficient of the first term on the righthand side becomes $[k + (4/3)\eta]$, instead of 3η, with the reduction to Navier's form when point centers of force are assumed.

As mentioned above, in the same year Poisson effected a solution to the Navier-Cauchy equation by assuming that the vector displacement $\vec{\boldsymbol{\epsilon}}$ could be resolved into two independent vector components $\vec{\mathbf{b}}$ and $\vec{\mathbf{c}}$, $\vec{\boldsymbol{\epsilon}} = \vec{\mathbf{b}} + \vec{\mathbf{c}}$, where $\vec{\mathbf{b}}$ is irrotational (i.e., not rotational) and therefore obeys the relation curl $\vec{\mathbf{b}} = 0$, and where $\vec{\mathbf{c}}$ is rotational (i.e., not irrotational) and obeys the relation div $\vec{\mathbf{c}} = 0$. By inserting this form of $\vec{\boldsymbol{\epsilon}}$ into Cauchy's equation and employing the conditions above, the complex form is resolved into two wave equations, one for $\vec{\mathbf{b}}$ only and one for $\vec{\mathbf{c}}$:

$$\rho \frac{\partial^2 \vec{\mathbf{b}}}{\partial t^2} = \left(k + \frac{4}{3}\eta\right) \text{div grad}\,\vec{\mathbf{b}}, \quad \text{with curl}\,\mathbf{b} = 0,$$

and

$$\rho \frac{\partial^2 \vec{\mathbf{c}}}{\partial t^2} = \eta\, \text{div grad}\,\vec{\mathbf{c}}, \quad \text{with div}\,\vec{\mathbf{c}} = 0.$$

As in the case of the vibrating string, these equations are analogous to Newton's second law in three dimensions and have sinusoidal or vibrating solutions in the form of traveling waves. Moreover, the solutions for the three components of $\vec{\mathbf{b}}$ represent longitudinal waves (i.e., vibrations along the direction of propagation), moving with a velocity $\{[k + (4/3)\eta]/\rho\}^{1/2}$. The solutions for the three components of $\vec{\mathbf{c}}$ represent transverse vibrating waves propagated with a velocity $\sqrt{\eta/p}$. Hidden in this development lies a general theorem which later becomes an important part of the vector analysis: any vector field $\vec{\mathbf{F}}$ can, in general, be represented as the sum of an irrotational component and a rotational component.

After extending the equations of Navier in 1828, Cauchy presented the culmination of these studies in 1830, in the form of a theory of the propagation of light in crystalline substances. In 1836 he proposed a second theory which gave the same results, although his boundary conditions in the second exposition were completely different from those employed in his earlier work in 1828. Both theories were encumbered by inconsistencies. For one thing, longitudinal waves were inherent in both. Cauchy believed that more refined experiments would bear this out. Another difficulty was that the boundary conditions were artificial, and no physical reasoning could be given for the choice of boundary conditions at the interface. As presented, these conditions were inconsistent with true boundary conditions given later by George Green (1793–1841). Much of the same criticism can be leveled at the theories of reflection and refraction discovered simultaneously by James MacCullagh (1809–1847) and Franz Neumann (1798–1895).

More significant than the failure of these theories of electromagnetism was the fact that they laid the foundation for a correct description of the problem. As often happens in the development of physical theories, details of the representation frequently went through several stages. The earlier work in many instances illustrated vividly those reasonable hypotheses, providing results that are inconsistent with physical observations. In this particular example, as with the Maxwell equations of electromagnetic theory, the boundary conditions imposed upon the solutions to the general equation were designed to produce a specific result—here, the Fresnel equations. To illustrate the use and meaning of boundary conditions, we may recall the equation for the vibrating string in one dimension:

$$\frac{\partial^2 y}{\partial x^2} - \frac{1}{c^2}\frac{\partial^2 y}{\partial t^2} = 0.$$

A simple substitution for $y(x, t)$ demonstrates that the function $y(x, t) = A \sin kx \cos(\omega t - \phi)$ (where ϕ is a constant phase, A is a constant, and $\omega = kc$) satisfies the partial differential equation. If the string is fixed at both ends and is stretched over a distance L, then the physical statement that the ends are fixed implies that $y(x, t)$ is *always* 0 at $x = 0$ and at $x = L$. These are boundary conditions. In order to satisfy these conditions the adjustable parameter k must be assigned *unique* values (later called eigenvalues) to conform to these conditions. Since we have chosen $\sin kx$ as the space function, the boundary condition that $y = 0$ at $x = 0$ is automatically taken care of. The statement that $y = 0$ at $x = L$ can be included by remembering that sin kx is zero whenever kx is an integral multiple of π. Thus by setting $kL = n\pi$ (where n is an integer) the solution is forced to take the form $\sin kx = \sin [(n\pi/L)x]$, which is 0 when $x = L$ and when $x = 0$.

The boundary conditions for vibrating elastic solids and electromagnetic waves can appear in more complicated forms; however, the idea is the same. For instance, a restriction that a function or that the normal or tangential component of a vector field be continuous at every point including an interface is equivalent in the string problem to the implied restriction that the string be continuous at every point of x in the interval between $x = 0$ and $x = L$. When the parameter k is fixed by the boundary conditions, an infinite denumerable set of solutions is available; i.e., there is a $k_n = n\pi/L$ for every integer n. The particular values of n which are used in a problem depend upon the initial conditions. Initial conditions represent a type of boundary condition imposed upon the time-varying function. As an example, if the string has the shape $\sin (\pi x/L)$ at $t = 0$ and starts from rest, the values of the constants A and ϕ are 1 and 0, respectively, while the only value of k employed corresponds to the value associated with $n = 1$. When more complicated configurations are assumed at $t = 0$, one must begin with a solution which is a sum of all the possible space functions involving different values of the integer n. This is the general Fourier series representation of the string problem:

$$y(x, t) = \sum_{n=1}^{\infty} A_n \sin \left(\frac{n\pi}{L} x\right) \cos \left(\frac{n\pi c}{L} t - \phi_n\right)$$
$$= A_1 \sin \left(\frac{\pi x}{L}\right) \cos \left(\frac{\pi ct}{L} - \phi_1\right) + A_2 \sin \left(\frac{2\pi x}{L}\right) \cos \left(\frac{2\pi ct}{L} - \phi_2\right) + \cdots.$$

Here the A_n, ϕ_n and the values of n allowed are determined from the initial conditions on $y(x, t)$ and on its derivative with respect to time—in other words, the initial velocity of the string.

By 1837 George Green had reformulated the solution to the elastic-solid equation for reflection, exhibiting superior insight in setting boundary conditions for reflection which must be satisfied. Green assumed that the three components of displacement and the three components of stress across an interface are equal in the two media, and although he is regarded as a lesser analyst than Cauchy, his contributions to mathematics have been of primary significance. The theorem bearing his name and the method for the inversion of a differential equation to an integral equation are but a few of his gifts to mathematical physics.

Stimulated by Green's work on the propagation of light in an elastic solid, Cauchy proposed a new relation between the elastic constants, which had been fashioned in such a way that the longitudinal component of a wave was reduced to zero. This theory of the aether was later known as the "contractile aether."

Cauchy's contributions to mathematics were of far higher importance than any of his endeavors in theoretical physics. Previous attempts to put the calculus and the power series expansion upon a rigorous basis had failed. More than one hundred years elapsed between the early attempts of Brook Taylor and the work of Cauchy. Cauchy, a graduate of the École Polytechnique, published a brilliant series of memoirs in 1811, which attracted the notice of both Lagrange and Laplace. He was appointed professor of mathematics at the École Polytechnique in 1816. His papers, which appeared in rapid succession, covered the theory of polyhedra, the theory of functions, differential equations, the rectification of curves and the quadrature of surfaces, the theory of probability and the calculus of variations.

In 1814 Cauchy began investigations in the theory of complex variables, the field for which he is best known today. His first book, *Cours d'analyse*, published in 1821, contained one of the first rigorous treatments of the convergence of an infinite series and provided tests which are still in use. Coupled with this was the first proof of Taylor's theorem accompanied by a form of the remainder. The concept of an integral as the limit of a sum was reintroduced by Cauchy, again in a rigorous form, and one result of this work, which is known to every student of mathematics and physics, was the Cauchy integral theorem. This theorem states that the closed path integral of a complex function $f(z)$, which is analytic everywhere within the region enclosed by a closed path C (or contour as it is called) and is analytic at every point on the boundary of the contour, is 0; i.e., $\oint_C f(z)\,dz = 0$, if $f(z)$ is analytic within and on C—here $z = x + iy$. The circular symbol superimposed on the integral sign merely states that the path of integration is closed, and the subscript C denotes the path, or contour.

Finite polynomials in z are familiar examples of such analytic functions. As a special example, consider $f(z) = z^n$, where n is a positive integer. For simplicity, let the path be a circle of radius 1 (a unit circle) with its center at the origin.

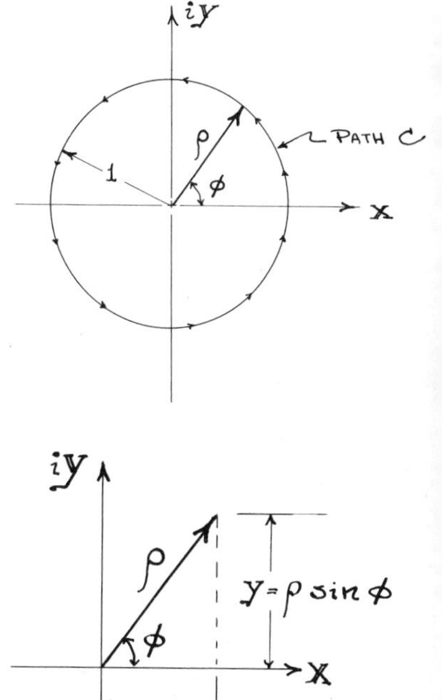

In this example $f(z) = z^n = (x + iy)^n$ and $dz = dx + idy$. This integral can be greatly simplified by writing the complex variable z in the polar form $\rho e^{i\phi}$, where ρ is the modulus (or magnitude) and $i\phi$ is the argument. We remind the reader of Euler's reduction of the exponential $e^{i\phi} = \cos\phi + i\sin\phi$. With this expansion one can write z as

$$z = x + iy = \sqrt{x^2 + y^2}\left(\frac{x}{\sqrt{x^2 + y^2}} + i\,\frac{y}{\sqrt{x^2 + y^2}}\right)$$
$$= \rho(\cos\phi + i\sin\phi) = \rho e^{i\phi}.$$

Applying the Pythagorean theorem, we observe that $\rho = \sqrt{x^2 + y^2}$ and $\tan\phi = y/x$.

The function $f(z) = z^n$ may be handled in a similar fashion:

$$f(z) = z^n = (\rho e^{i\phi})^n = \rho^n e^{in\phi} = \rho^n(\cos n\phi + i\sin n\phi).$$

By employing the unit circle as a contour, the differential dz reduces to a differential in ϕ because the modulus of a vector terminating on a unit circle is constant and equal to 1. On the circumference of the unit circle,

$$dz = dx + i\,dy = d(\rho e^{i\phi}) = e^{i\phi}\,d\rho + i\rho e^{i\phi}\,d\phi,$$

and when $\rho = 1$, a constant,

$$dz \xrightarrow[\rho=1]{} ie^{i\phi}\,d\phi.$$

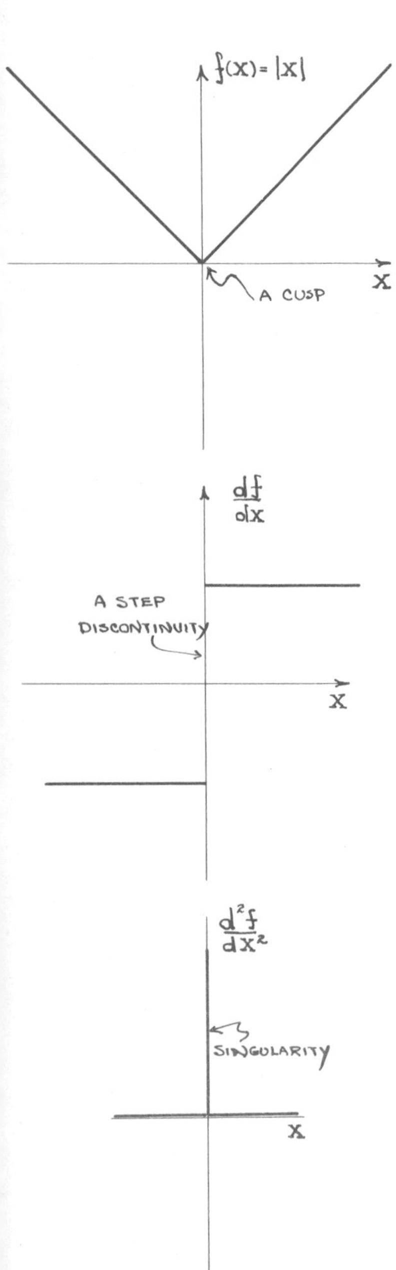

For this special problem, then, Cauchy's integral formula reduces to an integral over ϕ from 0 to 2π:

$$\int_{\substack{\text{unit}\\ \text{circle}}} z^n\, dz \xrightarrow[\rho=\text{const}]{} \int_0^{2\pi} \rho^n e^{in\phi} i\rho e^{i\phi}\, d\phi \xrightarrow[\rho=1]{} i\int_0^{2\pi} e^{i(n+1)\phi}\, d\phi = 0.$$

Although this is a special case, it illustrates the theorem quite well.

During this development, the term "analytic function" was stressed. The concept of analyticity is analogous to the concept of continuity of a function of one variable. The type of behavior which would be of interest in a function of one variable is the value of the function and its derivatives at every value of the independent variable. This is particularly obvious if one regards the Taylor's expansion of a function $f(x)$ expanded about a specific value of x, call it x_0:

$$f(x) = f(x_0) + \frac{1}{1!}\left[\frac{df}{dx}\right]_{x=x_0}(x - x_0) + \frac{1}{2!}\left[\frac{d^2f}{dx^2}\right]_{x=x_0}(x - x_0)^2 + \cdots$$

$$= \sum_{n=0}^{\infty} \frac{1}{n!}\left[\frac{d^nf}{dx^n}\right]_{x=x_0}(x - x_0)^n.$$

From this equation one observes that there is a special class of functions for which $f(x)$ and all of the derivatives of $f(x)$ must be finite. If one of the derivatives is undefined, this expression is not meant to imply that $f(x)$ is undefined. In fact, one can construct very simple cases for which $f(x)$ is defined at a point but for which one or more of the derivatives is undefined. Consider for example $f(x) = |x|$; the function takes on positive values for both positive and negative x. From the diagram one observes that although $f(x) = |x|$ is defined at $x = 0$, the first derivative has a step discontinuity at $x = 0$, and the second derivative is undefined at that point. Thus for a real variable analyticity is associated with the existence of *all* of the derivatives. In the case of a function of a complex variable, the condition of analyticity requires that all of the derivatives with respect to z exist; further, because the differentiation takes place in a two-dimensional space, it requires that the derivative be independent of the direction of approach to the point where the derivative is taken.

The theory of substitution begun by Cauchy produced the theory of finite groups. One can, therefore, regard Cauchy, along with Lagrange and others, as one of the founders of the theory of finite groups. As Gauss had done before him, Cauchy gave another proof of the fundamental theorem of algebra. In one memoir* among those dealing with radical polynomials,† he anticipated the consideration of higher spaces: for the first time the concept of higher dimensional points, loci, lines, and distances was introduced. Although Cauchy intended these ideas only as mechanisms for clearing delicate questions referring to the theory of radical polynomials, his definitions were suggestive of the studies of higher spaces which were to come later.

* *Comptes Rendus*, 24 (1847):885.

† A radical polynomial is a polynomial $\alpha + \beta\rho + \gamma\rho^2 + \cdots + \eta\rho^{n-1}$ where ρ is a primitive root of the equation $x^n = 1$. Thus it stems from the formalism of Lagrange resolvants.

Beginning with Gauss, mathematicians began more and more to question the basic postulates of various fields of mathematics. After 1800 there was a steady trend toward the abstract and toward greater generality, and by the middle of the nineteenth century the spirit of mathematics had undergone a complete change. Accompanying these changes was a steady loss of interest in the theories of physics, although physics had played a major role as a stimulus in the development of mathematics. Before 1850 most first-rate mathematicians had contributed, in some degree, to the formulation of physical theory. The separation of interests was not altogether complete, for as late as the twentieth century advances in physics would renew interest in such areas as tensor analysis and differential geometry.

The axiomatic method of mathematics first appeared with the Greeks, and most students gain some familiarity with it in Euclid. Theorems are *proved* by demonstrating that the theorem is a logical consequence of some propositions proved previously. The process is not infinite in regression; therefore, at some initial point a number of statements called postulates or axioms are accepted as true without proof. The choice of axioms with which to start is by and large arbitrary. Experience has shown that the axiomatic method is not particularly useful unless the initial axioms are simple and few in number. Furthermore, the axioms must be consistent and complete—consistent in that no two theorems deduced from them are contradictory, and complete in that every theorem of a system is derivable from them. Initial postulates should also be independent for reasons of economy, implying that no one of the initial axioms is a logical consequence of the others.

Modern mathematics, with its emphasis on structure, is deeply involved with questions of completeness and consistency. As a result, there are two different fundamental philosophical attitudes toward the manner by which mathematics should be done. The Kantian point of view is that mathematical objects are considered as substantial objects in a realm of "pure intuition." There can be no ultimate contradictions because mathematical statements are objective descriptions of existing reality. Intuition is not allowed too great a control; for instance, modern intuitionists would not assert that the number continuum is intuitively obvious. On the other hand, the denumerably infinite *is* accepted as intuitively obvious.

The modern formalists constitute the second approach to mathematics. Intuitive reality is not attributed to mathematical objects, nor are axioms regarded as obvious truths. The formalist is concerned only with the logical procedures of reasoning based upon abstract postulates. This approach has great advantages in that it provides a great deal of freedom to construct a multitude of mathematical systems. Coupled with this freedom, however, is the strict necessity of proving that these axioms, which appear as arbitrary creations of the human mind, cannot possibly lead to a contradiction. Much recent effort has centered about the search for such consistency proofs for the axioms of arithmetic and algebra.

Recent results suggest that such efforts cannot be completely successful within a closed system of concepts. Formalization certainly is the appropriate method for understanding and deciphering the network of interconnections between various facts and for exhibiting an essential logical structure. Generalization is facilitated by this technique, although it is thought that the most creative mathematics will not be produced by formalist methods.

The very close ties between physics and mathematics were present in those periods of mathematics when the intuitive approach was preeminent. In the mid-nineteenth century, as the formalist approach became the primary basis of mathematics, the two fields diverged. At present mathematicians not only are unconcerned with the problems of physics but assiduously avoid physically motivated problems.

Two great changes in the philosophical approach to geometry occurred in the nineteenth century, the creation of topology and of non-Euclidean geometry. Topology, or *analysis situs*, has as its object the study of the properties of geometric figures which persist even when the figures are subjected to deformations so drastic that all of their metric and projective properties are lost.* August Ferdinand Moebius (1790–1868) introduced a highly original work in 1827, the baryocentric calculus, which initiated the use of synthetic methods in geometry as opposed to those methods which rely in great part on the use of some coordinate reference. Moebius by career was an astronomer of some insignificance. At the age of sixty-eight he submitted his famous memoir on "one-sided" surfaces, sometimes called "nonorientable" surfaces. One of the simplest examples of such a surface is the Moebius strip. It has only one edge as its boundary. If the ends of a rectangle are pasted together without twisting, one achieves a ring or loop with two boundary curves and two sides. When this two-sided figure is cut along its center line, it falls into two different loops of the same kind. By inserting a 180° twist in the rectangle before pasting the ends together, one obtains the Moebius strip with one boundary. Cutting a Moebius strip along the center line does not give two separate pieces but rather another single object similar to the original strip with the exception that two twists are introduced between ends. A second cut along the center line of the new surface results in two separate pieces which are linked. These topological surfaces have fascinating general properties. For instance, because the boundary of the Moebius strip is an unknotted closed curve, it can be continuously deformed into a flat curve such as a circle. To accomplish this, one must allow the strip to intersect itself to form a one-sided, self-intersecting circle. Intersections do not obstruct the topological properties of the strip because a one-sided surface cannot be continuously transformed into a two-sided surface. Closed one-sided surfaces are not outside the field of serious mathematics. Rather, these discoveries were but another advance in geometric thought, which progressed to the concepts of analytic and projective geometry in the

* Courant and Robbins, *What Is Mathematics?*, p. 235.

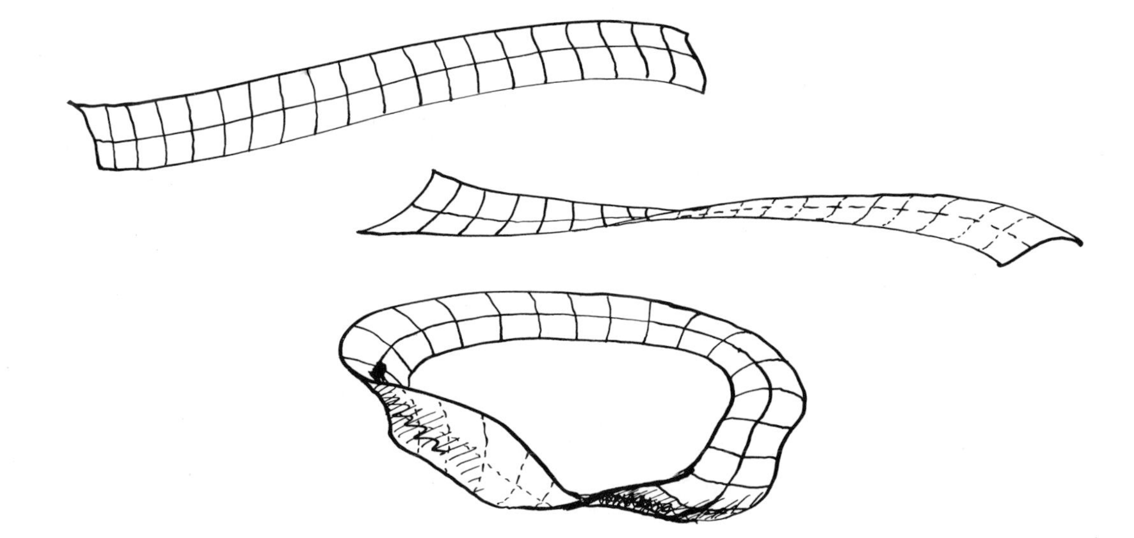

sixteenth and seventeenth centuries. Projective geometry provided the transition from the ordinary plane to the projective plane, wherein the improper or infinitely distant points are introduced. Moebius' one-sided surfaces present even further distortions of geometry where the adjacency of the various parts of a given figure are fundamental topological concepts. Topological transformations, then, are transformations which preserve the relations of adjacency.

During this period of indecision regarding the ultimate form of geometry, the fundamental axioms of Euclid seemed to provide a stable and satisfying basis. This appearance was illusory. Because of the five fundamental axioms, the fifth postulate, to the effect that "through a point not lying on a given line not more than one line parallel to the given line can be drawn," had attracted the attention of geometers for eighteen hundred years. Much of the effort of such men as Proclus (fifth century C.E.), Nasir-Eddin of Tus (thirteenth century), John Wallis, G. Saccheri (Italian, 1667–1733), Johann Heinrich Lambert (German, 1728–1777), and Legendre was aimed at eliminating the fifth postulate by deducing it as a theorem from the other basic axioms, to no avail. In each attempt, the proofs relied on some proposition which did not follow logically from the other premises of geometry, and the fifth postulate was invariably replaced by an independent statement which would itself have required proof.

The attempt of G. Saccheri, a Jesuit mathematician, in 1733 was a brilliant failure, for in attempting to show the absurdity of any alternative to the parallel postulate, Saccheri demonstrated that a proper non-Euclidean geometry could be constructed. Gauss undertook to construct a consistent geometry without the parallel postulate and was in possession of the main consequences of hyperbolic geometry in the early part of the nineteenth century. Unfortunately, his results were discovered in his notebooks only after his death in 1855. Nicolai Ivanovitch Lobachevski (Russian, 1793–1856) in 1826–1829 and János Bolyai (Hungarian, 1802–

1860) in 1833 settled the question by establishing a geometry in which the parallel postulate did not hold. Interestingly enough, Bolyai submitted his manuscript to Gauss, who had suppressed his own work on the subject.

Let us examine the approach to the fifth postulate which led Lobachevski and Bolyai to the new geometry. The crux of the problem lies in our concept of space in the large, and ultimately involves the intuitive concept of infinity. Because the fifth postulate is not provable from the other axioms of geometry, one can specify any independent fifth postulate without introducing a contradiction to the first four. Lobachevski for this reason suggested the contrary axiom and therefore developed a new geometry that was as logical and as rich as the Euclidean. Consequently, one must conclude that more than one geometry is logically conceivable.

According to Lobachevski, geometric truth, like physical laws, can only be verified by experiment. This assertion implies that truth represents some sort of correspondence between abstract representation and our observations of the real world. Euclidean geometry had proven adequate for a local space, but there had always been questions about some of its postulates when applied to space in the large. As an instance, Gauss suspected the closure principle for plane triangles and proposed that very accurate measurements should be made over great distances to check it. It is ironic that this movement, which originally focused on the necessity for experimental testing, led to a complete abstraction of the problem and to a realization that concern with logical structure represented an end in itself. Mathematical truth ultimately *does not* necessarily involve a correspondence between observations of the real world and an abstract representation. The internal consistency of the logical structure of mathematics represents the truth of a mathematical form.

To grasp the initial approach to non-Euclidean geometry, it is instructive to observe the method by which one can change the theory of parallel lines. Suppose that a line a and a point A (not on a) are given. The fundamental assumption of the contrary postulate is "through a point not lying on a given line, *at least two* lines parallel to the given line can be drawn." We drop the perpendicular AB from A to a. Notice the anomaly that a unique perpendicular is permitted. This is not surprising because the perpendicular segment is finite in extension and exists only in the local space of A.

By the fundamental axiom at least two lines exist which pass through A without intersecting a. As a result, every line between one of these and the extension of the other also does not intersect a. If one resorts to a diagram as shown, it would appear that the two lines through A, b and b', if produced far enough would intersect a. One must now keep in mind that Lobachevski did not argue from figures as we draw them in an ordinary local space, and, further, that our basic assumption, that extensions of the plane local spaces necessarily maintain the axiom of parallelism is subject to argument. With respect to this last observation, one must remember that in all cases the introduction of the concept of infinity into ordinary Euclidean geometry lends to the final formalism an element of uncertainty.

Returning to the line a and the point A, through A we construct a half line x that does not intersect a. By rotating x about A, we find the least position of x which does not bring it into intersection with a. In this least position, x makes a limiting angle ϕ_0 less than $\pi/2$ relative to the perpendicular AB. In the limiting position, call the half line c. Every half line making an angle ϕ larger than ϕ_0 (and less than $\pi - \phi_0$) relative to AB also does not intersect. The argument here is this: suppose the limiting value c does intersect a at C. If this were the case, then one should be able to take a point C' such that $BC' > BC$, where a second line c' intersects a. Such a situation, however, violates the condition that c is a limiting half line.

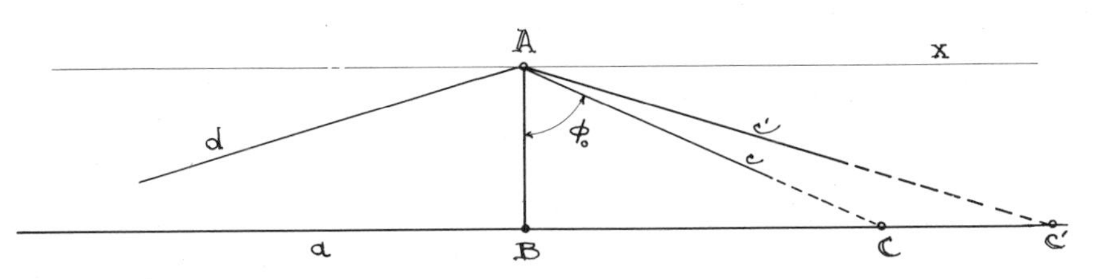

Therefore, c does not intersect a, and, moreover, it is the extreme half line passing through A and not intersecting a. By symmetry there then exists another limiting half line d which does not intersect a to the left of B. If c and d were continuations of each other, they would form a single line $c + d$, which would be the unique parallel through A. Once it has been assumed that the parallel is not unique, c and d *cannot* be the continuation of each other. When the half lines c and d are produced, we obtain the two limiting lines not intersecting a. The angle β between c and d is called the angle of parallelism, and is less than two right angles.

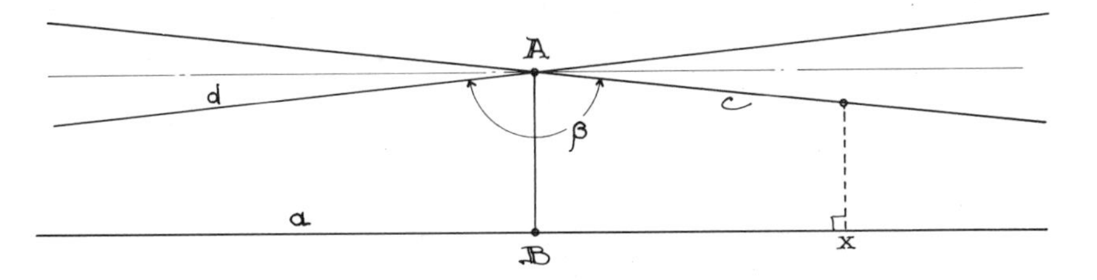

The results of these arguments are startling. If perpendiculars are dropped to a from points on c and d at increasing distances from A, the lengths of the perpendiculars decrease and actually tend toward zero as the point of intersection, say x, tends to infinity. Thus, the Lobachevski parallels c and d converge asymptotically to a. One can contrast this result with Euclidean geometry, where the perpendiculars remain fixed in length and do not converge. There is some indefiniteness in this concept when the lines have infinite extension. In the Lobachevski space, a line which maintains a constant distance from another is not straight but is a curve known as an equidistant. Most of the negative reaction to this geometry stems from our reliance upon our visual intuition. Accordingly, once the distinction between a local space and a complete space is accepted, one recognizes that visual intuition, which applies only to a local space, should not lead us to a narrow interpretation of space in the large. More important, discrepancies between the Lobachevski space and our concept of a local space cannot be used as a valid argument when the new geometry is viewed as an abstract structure which has intrinsic merit as such.

Many interesting variations arise when non-Euclidean geometry is compared with Euclidean. For instance, the angle of parallelism β is a function of the distance between the point A and the line a. The further a point is from the line, the smaller the angle of parallelism becomes. This last statement is more or less obvious when one considers two lines c and e, each parallel to a but at different distances.

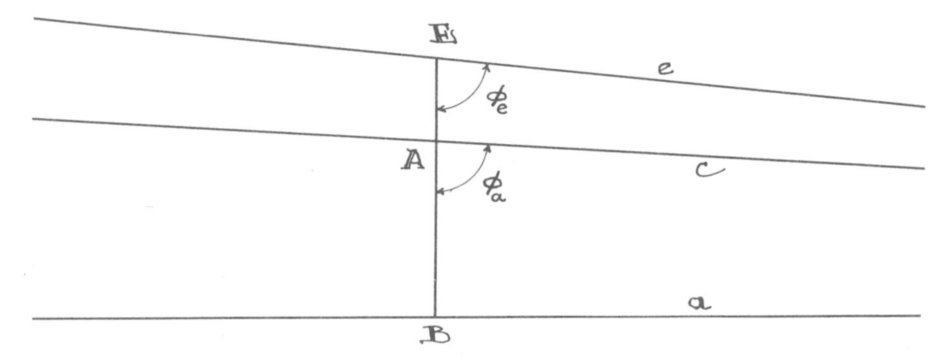

If c and e are both parallel to a and thus parallel to each other, c and e must converge asymptotically or else one or the other will be an equidistant curve, violating the assumption of parallelism. There are other interesting results.

1. If two lines a and b have a common perpendicular, then two perpendiculars c and d can be drawn to a which are parallel in the sense of Lobachevski to b, and the whole line b lies in the strip between c and d.

2. The limit of a circle of infinitely increasing radius is not a line but a particular curve called a limiting circle. Although it is not always possible to draw a circle through three points not on a line, either a circle, a limiting circle, or an equidistant can be drawn through the three points.

3. The sum of the interior angles of a triangle is less than two right angles. If the dimensions of a triangle become arbitrarily large, the three angles tend toward zero. Thus, there are no triangles of an arbitrarily large area.

4. The length l of the circumference of a circle is not proportional to the radius r but grows more rapidly:

$$l = \pi k(e^{r/k} - e^{-r/k}) = 2\pi k \sinh r/k,$$

and

$$l = 2\pi r\left(1 + \frac{1}{6}\frac{r^2}{k^2} + \cdots\right),$$

where k is a constant depending upon the unit of length. When r is much less than k, the Euclidean result $l = 2\pi r$ is approximately valid.

When compared with Euclidean geometry in a sufficiently small domain, hyperbolic geometry differs but little from the classical geometry. Therefore, in this sense we can claim that Euclidean geometry is just a limiting case of the hyperbolic geometry. It was not until 1868 that an intuitive interpretation of the new geometry was given by Eugenio Beltrami and later by Felix Klein. Beltrami noted that all the geometric relations on the surface of a pseudosphere were those of hyperbolic geometry, where the straight line on a plane is interpreted as a geodesic on the pseudosphere (the geodesic connecting two points on a surface is the path having the shortest, or longest, length).

The pseudosphere of Beltrami is formed by rotating a *tractrix* about its asymptote, the tractrix being a curve which has the property that the length of the segment of a tangent between the point of tangency to the intersection, with the axis corresponding to the asymptote, is always the same.

As the internal structure of mathematics was revealed, it began to change in the direction of greater generality and abstraction. The invention of non-Euclidean geometry serves as a signpost to the future.

During this same period, physics was advancing primarily in the area of electromagnetism. There is little similarity between the abstract changes in geometry and the basic discoveries in magnetism, except that

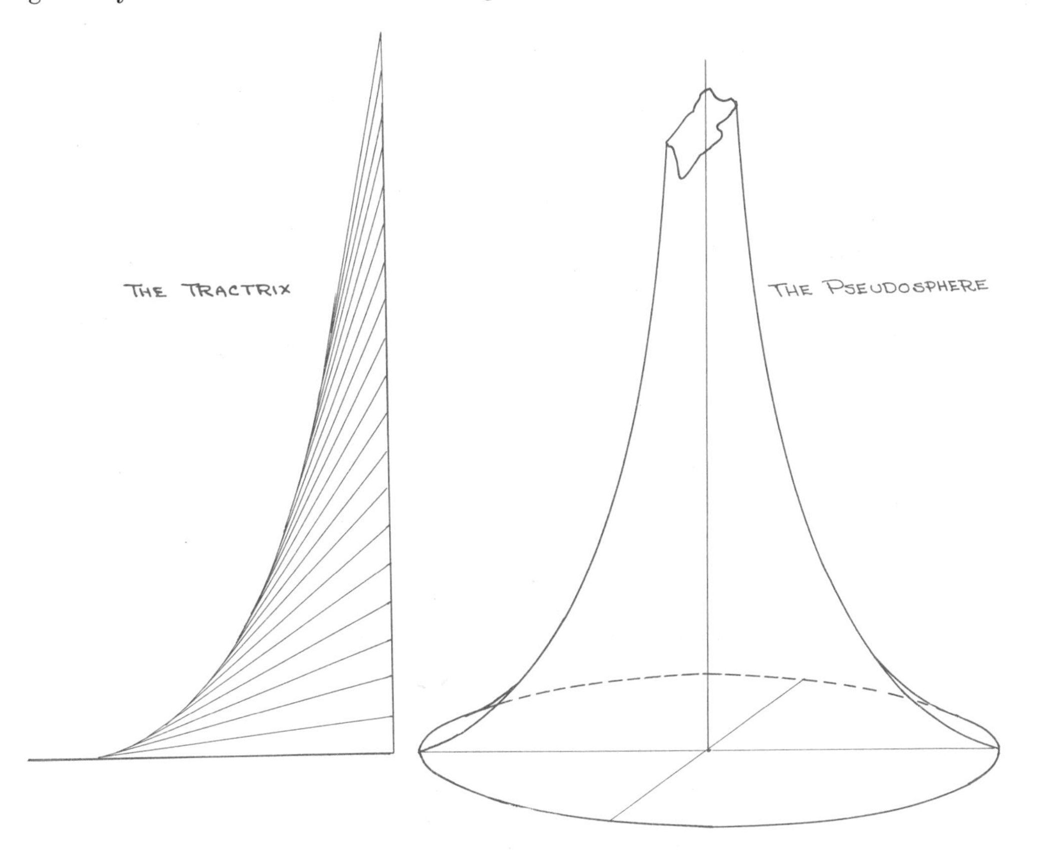

both represented an advance to the frontiers of the respective fields. The fundamental breakthrough in electromagnetism centered about the efforts of Michael Faraday (1791–1867), a man of little, if any, formal education but endowed with both genius and ambition. Apprenticed to a bookbinder as a young man, Faraday seized the opportunity to read much of the scientific literature which passed through his hands. In 1810 and 1811 he attended public lectures on natural philosophy, and the following year he was present at a series of lectures given at the Royal Institution by Sir Humphry Davy. At this point Faraday made up his mind to become a scientist. He submitted his plea and personal notes, and was accepted as Davy's laboratory assistant. He began his assistantship by traveling through Europe with Davy, and upon his return to England he began lecturing in chemistry.

The rise of this man was truly astounding. In 1821 Faraday wrote a review of electricity and magnetism after having repeated all of the experiments known at the time. By 1830 he had made three notable discoveries in chemistry and had published sixty original papers. Faraday was convinced that the analogue of electrostatic induction existed for dynamic charge or currents. Initially, Faraday thought that the mere proximity of two currents would produce some change in them, and he assumed that this effect would be dependent upon their separation. However, experiments based upon these assumptions showed no effect.

While Faraday was carrying out these researches, Joseph Henry, in the United States, and William Jenkins, in England, noted that an electric shock was produced when an electric circuit carrying a current was broken. Faraday pursued this avenue of investigation and showed that large currents were induced in a circuit, or in a nearby circuit, when the initial current-carrying circuit was interrupted. Thus, the induced effects were associated with the time rate of change in the initial current. In addition, Faraday found that currents could be induced by the motion of a permanent magnet nearby.

Faraday lacked mathematical skills; pictorial intuition alone enabled him to create a number of fundamental concepts which later became the basis of the mathematical formulation of the laws of electromagnetism. He constantly thought in terms of lines of force: a tangent to a line of force at any point gave the direction of the force, and the density of lines in the vicinity gave the strength of the force at the point in question. Faraday's concept of lines of force turned out to be consistent with the vector field representation which evolved toward the end of the century. With these concepts, Faraday was able to formulate the law of electromagnetic induction in terms of his lines of the magnetic force field. His further experiments showed that the induction is independent of the nature of the wire, depending only upon the magnetic flux linked by the wire. Induction was interpreted as an induced electromotive force ("emf") which is the equivalent of a battery potential. The emf was found to be proportional to the time rate of change of the total number of magnetic lines of flux linking a given circuit. By pictorializing the mathematical problem of the magnetic field, Faraday created a proper law for magnetic induction. His reasoning was relatively simple. A force field is thought of as a flux of lines, i.e., a given number of lines per unit area where the area is taken perpendicular to the lines. The total flux linkage of a loop which circumscribes an area A is the flux density times the elements of area summed over the circumscribed area, i.e., the total lines encompassed.

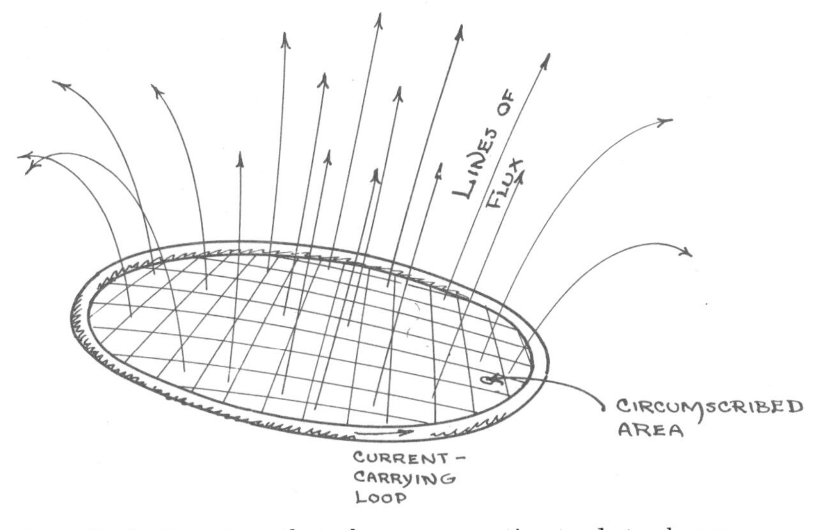

The law of induction states that when any operation tends to change the total magnetic flux linked by a conducting loop, an electromotive force is set up in such a direction as to create a current opposing the change. If the current in a loop is reduced by dropping the driving voltage or inserting a resistance, the back emf of induction tends to oppose the reduction, with the net result that the current drops more slowly than one might have anticipated initially.

Because the magnetic field $\vec{\mathbf{B}}$ of a current-carrying loop is of the same nature as the magnetic field of a permanent magnet, the motion of a permanent magnet near a conducting loop will cause an induced emf with an associated current to build up again in the circuit in a direction that will oppose the change in $\vec{\mathbf{B}}$.

The transformer configuration is similar to the example of two loops. A primary loop carrying a current I through the action of a battery E is shown in the diagram.

Initially, at times less than t_0 (t_0 is the time when the switch is thrown) when the primary circuit is constant, the secondary loop carries no current. If the primary circuit is altered by switching the battery out of the circuit, the initial magnetic field of the primary will tend to drop. Two things will then happen: first, there will be an induced current in the primary which tends to maintain the initial field, and second, there will be an induced emf in the secondary loop, producing a current which tends to maintain the initial magnetic field originally linked by the secondary. Because the compensation is not complete, the currents damp out in time.

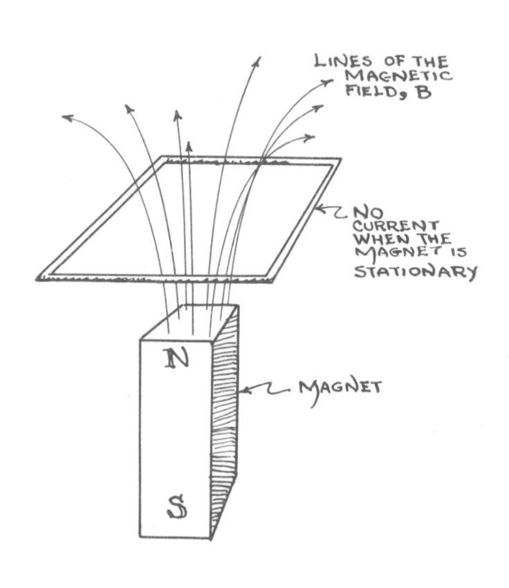

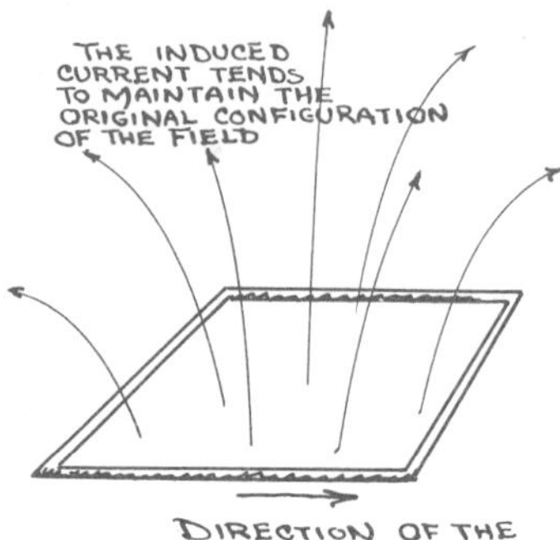

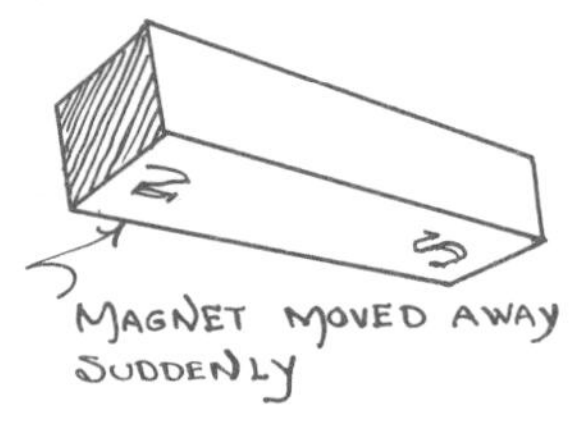

In mathematical language, if Φ is the total flux linked by a conducting loop $\mathscr{L}$ having a surface area A, then the induced emf E is equal to the negative of the time rate of change of Φ: $E = -d\Phi/dt$, where

$$\Phi = \iint_A \vec{\mathbf{B}} \cdot \vec{\mathbf{n}}\, dA = \int_\Phi d\Phi .$$

The last expression implies that the flux through an oriented surface element $\vec{\mathbf{n}}\, dA$ (where $\vec{\mathbf{n}}$ is a unit normal perpendicular to the surface) is the

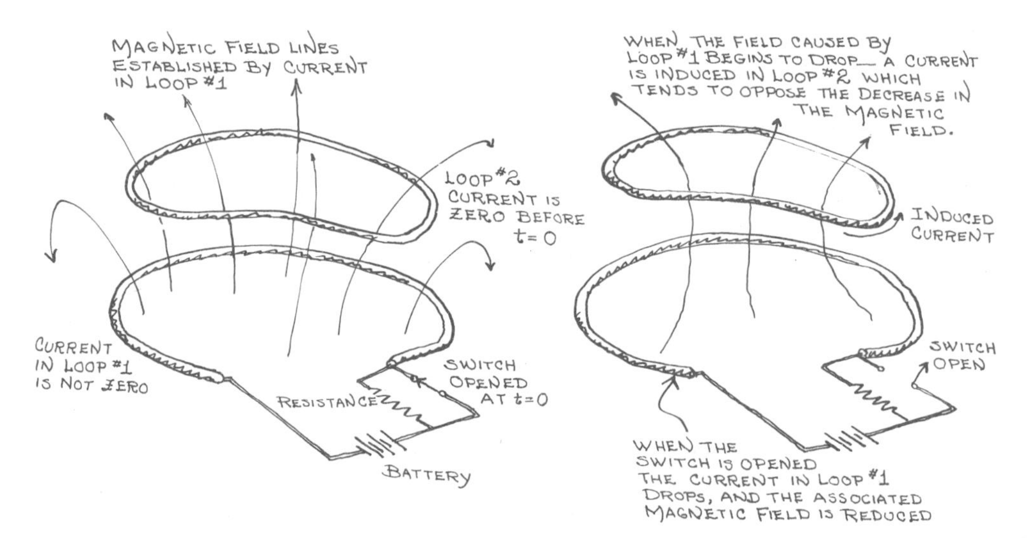

projection of the magnetic field $\vec{\mathbf{B}}$ on $\vec{\mathbf{n}}$ times dA, or $d\Phi = (\vec{\mathbf{B}} \cdot \vec{\mathbf{n}})\, dA$. Thus, the total number of lines or the total flux is just the sum of all of the $d\Phi$'s in A. The minus sign in the expression for E indicates that the induced emf opposes the change.

The practical importance of this discovery was immeasurable—in a short space of time the electric generator, the electric motor, and the transformer were invented, all based on Faraday's work. The discovery of electromagnetic induction would itself have singled out Michael Faraday as one of the great physicists. In this period he also showed that charge flow from a condenser was of the same nature as that from a voltaic pile.

For his second phase of research, Faraday attacked the question of the decomposition of a solution when a current is passed through it. By assuming that materials in solution decompose into ions of the opposite sign, he escaped many of the encumbrances of older theories. From his work came the names "electrolyte," "electrode," "cathode," "anode," "anion," and "cation." Further, he found that the total mass of any ion liberated by a fixed amount of charge passed through a solution was proportional to its chemical equivalent mass. A microscopic concept of matter was evolved when he postulated that there is an absolute quantity of electricity associated with each atom of matter. The amount of charge necessary to liberate one gram atomic weight of any substance is called the faraday, and is 96,580 coulombs (this is a modern figure). This discovery provided a mechanical means of determining absolute charge measurements.

Faraday's experiments in 1837 on the decomposition of liquids led him to reflect upon the behavior of insulators or dielectrics. When the plates of a condenser are maintained at a constant electrostatic potential, the admission of a dielectric into the space between the plates of the condenser causes an increase in the total charge on the plates; the amount of increase varies from substance to substance. Faraday assigned to each insulator a specific inductive capacity, today called the dielectric constant.

Consider a parallel plate capacitor having a vacuum in the separation between plates with a capacitance C_0, where the magnitude of the charge on either plate, Q, is given by $Q = C_0 V$. Electrostatic analysis demonstrates that C_0 is determined by the geometry of the two conductors which form the condenser. In the case of a parallel plate capacitor, C_0 is proportional to the area of the plates and inversely proportional to the separation of the plates. When a dielectric is inserted between the plates, the capacitance C is proportional to the vacuum capacitance C_0, $C = KC_0$, where K is a number greater than or equal to unity. If the dielectric completely fills the gap or the field region of the condenser, K is the dielectric constant of the dielectric. Because the charge under a fixed voltage V tends to increase when a dielectric is introduced into the condenser, the dielectric constant is written as $K = (1 + \chi)$, where χ is called the electric susceptibility and is positive—thus K is greater than or equal to 1. Our original equation for the charge, Q, then shows that $\chi C_0 V$ is the increase in charge caused by the presence of the dielectric.

Faraday modeled his description of a dielectric after the electrolyte except for the assumption that the dielectric could sustain stress without decomposing. He supposed that in the ordinary or unpolarized condition of a body the structure consisted of atoms bound by electrical forces. By introducing lines of electric force in a direct analogy to the magnetic force field, Faraday viewed the action of a dielectric in an electric field as a polarization of the atomic charge. Here again the analogue with magnetic materials was employed. With unfailing insight, Faraday conceived of a volume of small conducting spheres whose dimensions were small compared with their separation. Thus, when an electric field was imposed upon these small spheres they polarized in the manner of a dielectric, with a consequent reduction of the internal field.

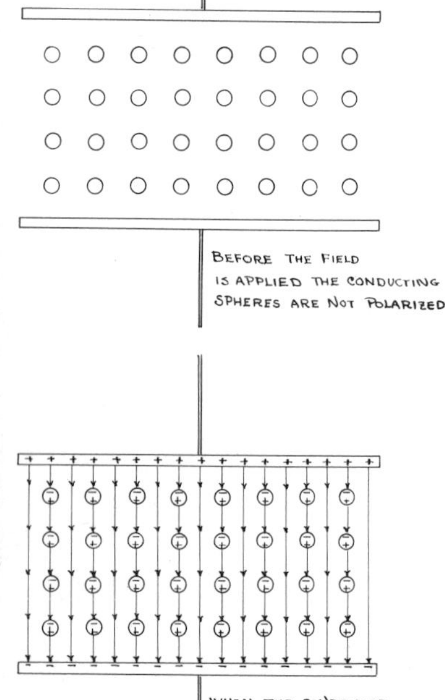

Although Faraday was incapable of participating in sophisticated mathematical descriptions of the physical world, with rare insight he prepared the way for much of the mathematical theory of electromagnetism which followed. Faraday's geometric and pictorial models were essential to the subsequent theories—in this sense, these descriptions *are* theories.

Many investigators in the early nineteenth century had sought in vain for some connection between magnetism and light. Sir John Herschel suggested that there should be a connection between the plane of polarization of transmitted light in a material and the state of magnetization. In September of 1845 Faraday discovered that when light was transmitted through heavy glass in a strong magnetic field in such a way that it was propagated parallel to the field, the plane of polarization was rotated. In 1846 Sir George Biddell Airy gave a special mathematical description of this effect in quartz. The basis of the mathematical description is similar to the model used for optically active materials. Assuming that plane polarized light is composed of two circularly polarized components, one rotating clockwise and the other counterclockwise, the rotation is achieved because the two circularly polarized components travel with different velocities of propagation along the magnetic field lines. After transmission, the ray emerges with a relative phase shift between the two circular components, resulting in a rotation of the initial plane of polarization. In 1881 Marie Alfred Cornu showed that the mean velocities of the righthand and lefthand waves in the material were equal to the velocity of light in the medium.

A few weeks after the discovery of the magnetic rotation of light, Faraday noticed that a bar of heavy glass suspended between the poles of the magnet tended to set itself across a line joining the poles, minimizing the volume of glass penetrated by the magnetic field. Many other materials were found, such as bismuth, which exhibited repulsion to the magnetic field. These materials were termed diamagnetic. Faraday's analysis of this phenomenon resulted in a description wherein diamagnetism is similar in all respects to ordinary induced magnetism, except that the direction of the induced magnetic polarization is reversed. More modern theories in fact show that all materials are diamagnetic, although the diamagnetic effects are usually weak. Paramagnetism and ferromagnetism, when present, are strong effects which mask the diamagnetic effect. Modern treatments of diamagnetism indicate that this effect can be accounted for as an electromagnetic reaction or as an induced electromagnetism resulting from the motion of electrons in the presence of a magnetic field. Thus the electromagnetic induction of Faraday became the ultimate explanation of diamagnetism.

While Faraday was discovering the law of induced currents, utilizing his own pictorial geometric models, Franz Neumann (1798–1895) developed what appeared to be an equivalent relationship using Ampère's law and the law of conservation of energy. In the final analysis this derivation has serious flaws; however, the approach did serve to emphasize the importance of the energy stored in the electromagnetic field. According to Ampère's law, a closed current loop gives rise to a field $\vec{\mathbf{B}}$. The law of force discovered by Ampère specifies that a current element $I\,\Delta\vec{\mathbf{l}}$ interacts with this field and experiences a force $\Delta\vec{\mathbf{F}}$ directed at right angles to $\vec{\mathbf{B}}$ and $\Delta\vec{\mathbf{l}}$:

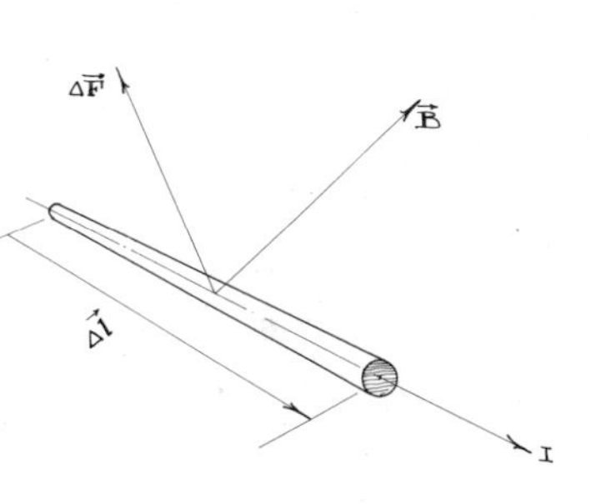

$$\Delta\vec{\mathbf{F}} = I\,\Delta\vec{\mathbf{l}} \times \vec{\mathbf{B}}.$$

With these forces one may associate a potential energy. If the second loop, of which $I\,\Delta\vec{\mathbf{l}}$ is a portion, is varied in position, the net force on the loop times the change in the position involves work which would be negative of the change in potential energy.

Mathematical development of the expression for the work done shows that the potential energy of a current loop in a magnetic field is equal to the current in the loop times the total magnetic flux Φ linking the loop; Φ, as mentioned before, is just the sum (or integral) of all the lines of $\vec{\mathbf{B}}$ linking the loop. The total energy U for two loops interacting with one another is then $U = I_2\Phi_2$, or $U = I_1\Phi_1$. Because the magnetic interactions between loops always occur between pairs, the total energy of a pair is divided equally between the two.

The flux Φ_2 due to a primary current I_1 is always a geometric factor L_{21} times the current I_1. Thus, the potential energy of loop L_2 in the field of a loop L_1 carrying a current I_1 is

$$U_2 = (1/2)I_2\Phi_2 = (1/2)I_2L_{21}I_1.$$

The geometric factor L_{21} between two separate loops is known as the mutual inductance. If one thinks of the current of a single loop as being made of a set of current filaments, one concludes that the filaments of a single loop interact, giving a self-energy and an associated geometric factor L, called a self-inductance. Thus, the self-energy of a single loop would have the form $U = (1/2)LI^2$, where I is the current of the loop. These energies associated with current distributions can also be regarded as energies stored in the extended magnetic fields of distributions, and such considerations lead to the same results.

When one considers extended current distributions in space, interacting with extended fields, these relatively simple forms for the stored energy can be represented in a much more sophisticated form by the introduction of a vector magnetic potential, $\vec{\mathbf{A}}$, where $\vec{\mathbf{A}}$ is defined by $\vec{\mathbf{B}} = \text{curl}\,\vec{\mathbf{A}}$, or $\vec{\mathbf{B}}$ equals the circulation of another field $\vec{\mathbf{A}}$. Such a form gives the energy stored as the volume integral over all space of one half the projection of the current density $\vec{\mathbf{J}}$ (current per unit of cross sectional area) on $\vec{\mathbf{A}}$:

$$U = \frac{1}{2} \iiint_{\substack{\text{all}\\\text{space}}} \vec{\mathbf{J}} \cdot \vec{\mathbf{A}}\, d(\text{volume}).$$

Consider a very long solenoid having N_1 turns of wire per unit length (measured along the axis), each turn carrying a current I and circumscribing a circle of radius R. A simple calculation shows that the field inside the long solenoid is approximately $B = \mu_0 N_1 I_1$, where μ_0 is a constant equal to $4\pi(10)^{-7}$ if N_1 is in turns per meter and I_1 is in amperes, giving B in webers per square meter. By winding one secondary turn of radius R about the solenoid, the flux linkage is $\Phi_2 = (\pi R^2)B = (\pi R^2)\mu_0 N_1 I_1$. Then, the energy stored in the secondary, if the current in the second single loop is I_2, is

$$U_2 = (1/2)I_2\Phi_2 = (1/2)I_2(\pi R^2)\mu_0 N_1 I_1 = (1/2)I_2 L_{21} I_1.$$

This is not the total stored energy because we have neglected the self-energy of the primary solenoid. The expression above shows that the mutual inductance between the primary solenoid and the secondary loop is truly geometric with $L_{21} = \mu_0 N_1 \pi R^2$.

Contemporary with Neumann was Wilhelm Weber (1804–1891), a friend and colleague of Gauss at Göttingen. Using some assumptions of G. T. Fechner, Weber viewed electric currents as composed of streaming electrical charges. Modifications of the definition of current then allowed him to formulate the Ampèrian force laws in terms of the charge on a particle. The results of Weber's efforts were unimpressive; however, his work was of the greatest significance in initiating the electron theories of matter, the point of view that electrodynamics was to be constructed from the properties of moving electric charges, providing forces which not only depend upon position but upon velocity.

Using his concept of electrons, in 1847 Weber attacked the question of diamagnetic bodies. He assumed, as Ampère had, that molecular currents exist which flow without resistance and, therefore, without dissipation of energy, and that, when a magnetic field is applied, currents would be induced in the molecular circuits which would oppose the applied field, producing a diamagnetic effect. His inference then, which was quite correct, was that all bodies, without exception, would be diamagnetic. Weber concluded that iron and substances exhibiting paramagnetism owed their magnetic properties to the presence of strong permanent molecular currents which are superimposed upon the ordinary weak diamagnetic effect, thereby masking the diamagnetism. These conjectures were essentially correct and were confirmed fifty years later through experiments and associated theories of magnetism. Working in areas well ahead of his time, Weber's record is remarkable. Because of his ideas of velocity-dependent forces, astronomers began attempting to modify Newton's laws of motion in order to account for the anomalous precession of the perihelion of Mercury. Their attempts, though unsuccessful, helped to awaken interest in the problem.

Little attention has been directed thus far to the subject of continuum thermodynamics. Thermometry had been a central interest of scientists since the early seventeenth century. Galileo's student Viviani gives 1593 as the date of the invention of the thermometer. The modern theory of heat was anticipated by Descartes, Boyle, Francis Bacon, Newton, and Hooke, who viewed heat as some form of internal motion of a body. In 1702 Galileo's air thermometer was improved by Guillaume Amontons (1663–1705), and from an extrapolation of Amontons' data, the absolute zero of temperature was found. These researches stimulated the invention of the alcohol and mercury thermometers, culminating in the invention of the centigrade scale in 1737. As often happens in the evolution of a proper theory, the theory of heat regressed in the eighteenth century from the early concepts based on motion to a conception of heat as a material fluid known as "caloric." In spite of these erroneous theories, specific heat capacity, the latent heat of vaporization, and the heat of fusion were all defined. In 1705 the first device for the application of steam power was produced. The invention of the steam engine between 1736 and 1760 led to the important concept of the efficiency of a thermodynamic engine.

The caloric theory of heat was disproved by Benjamin Thompson, Count Rumford (1753–1814), who first made quantitative estimates of the relationship between mechanical energy in terms of friction and the heat of water. Rumford's view of heat as energy was confirmed in 1799 by Sir Humphry Davy, and the propagation of heat in solid bodies was analyzed in 1822 by Fourier. The Fourier analysis was applied quite successfully to this problem; in fact, the success of practical applications of his mathematics led Fourier to assume that the technique was correct.

During the early part of the nineteenth century the thermometer was perfected to a high degree. Increments of temperature could be measured quite accurately, but a uniform temperature scale independent of the particular properties of materials was needed. The thermodynamic scale based upon the Carnot heat cycle was developed by Lord Kelvin in 1848 and remains the ultimate scale of reference. J. A. Charles (1746–1823), J. L. Gay-Lussac (1778–1850), and John Dalton (1766–1844) carried out careful researches on the properties of gases, measuring coefficients of expansion, latent heats, etc. These works culminated in the ideal gas law and the necessary modifications for real gases.

Thermodynamic engines, first steam and later gasoline and diesel, focused interest upon the mathematical problem of determining the amount of work which could be extracted from an engine moving through a complete cycle. Nicolas Léonard Sadi Carnot (1796–1832) introduced the consideration of the ideal thermodynamic cycle, in which a working substance is brought back to its initial state after a series of thermodynamic changes. These changes generally involve a series of expansions, during a part of which the ideal engine may absorb heat. The expansions are followed by a series of compressions, accompanied again in part by the loss of heat from the working substance to a condenser. At the end of the compression interval the working substance is returned to its initial state. Carnot invented the image of the ideal engine, in which the work extracted would be just sufficient to return to the high temperature reservoir the heat lost to the condenser. This quality of the ideal engine is known as reversibility: it assumes that perpetual motion is impossible, according to Carnot, and that an ideal engine possesses the maximum efficiency possible for any engine working between two limiting temperatures T_1 and T_2.

These concepts were later to be formulated as the second law of thermodynamics. One way of stating the second law is to say that heat energy cannot be transferred into work with 100 per cent efficiency by a heat engine which carries a working substance through a cyclical process. The idea of conservation of energy anticipates a general trend in the physics of the nineteenth century. Energy and its conservation were to become the dominant physical quantities during this century in thermodynamics, in electricity and electromagnetism, and in mechanics, although the invariance of the total mechanical energy of certain isolated systems had been established as an important concept earlier. Changelessness in the midst of change and persistence of configuration in the presence of transformation also characterized an important concept for the mathematicians of the late nineteenth century.

In the field of elementary algebra Lagrange had noted in 1773 that a particular linear transformation of x and y in the bilinear form $ax^2 + 2bxy + cy^2$ left the discriminant, $b^2 - ac$, unchanged. His discovery can be demonstrated quite briefly. Let $x = mx' + ny'$, and $y = px' + sy'$, where m, n, p, and s are numbers representing a linear transformation of x and y to x' and y'. Substituting these forms for x and y into $ax^2 + 2bxy + cy^2$ provides a new bilinear form in x' and y' of the form $Ax'^2 + 2Bx'y' + Cy'^2$, where A, B, and C are constituted from a, b, c, m, n, p, and s. The discriminant of this second quadratic is $B^2 - AC$. Carrying out the algebra, one discovers that $B^2 - AC = (ms - np)^2(b^2 - ac)$. In other words, the linear transformation merely scales the discriminant by $(np - ms)^2$. Lagrange actually used the specific values $m = 1$, $n = \lambda$, $p = 0$, and $s = 1$, so that $(np - ms)^2 = (1)^2$, and this discriminant remained invariant under this specific linear transformation. Lagrange's discovery is, therefore, a special case of a general theorem in algebraic invariants. The general theorem states that a linear transformation of the variables in a quadratic or bilinear form converts the form into another form whose discriminant is equal to that of the original form multiplied by a factor depending only upon the coefficients (m, n, p, s) of the transformation. This general theory was discovered in 1842 by George Boole (1815–1864), a self-taught mathematician considered to be one of the founders of symbolic logic. His contributions to mathematics were much more extensive than is generally recognized.

The algebraic polynomial represents one of the simplest of functions, and, as such, it received primary attention for centuries. The general cubic and quartic equations were solved in the sixteenth century, and the only major contribution for the next two centuries was the generalization of the solution to the quartic by Lagrange in 1770. Lagrange initiated the methods later expanded into the theory of groups, in that he obtained the equivalents for some of the simpler properties of permutation groups. In modern terminology, one of his results states that the order of a subgroup of a finite group divides the order of the group.

Niels H. Abel (1802–1829) published at his own expense a very complex proof that equations of degree higher than four could not be solved in terms of rational radicals.* In spite of minor defects, Abel's proof, along with similar work by P. Ruffini, was the first proof of impossibility. Encouraged by this proof, Évariste Galois (1811–1832) searched for a more fundamental condition of solvability. He found that an algebraic equation is solvable by rational radicals if and only if its group for the field of its coefficients is solvable. Group theory was the creation of a number of mathematicians, starting with Lagrange, continuing with Abel, and culminating in the inspired attack on the problem by Galois.

Before describing Galois' contributions it is useful to inspect some of the basic definitions of group theory and their consequences. Many details were introduced later, such as Cayley's design of the group multiplication table. The essence of the group concept is the notion of structure or pattern in a set of objects. The two most important features of a group are a set of elements and a binary operation. As an example, we can consider the set of all integers— $-3, -2, -1, 0, 1, 2, 3, \cdots$—and the binary operation of addition. One notices that when the binary operation is performed upon any two elements of the set, the result is an element of the set. As a second example, one could employ the set of all positive rational numbers in the binary operation of multiplication. Again, it is clear that the combination of two elements under the binary operation results in an element of the set. It is important to note that a binary operation on a set is a correspondence which assigns to each ordered pair of elements of the set a uniquely determined element of the set. If in a group G, r and s are any two elements of the set, the binary operation on the ordered pair, designated by the symbol $\otimes$, produces t: $r \otimes s = t$.

The symbol $\otimes$ may represent any one of a number of combinative operations, such as addition, multiplication, subtraction, division, etc. In practice, the operation can be something more complex than these. For instance, if the elements are the matching positions of an equilateral

* This proof implies that roots cannot be generated by taking rational roots of quantities formed from the number field of the coefficients. The number field of the coefficients constitutes all of the numbers which can be produced by the multiplication, division, addition, and subtraction of the coefficients of the polynomial.

triangle when rotated about an axis through its centroid, the operation may consist of rotations through angles which are multiples of 120°.

Because it may produce elements outside of the set, the binary operation must be somewhat restricted: in some cases the set would not be closed under the operation specified, an example being the set of positive integers under the operation of division. Subsets may also be defined within a set: if the binary operation upon any two elements of the subset gives an element in the subset, the subset is said to be closed under the binary operation. The *closure property* of a subset under a binary operation plays an important role in defining subgroups.

Throughout this discussion the qualification "ordered pair" has been used. This distinction is important because the binary operation $\otimes$ is not necessarily commutative; i.e., in general $r \otimes s$ is not necessarily equal to $s \otimes r$. In fact, if the reader considers the set of positive rational numbers under the binary operation of division, it is obvious that the operation is not necessarily commutative: $(a/b) \div (c/d) \neq (c/d) \div (a/b)$ always. In those special cases where $r \otimes s = s \otimes r$, r and s commute.

If a set with a binary operation is to form a group, one must postulate that the binary operation possesses certain properties relative to the elements. The three required postulates are the group axioms.

1. *Associativity:* for any elements r, s, t of G, $r \otimes (s \otimes t) = (r \otimes s) \otimes t$; in other words, for more than one operation the grouping of the operations should not change the result as long as the order is maintained.

2. *The identity element:* there is a unique element I in G such that for every r of G, $r \otimes I = I \otimes r = r$.

3. *The inverse:* for any element r of G there exists a unique element r^{-1} of G, called the inverse of r, such that $r \otimes r^{-1} = r^{-1} \otimes r = I$.

The inverse and identity elements do not always assume the special form which we encounter when multiplying or dividing ordinary numbers. As an example, again consider the set of integers under the operation of addition. Here the identity element is 0: $r + 0 = 0 + r = r$. For this group the generalized inverse turns out to be the negative of r in order that the law $r \otimes r^{-1} = I$ be maintained (remember that 0 had just been shown to be the identity element for this group): $r + (-r) = (-r) + r = 0$.

As mentioned, a specific group is customarily defined by writing down the multiplication table or, in better language, a binary combination table. Cayley did not introduce such a table until 1854. Just as useful are the generators of the group. Consider a simple example: if, for instance, all of the elements of a group can be expressed as binary combinations of a and b together with their inverses, one calls a and b the generators of the group. This concept can be generalized to sets of more than two elements. If S is a subset of elements of a group G, and if all elements of G can be expressed as binary combinations of elements of S together with their inverses, we then call the elements of S the generators of group G.

As a simple but important case, consider the situation in which G is a group generated by the single element r, and, further, that G is cyclic with a period 3 such that $r^3 = I$, with the stipulation that the single relation

$r^3 = I$ constitutes a complete set of defining relations for G; here $r^2 = r \otimes r$ and $r^3 = r \otimes (r \otimes r)$. With this definition one observes that the elements of G—$r, r^2, r^3, r^4, r^5, \cdots, r^n$—can be written as r, r^2, I, r, r^2, I. Therefore, there is a cyclical repetition of the basic pattern r, r^2, I, giving rise to the qualification that this group is a cycle group of order 3.

An important concept in Galois' theory is that of the subgroup. Most groups exhibit their internal structure in terms of subgroups, which are groups within a group. H is said to be a subgroup of G if every element of H is an element of G and if H is a group under the binary operation of G. The proof of the group property of the subgroup rests upon whether the three group axioms hold; that is, (1) the operation $\otimes$ is associative (this is automatically satisfied); (2) the inverse of each element h in H is in H; and (3) the identity element of G is in H. A small point that must be made is that every group G contains two special subgroups, G and I, and any subgroup that is not one of these is shown as a proper subgroup. The order of a group measures the number of elements, and consequently groups may be infinite or finite. In 1771 Lagrange formulated the important theorem that the order of any *finite* group is a multiple of the order of any subgroup. As a result a finite group has no subgroups if and only if the order of the group is a prime number.

The co-sets of a group G can be manufactured from a subgroup H. If b is an element of G but not an element h_j of H, then b, operating from the left on the elements of H, forms the left co-sets of G. In the same manner, if the elements of H each operate from the left on b, the right co-sets of G, h_jb, are formed. The left and right co-sets of G with respect to a subgroup H may in general yield different decompositions of G.

Closely associated with the group concept is the concept of mapping. Mathematical mapping is abstracted from the customary idea of mapping. To map an ordered set of objects A into an ordered set of objects B, one must maintain a one-to-one correspondence between an object in one set and an object in the second set. Consider the objects a, b, and c ordered in numbered columns,

$$\begin{pmatrix} 1 & 2 & 3 \\ a & b & c \end{pmatrix}.$$

A specific mapping of this set could result in the array

$$\begin{pmatrix} 1 & 2 & 3 \\ b & c & a \end{pmatrix},$$

where a, initially in position 1, has been mapped to position 3, while b has gone from 2 to 1, etc. A group binary operation can be thought of as a mapping: $r \otimes s = t$. Here the ordered pair of group elements has been mapped into the group. In the first bracket notation above, one can think of 1 mapped into a: $1 \to a$, $2 \to b$, $3 \to c$. A set may be mapped into itself:

$$\begin{pmatrix} a & b & c \\ b & a & c \end{pmatrix}.$$

Then multiple mapping can occur. The elements in the lower row pick up corresponding elements in the succeeding upper row at the right to insert the final term in its appropriate position. For instance,

$$\begin{pmatrix} a & b & c \\ b & a & c \end{pmatrix} \begin{pmatrix} a & b & c \\ b & a & c \end{pmatrix} = \begin{pmatrix} a & b & c \\ a & b & c \end{pmatrix}.$$

Mappings of one set onto itself, in fact, can be elements of a group. This fact introduces a new term. A "homomorphic" mapping of a group G into a group H has the characteristic property that if a and b are elements of G, their product or combination $a \otimes b$ is mapped into the element $f(a) \otimes f(b)$ in H, where $f(a)$ is the operation which maps a of G into H. When a homomorphic mapping of one group onto another is also one-to-one, it is called an "isomorphic" mapping. The class of groups known as permutation groups are significant in the theory. Every finite group is isomorphic to some permutation group. All mappings of a set onto itself constitute a group of mappings and are permutation groups.

In the general area of the mapping on the subgroups of a group, one encounters the so-called normal subgroups, which form an important part of Galois' theory. A normal subgroup K of a group G has the property that if y is any element of G the left co-set of K with y is equal to the right co-set of K with y. This is written $yK = Ky$, meaning that each element of yK (i.e., $yk_1, yk_2, \cdots$) is in the set Ky. The simplified forms yk_j again imply the presence of the binary operation of the group $y \otimes k_j$.

The Abelian group, named after Abel, is that in which any two elements commute. Thus any subgroup of an Abelian group is normal—in his work with algebraic equations, Abel employed the concept of commutative groups. The converse theorem is that given a normal subgroup K or a group G, there exists a group H and a homomorphic mapping of G onto H such that the elements of K are those precise elements of G that are mapped onto the identity of H. Galois first demonstrated that the co-sets of a group G with respect to a normal subgroup K of G form a group called the factor group. To discuss this theorem one must define the binary combination of two ordered co-sets R and S as the set of all combinations $r \otimes s$, where r is an element of R and s is an element of S.

Galois' proof of the general insolubility of algebraic equations of degree higher than four rests upon the ability, starting with symmetric functions of the roots of the equation of order $n!$ (where n is the degree of the original equation) belonging to a group G, to find a sequence of normal subgroups $K_1, K_2, \ldots, K_r = I$, ending in a subgroup of order one consisting of the identity. Each subgroup K_j is an invariant subgroup of the preceding subgroup K_{j-1}. Only when such a sequence of subgroups can be obtained is the general nth-degree equation solvable in rational radicals. Complex as this problem is, some insight into it can be had by considering the cubic equation $x^3 + ax^2 + bx + c = 0$, with roots α, β, and γ; such that

$$(x - \alpha)(x - \beta)(x - \gamma) = x^3 + (-\alpha - \beta - \gamma)x^2 + (\alpha\beta + \beta\gamma + \gamma\alpha)x - \alpha\beta\gamma = 0.$$

The roots form a symmetric group of order 3!. The binary operation of the group is the permutation operation. The operations of the group on certain functions of the roots (such as $\alpha + \beta + \gamma$, or $\alpha\beta\gamma$) leave the function unchanged. Such functions are called symmetric functions, and any symmetric function of the roots can be expressed directly in terms of the coefficients of the equation. Here $\alpha + \beta + \gamma = -a$, and $\alpha\beta\gamma = -c$.

It happens that some functions are changed by one given permutation but are unchanged by others. The permutations which leave a function unchanged will form a subgroup of the symmetric group, and the function is said to belong to this subgroup. For instance, consider the subgroup of order 2 which contains the identity and the interchange (α, β). Functions belonging to this subgroup are, for example, $(\alpha^2 + \beta^2)$, $(\alpha + \beta)$, γ, $(\alpha\gamma + \beta\gamma)$, etc. These, it turns out, can also be expressed in terms of a, b, and c.

For the symmetric group of order 3! there are three subgroups of order 2* involving the permutations $G_1, I, (\alpha, \beta)$; $G_2, I, (\beta, \gamma)$; and $G_3, I, (\gamma, \alpha)$. Another subgroup is that of order 3 involving the elements $G, I, (\alpha, \beta, \gamma)$, (α, γ, β).

The general solution of an algebraic equation of degree greater than 1 involves irrational numbers, the most familiar of which is a rational radical. Now, if those functions of the roots of an equation which correspond to a given group G are known, and if the group G has a subgroup H of index r, then the functions belonging to the subgroup H can be obtained by solving an equation of degree r. Further, if H is a self-conjugate subgroup of G, and r is a prime number, the equation will appear in the form $x^r =$ constant. Finally, if G is the symmetric group on n symbols, we can solve the nth-degree equation, that is, if we find a sequence of subgroups $G, H_1, H_2, \ldots, H_r = \mathbf{I}$ terminating in a group of order 1 (or the identity subgroup). Each subgroup is an invariant subgroup of the preceding group. Only when such a sequence occurs is the general nth-degree equation solvable.

In the example of the cubic equation, G is the symmetric group and H is the invariant subgroup of order 3. For the quartic equation with roots $\alpha, \beta, \gamma, \delta$ the appropriate group is the symmetric group of order $24 = 4!$. The 24 operations of this group separate into five classes. This symmetric group has a self-conjugate subgroup of order 12 followed by a self-conjugate subgroup of order 4 composed of $\mathbf{I}$, $(\alpha, \beta)(\gamma, \delta)(\alpha, \gamma)(\beta, \delta)(\alpha, \delta)(\beta, \gamma)$. Because these elements are self-conjugate, there are self-conjugate subgroups of order 2.

When Cardano's rule is employed to solve the quartic, one proceeds from the symmetric group to a group of order 8 of index 3, which is not self-conjugate. This last group gives rise to a general cubic equation (not of the form $y^3 = k$), which is solvable.

The appropriate group for the fifth-degree equation is the symmetric group of order $5! = 120$. This group has a self-conjugate subgroup of order 60. It can then be shown that this subgroup has no invariant subgroup. Hence the Galois criterion is violated, and it is impossible to solve the general quintic by the repeated extraction of rational radical roots. Of course, special quintics and special higher-degree equations can be solved in terms of rational radicals. The question originally, however, was directed to a general solution.

Generalization and abstraction now appeared in most active areas of mathematics. Algebra had first been conceived as an abstract hypothetico-deductive science by George Peacock (1791–1858); for example, the algebra

* Any one of these subgroups can be transformed into any other by an operation of the symmetric group. Thus they are known as conjugate subgroups.

of complex numbers was reduced to an abstract algebra of number couples by Gauss and later by Augustus De Morgan (1806–1871) and William Rowan Hamilton (1805–1865). Number couples with appropriate binary combination rules allowed representation of rational fractions and negative numbers in an arithmetized algebra. One is immediately impressed by the similarities between the theories of groups and the developments in abstract algebra. An important aspect of the new movement in algebra, which also appeared in group theory, was the suppression of the commutative law in binary combinations; in other words, the relationship between the products ab and ba was reexamined on the principle that the equality of the commuted products, in some instances, may be overrestrictive.

By the age of thirty-eight Hamilton had developed the algebra of quaternions—an algebra concerned with the ordered arrays of four elements a, b, c, and d, which we shall write as a, b, c, d or as $a\mathbf{1} + b\mathbf{i} + c\mathbf{j} + d\mathbf{k}$. The parameters **i**, **j**, and **k** played a role analogous to the imaginary $\sqrt{-1}$ in complex numbers, which, as we have noted, could be handled as binary arrays. To achieve multiplication, Hamilton specified the manner by which the parameters **1**, **i**, **j**, and **k** could be combined. The **1** multiplied by **1**, **i**, **j**, or **k** left the parameter the same as in ordinary multiplication. Products between the last three parameters were somewhat more involved. When **i**, **j**, or **k** is multiplied by itself, the product is -1, while the product of unlike quantities in this set satisfy anti-commutation rules of the type $(\mathbf{ij} + \mathbf{ji}) = 0$. To gather all of these rules into one table, we anticipate the device invented by Cayley in 1854, and write a multiplication table for all of the parameters. Any box in the table presents the product of the parameter, which labels the row times the parameter labeling the respective column. The order in which parameters are multiplied is significant, and this is apparent from the anti-commutation relations of the form $\mathbf{ij} = -\mathbf{ji}$:

	1	**i**	**j**	**k**
1	**1**	$\mathbf{1i} = \mathbf{i}$	$\mathbf{1j} = \mathbf{j}$	$\mathbf{1k} = \mathbf{k}$
i	$\mathbf{i1} = \mathbf{i}$	$\mathbf{ii} = -\mathbf{1}$	$\mathbf{ij} = \mathbf{k}$	$\mathbf{ik} = -\mathbf{j}$
j	$\mathbf{j1} = \mathbf{j}$	$\mathbf{ji} = -\mathbf{k}$	$\mathbf{jj} = -\mathbf{1}$	$\mathbf{jk} = \mathbf{i}$
k	$\mathbf{k1} = \mathbf{k}$	$\mathbf{ki} = \mathbf{j}$	$\mathbf{kj} = -\mathbf{i}$	$\mathbf{kk} = -\mathbf{1}$

To illustrate the product of two quaternions, take the two arrays

$$A = a\mathbf{1} + b\mathbf{i} + c\mathbf{j} + d\mathbf{k},$$

or $A = [a, b, c, d]$; and

$$E = e\mathbf{1} + f\mathbf{i} + g\mathbf{j} + h\mathbf{k},$$

or $E = [e, f, g, h]$. Then,

$$AE = [a, b, c, d][e, f, g, h]$$
$$\begin{aligned} AE = {} & ae\mathbf{1} + af\mathbf{i} + ag\mathbf{j} + ah\mathbf{k} \\ & + be\mathbf{i} + bf\mathbf{ii} + bg\mathbf{ij} + bh\mathbf{ik} \\ & + ce\mathbf{j} + cf\mathbf{ji} + cg\mathbf{jj} + ch\mathbf{jk} \\ & + de\mathbf{k} + df\mathbf{ki} + dg\mathbf{kj} + dh\mathbf{kk}. \end{aligned}$$

With the multiplication rules for the **1**, **i**, **j**, and **k** this becomes

$$\begin{aligned} AE = {} & (ae - bf - cg - dh)\mathbf{1} + a(f\mathbf{i} + g\mathbf{j} + h\mathbf{k}) \\ & + e(b\mathbf{i} + c\mathbf{j} + d\mathbf{k}) + (ch - dg)\mathbf{i} + (df - hb)\mathbf{j} + (bg - cf)\mathbf{k}. \end{aligned}$$

The coefficients of each array (i.e., a, b, c, d, e, f, g, h) are treated as scalars and play no role in forming the position of a product. Those familiar with vector analysis will recognize that the quaternian product illustrated above is a catch-all. The quaternian is the linear combination of a scalar, a vector, and a pseudovector. The product of two quarternians in the notation of vector analysis gives the product of the scalar terms (i.e., as above), the negative of the inner product of the vectors, the scalar of each times the vector of the other, and the vector or outer product of the two vector terms. If we write $A = a\mathbf{1} + \vec{\mathbf{b}}$, where $\vec{\mathbf{b}} = b\mathbf{i} + c\mathbf{j} + d\mathbf{k}$; and $E = e\mathbf{1} + \vec{\mathbf{f}}$, where $\vec{\mathbf{f}} = f\mathbf{i} + g\mathbf{j} + h\mathbf{k}$; then $AE = ae - \vec{\mathbf{b}}\cdot\vec{\mathbf{f}} + a\vec{\mathbf{f}} + e\vec{\mathbf{b}} + \vec{\mathbf{b}} \times \vec{\mathbf{f}}$. Here the dot indicates an inner product and $\times$ the outer product.

This formulation was not without practical application in that it accounted for rotations and stretches in a three-dimensional space and, as such, contained in a rather obscure form elements of vector algebra. Gauss in 1827 and Moebius in 1823 had attempted algebras of four fundamental units. In each instance the results were shallow because the commutative law of multiplication was assumed to hold. Today, when it is natural to presume that any given fundamental axiom can be suppressed, it is difficult to appreciate the insight of men like Hamilton, who swept away centuries of tradition.

At the age of thirty-five, Hamilton completed a comprehensive theory of light rays and established the most powerful representation of classical mechanics, a system which was to become the foundation of modern quantum theory. Hamilton assigned to every mechanical system a function analogous to the Lagrangian, known today as the Hamiltonian function, H. The Hamiltonian was defined in such a way that this function contained only position components, generalized momentum components, and time as the independent variables of a system. In Lagrangian mechanics the generalized momentum components were defined as the derivative of the Lagrangian function, L, with respect to the appropriate velocity component $\dot{q}_j$; i.e., p_j = the jth component of momentum $= \partial L/\partial \dot{q}_j$. One refers to this term as a generalized momentum in view of the fact that it conforms to our simpler definition of momentum as mass times velocity when the system is of an elementary nature. When the system contains coupling terms or further complications, the generalized momentum conforms in spirit to Newton's laws in the final result.

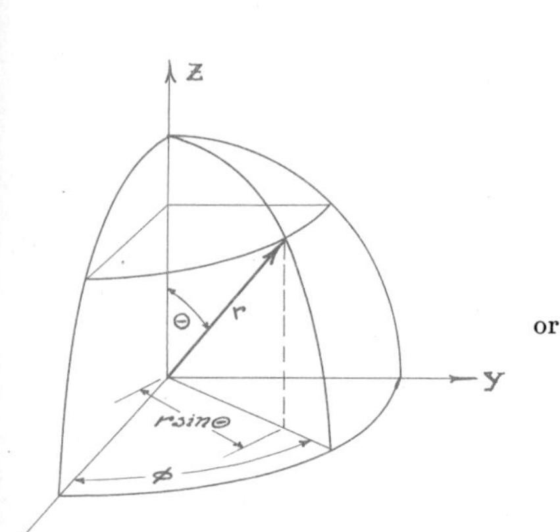

To illustrate one elementary advantage of the generalized momentum definition, consider a mass point moving in spherical coordinates. Although the resolved velocity vectors and the resolved acceleration vectors have quite complex forms in this set of coordinates, the square of the velocity has a relatively simple form:

$$v^2 = \left(\frac{dr}{dt}\right)^2 + r^2\left(\frac{d\theta}{dt}\right)^2 + r^2 \sin^2 \theta\left(\frac{d\phi}{dt}\right)^2,$$

or

$$v^2 = \dot{r}^2 + r^2\dot{\theta}^2 + r^2 \sin^2 \theta\dot{\phi}^2.*$$

The Lagrangian for a single particle in this representation is then $L = (1/2)mv^2 - V(r)$. If $V(r)$, the potential energy function, is not a function of the time derivatives $\dot{r}$, $\dot{\theta}$, or $\dot{\phi}$, one observes that the generalized momenta p_r, p_θ, and $p\phi$ can be easily computed: p_r is the derivative of L with respect to $\dot{r}$, p_θ is the derivative of L with respect to $\dot{\theta}$, and $p\phi$ is the derivative of L with respect to $\dot{\phi}$: Then

$$p_r = \frac{\partial L}{\partial \dot{r}} = m\dot{r} = \text{the radial momentum;}$$

$$p_\theta = \frac{\partial L}{\partial \dot{\theta}} = mr^2\dot{\theta} = \text{the component of angular momentum in } \dot{\theta};$$

and

$$p_\phi = \frac{\partial L}{\partial \dot{\phi}} = mr^2 \sin^2 \theta\dot{\phi} = \text{the second component of angular momentum in } \dot{\phi}.$$

Although the first momentum p_r is elementary, the last momentum component has a more complex form.

For systems which are more complex than this, the Lagrangian definition of the generalized momenta is essential. Hamiltonian mechanics is based upon the function H, according to

$$H = \sum_{\text{all } j} p_j\dot{q}_j - L(q_j, \dot{q}_j; t),$$

where p_j is the generalized momentum component for the jth coordinate and q_j is the time derivative of the jth coordinate. When the system is relatively simple and time-independent, the Hamiltonian takes the form of the total energy: $H \to T + V$, where T is the total kinetic energy of the system and V is the total potential energy. For our example of the single particle of mass m moving in spherical coordinates under the action of a potential $V(r, \theta, \phi)$, H is written,

$$H = \frac{p_r^2}{2m} + \frac{p_\theta^2}{2m} + \frac{p_\phi^2}{2m} + V(r, \theta, \phi).$$

The equivalent of Newton's second law is obtained from the canonical

* Here we employ the shorthand superior dot to indicate a time derivative, i.e., $dr/dt = \dot{r}$.

equations of Hamilton. These relate the derivatives of H to the time variation of the momenta p_i and of the coordinates q_i:

$$\dot{p}_i = \frac{dp_i}{dt} = -\frac{\partial H}{\partial q_i},$$

and

$$\dot{q}_i = \frac{dq_i}{dt} = \frac{\partial H}{\partial p_i}.$$

Application of the first equation to the single particle Hamiltonian shows that Newton's second law appears appropriate. To illustrate the complexity of the acceleration components, take p_θ as an example:

$$\begin{aligned}\dot{p}_\theta = \frac{dp_\theta}{dt} &= \frac{d}{dt}(mr^2 \sin^2 \theta\dot{\phi}) \\ &= m(2r\dot{r} \sin^2 \theta\dot{\phi} + r^2 \sin 2\theta\dot{\theta}\dot{\phi} + r^2 \sin^2 \theta\ddot{\phi}) \\ &= -\frac{\partial H}{\partial \theta} = -\frac{\partial V}{\partial \theta},\end{aligned}$$

a form which would be tedious to develop by Newtonian methods.

While quaternian algebra was being developed, in the years 1840–1844, Hermann Günther Grassmann (1809–1877), a Sanskrit scholar and an elementary school teacher, developed a broader algebra of n dimensions. He endowed his theory with the utmost generality, virtually overwhelming it in philosophical abstractions.* Unlike Hamilton, who sought fame and priority rights with fervor, Grassman wrote at his leisure and was in a position of complete obscurity for some forty years. It is to the credit of mathematicians and physicists that ultimately the power of Grassmann's methods was recognized. For eighteen years after the publication of his *Lineale Ausdehnungslehre* Grassmann remained unknown, and he was never recognized as a mathematician in his own lifetime.

The multiple algebra of Grassmann was sufficiently broad to encompass quaternians as a special case, an advantage which was not fully appreciated until the twentieth century. The starting point of the multiple algebra is similar to the barycentric calculus of Moebius, wherein scalar weights a, b, c, and d are associated with points A, B, C, and D to form a type of vector or hypercomplex number $aA + bB + cC + dD$. Grassmann associated with the array of terms $x_1, x_2, x_3, \cdots, x_n$ the hypercomplex number $x_1\boldsymbol{\epsilon}_1 + x_2\boldsymbol{\epsilon}_2 + x_3\boldsymbol{\epsilon}_3 + \cdots + x_n\boldsymbol{\epsilon}_n$, where the quantities $\boldsymbol{\epsilon}_1, \boldsymbol{\epsilon}_2, \boldsymbol{\epsilon}_3, \cdots, \boldsymbol{\epsilon}_n$ were the fundamental units or bases of the algebra. Addition of hypercomplex numbers (or vectors) was known long before this, having

* E. T. Bell, *The Development of Mathematics* (New York: McGraw-Hill Book Co., 1945), p. 200.

been mentioned with respect to a three-dimensional space by Lagrange and others. Two hypernumbers, call them **a** and **b**, where

$$\mathbf{a} = a_1\boldsymbol{\epsilon}_1 + a_2\boldsymbol{\epsilon}_2 + a_3\boldsymbol{\epsilon}_3 + \cdots + a_n\boldsymbol{\epsilon}_n = \sum_{j=1}^{n} a_n\boldsymbol{\epsilon}_n,$$

and

$$\mathbf{b} = b_1\boldsymbol{\epsilon}_1 + b_2\boldsymbol{\epsilon}_2 + b_3\boldsymbol{\epsilon}_3 + \cdots + b_n\boldsymbol{\epsilon}_n = \sum_{k=1}^{n} b_k\boldsymbol{\epsilon}_k,$$

add by adding components of like bases:

$$\mathbf{a} + \mathbf{b} = (a_1 + b_1)\boldsymbol{\epsilon}_1 + (a_2 + b_2)\boldsymbol{\epsilon}_2 + \cdots + (a_n + b_n)\boldsymbol{\epsilon}_n$$

or

$$\mathbf{a} + \mathbf{b} = \sum_{m=1}^{n} (a_m + b_m)\boldsymbol{\epsilon}_m.$$

In this formalism the rule for subtraction is contained in the rule for addition. Multiplication became an open question. The presence of the fundamental units of bases $\boldsymbol{\epsilon}_j$ permitted a large amount of freedom. For multiplication of hypercomplex numbers the associative law and distributive law were expected to hold; however, the commutative law was abandoned. Consider three hypercomplex numbers **a**, **b**, and **c**. The rules for combination according to our description become

$$\mathbf{a}(\mathbf{bc}) = (\mathbf{ab})\mathbf{c}, \quad \text{the associative law,}$$
$$\mathbf{a}(\mathbf{b} + \mathbf{c}) = \mathbf{ab} + \mathbf{ac}, \quad \text{the distributive law, with}$$
$$\mathbf{ab} \neq \mathbf{ba}, \quad \text{the commutative law, not holding in general.}$$

Because of the additional freedom of the multiple algebra arising from the presence of the bases, Grassmann, for generality, defined several types of multiplication. The first assumption was that the coefficients of the bases, the a_j, the b_j, etc., would multiply as scalars and would commute with each other ($a_jb_j = b_ja_j$), and would also commute with the bases ($a_j\boldsymbol{\epsilon}_j = \boldsymbol{\epsilon}_ja_j$). The differentiation between different products arose in the manner in which the base units $\boldsymbol{\epsilon}_k$ were combined into products. Regard the product of two hypercomplex numbers. To illustrate the generality of the notation, let $\mathbf{a} = a_1\boldsymbol{\epsilon}_1 + a_2\boldsymbol{\epsilon}_2 + \cdots + a_n\boldsymbol{\epsilon}_n$ and $\mathbf{b} = b_1\boldsymbol{\epsilon}_1 + b_2\boldsymbol{\epsilon}_2 + \cdots + b_n\boldsymbol{\epsilon}_n$, where a_j and b_j are the respective scalar coefficients. Then the product **ab** exhibits two possible combinations of the bases, those which have like elements (i.e., $\boldsymbol{\epsilon}_k\boldsymbol{\epsilon}_k$) and those which are composed of unlikes (i.e., $\boldsymbol{\epsilon}_i\boldsymbol{\epsilon}_j$, where $i \neq j$ [i is not equal to j]). All of the possible terms of a product are then (notice the order is maintained),

$$\begin{aligned}\mathbf{ab} = a_1\boldsymbol{\epsilon}_1b_1\boldsymbol{\epsilon}_1 + a_1\boldsymbol{\epsilon}_1b_2\boldsymbol{\epsilon}_2 + \cdots + a_1\boldsymbol{\epsilon}_1b_n\boldsymbol{\epsilon}_n + a_2\boldsymbol{\epsilon}_2b_1\boldsymbol{\epsilon}_1 \\ + a_2\boldsymbol{\epsilon}_2b_2\boldsymbol{\epsilon}_2 + \cdots + a_2\boldsymbol{\epsilon}_2b_n\boldsymbol{\epsilon}_n + a_3\boldsymbol{\epsilon}_3b_1\boldsymbol{\epsilon}_1 + \cdots \\ + a_n\boldsymbol{\epsilon}_nb_1\boldsymbol{\epsilon}_1 + a_n\boldsymbol{\epsilon}_nb_2\boldsymbol{\epsilon}_2 + \cdots + a_n\boldsymbol{\epsilon}_nb_n\boldsymbol{\epsilon}_n.\end{aligned}$$

Commuting the coefficients but not the bases, this becomes

$$\begin{aligned}\mathbf{ab} = (a_1b_1)\boldsymbol{\epsilon}_1\boldsymbol{\epsilon}_1 + (a_1b_2)\boldsymbol{\epsilon}_1\boldsymbol{\epsilon}_2 + \cdots + (a_1b_n)\boldsymbol{\epsilon}_1\boldsymbol{\epsilon}_n + (a_2b_1)\boldsymbol{\epsilon}_2\boldsymbol{\epsilon}_1 \\ + (a_2b_2)\boldsymbol{\epsilon}_2\boldsymbol{\epsilon}_2 + \cdots + (a_2b_n)\boldsymbol{\epsilon}_2\boldsymbol{\epsilon}_n + (a_3b_1)\boldsymbol{\epsilon}_3\boldsymbol{\epsilon}_1 + \cdots \\ + (a_nb_1)\boldsymbol{\epsilon}_n\boldsymbol{\epsilon}_1 + (a_nb_2)\boldsymbol{\epsilon}_n\boldsymbol{\epsilon}_2 + \cdots + (a_nb_n)\boldsymbol{\epsilon}_n\boldsymbol{\epsilon}_n.\end{aligned}$$

Grassmann created the *inner product* such that $\boldsymbol{\epsilon}_r\boldsymbol{\epsilon}_s$ equals 1 if $r = s$ and $\boldsymbol{\epsilon}_r\boldsymbol{\epsilon}_s = 0$ if $r \neq s$. Further, an outer product can be defined such that $\boldsymbol{\epsilon}_r\boldsymbol{\epsilon}_s = -\boldsymbol{\epsilon}_s\boldsymbol{\epsilon}_r$ for $r \neq s$, and 0 for all $s = r$. From these two products, Grassmann constructed others for arrangements containing more than two bases. For a consistent notation we denote the inner product with a boldface dot, i.e., $\boldsymbol{\epsilon}_r \cdot \boldsymbol{\epsilon}_s$.* The outer product can be labeled with a bracket about the base pair, such as $[\boldsymbol{\epsilon}_r\boldsymbol{\epsilon}_s]$.† With this notation, then, the product of three bases $\boldsymbol{\epsilon}_r$, $\boldsymbol{\epsilon}_s$, $\boldsymbol{\epsilon}_t$ may take on several forms: $\boldsymbol{\epsilon}_r \cdot [\boldsymbol{\epsilon}_s\boldsymbol{\epsilon}_t]$, $[\boldsymbol{\epsilon}_r\boldsymbol{\epsilon}_s] \cdot \boldsymbol{\epsilon}_t$, and $[[\boldsymbol{\epsilon}_r\boldsymbol{\epsilon}_s]\boldsymbol{\epsilon}_t]$. If one restricts the number of bases to three with the condition that the outer product between two produce plus or minus the third, the parts of quaternian theory can be represented as a special case of the hypercomplex number theory of Grassmann. As a general algebraic form it includes theories of vector analysis, determinants, matrices, and tensor algebra which were to appear later in the nineteenth century.

A third type of product Grassmann suggested was the open or indeterminant product, in which the pair $\boldsymbol{\epsilon}_r\boldsymbol{\epsilon}_s$ is assigned a position rs to the associated coefficient. This open product contained the basic elements which imply the theory of dyadics and matrices. One need only think of a square array with sequentially numbered rows and columns. By placing the coefficient of $\boldsymbol{\epsilon}_r\boldsymbol{\epsilon}_s$ in a position corresponding to the rth row and sth column, the entire open product assumes the form of a square array of terms:

$$\mathbf{ab} \xrightarrow[\text{open product}]{} \begin{pmatrix} a_1b_1 & a_1b_2 & a_1b_3 & \cdots & a_1b_n \\ a_2b_1 & a_2b_2 & a_2b_3 & \cdots & a_2b_n \\ \vdots & & & & \\ a_nb_1 & a_nb_2 & a_nb_3 & \cdots & a_nb_n \end{pmatrix}.$$

Number theory, although an ancient and established interest of mathematicians, was just as vital in this period of spectacular innovation in geometry and algebra. The concept of the infinite had been a subject of investigation and controversy in the days of Zeno; Galileo showed some insight into the dangers of countability in the realm of arbitrarily large numbers. The problem of countability of the real numbers is tied to the presence of transcendental numbers in the set of all real numbers. The first to prove the existence of transcendental numbers was Joseph Liouville (1809–1882). Because the set of real numbers is composed of the two subsets consisting of all of the algebraic numbers and all of the transcendental numbers, Liouville's proof of existence is based upon the construction

* The boldface dot is modern notation. Originally, the inner product was written as $\boldsymbol{\epsilon}_r|\boldsymbol{\epsilon}_s$.

† In modern vector notation the outer product is written with a large $\times$ between bases. Sometimes in three dimensions this is called a cross-product.

of a number and the proof that this particular number is not algebraic. The Liouville number is alpha, where

$$\alpha = \sum_{n=1}^{\infty} (10)^{-n!} = 10^{-1!} + 10^{-2!} + 10^{-3!} + \cdots + 10^{-k!} + \cdots$$

$$= 10^{-1} + 10^{-2} + 10^{-6} + \cdots + 10^{-k!} + \cdots$$

$$= 0.110001000000\cdots,$$

where $n!$ is n factorial, i.e., $n! = n(n-1)(n-2)\cdots 3\cdot 2\cdot 1$. Proof is achieved by considering a truncated α, call it β, where β is known to be algebraic. By truncation we mean that the infinite series is terminated at a finite number, say N:

$$\beta = \sum_{n=1}^{N} (10)^{-n!} = 10^{-1} + 10^{-2!} + 10^{-3!} + \cdots + 10^{-N!}.$$

The proof is further carried out by assuming that α is algebraic and therefore is the solution of a polynomial with integral coefficients $f(x) = 0$. Following this one can show that $f(\beta)$ is not zero. Then, as a first step, it can be demonstrated that $|f(\alpha) - f(\beta)|$ is less than a specific number K times $(\alpha - \beta)$, where $(\alpha - \beta)$ is of the order of $10^{-(N+1)}$. Next it can be shown that $|f(\alpha) - f(\beta)|$ is larger than this estimate, leading to a contradiction with the conclusion that $f(\alpha) = 0$ is impossible and that α is not algebraic.

Liouville made extensive contributions to classical analysis. He is remembered for his solution to the question of the meaning of $d^n y/dx^n$ when n is not an integer. In the field of complex analysis, in 1847 he applied Cauchy's analysis to elliptic functions, and produced the theorem known by his name. The Liouville theorem states that if $f(z)$ is analytic for all values of z, and if $|f(z)|$ is bounded as $|z|$ tends to infinity, then $f(z)$ is necessarily everywhere a constant. This very simple theorem can be employed to prove the fundamental theorem of algebra discussed earlier by considering the inverse of a polynomial $P(z)$; i.e., $1/P(z)$. The reader should think about this in terms of Liouville's theorem and attempt to prove that $P(z)$ has at least one root.

Liouville and J. C. F. Sturm (1803–1855) applied themselves to the special problem of linear differential equations, with particular attention to the effect of the numerous types of boundary conditions. The result was a unified theory known as the Sturm–Liouville theory (1830). The generalized differential equations which they systemized had the form:

$$\frac{d}{dz}\left\{p(z)\frac{du(z)}{dz}\right\} + q(z)u(z) - \lambda w(z)u(z) = 0,$$

defining a set of solutions $u_\lambda(z)$ with a corresponding parameter λ called the eigenvalue of the differential equation; $p(z)$, $q(z)$, and $w(z)$ are functions of the independent variable z. In particular, the last function $w(z)$ is known as the weight function, and it defines the space of which z is the variable; for example, $w(z) = 1$ if the space is Cartesian, $w(z) = z$ if the space is cylindrical, and $w(z) = z^2$ if the space is spherical. This class of equations became known as the eigenvalue problem and had a profound effect upon later developments in mathematical physics. The introduction of the theory of integral equations was influenced by this formalism. From the sequence of solutions $u_\lambda(z)$ to these equations, sets of functions are developed which may be employed in a generalized Fourier representation of any arbitrary function $f(z)$ which satisfies the boundary conditions, is quadratically integrable (i.e., that $\int_a^b |f|^2 w(z)\, dz$ is finite), has a finite number of maxima and minima, and has a finite number of jump discontinuities in the interval a to b. These qualities refer to a generalized function space of the $u_\lambda(z)$, known later as a Hilbert space. With the advent of quantum mechanics and the associated Schroedinger's equation in 1926, the eigenvalue problem became the foundation of quantum physics. Almost one hundred years after the first unified treatment, the eigenvalue was discovered to have important physical interpretation in the applications of quantum theory, yet it should be remembered that the eigenvalue problem was an integral part of the classical theory of waves.

Although analysis had expanded into a productive field with large numbers of what one might call right results, it was clear even after the work of Cauchy that much of the basic reasoning about the calculus was unsound. As in the initial attempts of Leibniz and Newton, the concepts of limits, infinitesimals, and infinity were very shaky indeed, Some help came from the attack on the real number system led by Weierstrass (1815–1897), and from the work of Dedekind, Meray, and Cantor. Weierstrass, like Cauchy, had a profound influence upon the theory of a complex variable. The power series expansion, along with a reconstruction of the real number system, became the basis of his arithmetical theory of functions. Through his work the radius of convergence of a power series was introduced, along with the theory of analytic continuation. Wide classes of functions were characterized in terms of their singularities. A polynomial $f(z)$, for example, has no singularities for values of the independent variable in the finite part of the complex plane. However, it is singular or undefined for points at infinity. From this, one proceeds to a class of functions called integral transcendental (or entire), with an infinite radius of convergence and an essential singularity at infinity. These are defined by the infinite series $\sum_{n=0}^{\infty} a_n z^n$, with simple examples being $\sin z$, e^z, etc.

Cauchy's theory of a complex variable rested upon the representation of a function $f(z)$, analytic in the neighborhood of a point z_0 as the power series, $\sum_{n=0}^{\infty} a_n(z - z_0)^n$. The circle of convergence is centered at z_0 and extends to the first singular point of $f(z)$ nearest z_0 without passing through z_0. Unless the singularities of $f(z)$ are everywhere dense upon the circumference of the circle of convergence, the function as defined by some power series may be continued by expanding about another point in the original circle, not at the center. As long as the singularities of $f(z)$ are avoided, this process may be continued, as illustrated in the diagram. Here we have assumed but one singular point for $f(z)$ at P.

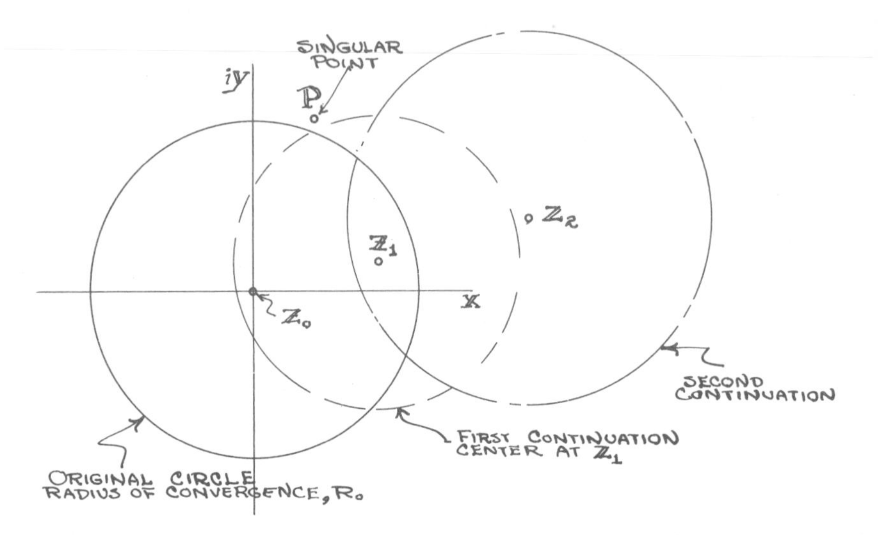

A subtle and important flaw of analysis had been exposed by the work of Fourier. In spite of his unwillingness to admit the lack of rigor in his work, Fourier developed a field of mathematics which was to have a profound influence upon the thinking of the formalists in mathematics. Physical and graphic intuition replaced mathematical understanding, and Fourier's major innovation was the introduction of functions which were discontinuous and arbitrary. It was obvious that much of formal theory, such as the Taylor's expansion with its associated hierarchy of existing derivatives, was inadequate in the presence of such functions. Consequently, the formalists came to realize that many of the concepts of continuity, arbitrary functions, and real numbers were in need of re-examination. The result of these considerations was the arithmetization of analysis, and one of the founders of this movement was Weierstrass. In Eudoxus' early use of the "method of exhaustion" compared quantities were areas, lengths of curves, and, in general, geometric entities. Little of this approach was changed in the early intuitionist years of analysis, when such concepts as points moving continuously on a curve were an important part of the teaching. In the formal program of Weierstrass, Dedekind, and Cantor, geometric magnitudes were replaced by real numbers, and in this connection Zeno's original objections to the methods of exhaustion achieved more meaning. The classical ϵ and δ technique of nineteenth-century analysts in defining the calculus is an inheritance from Weierstrass.

The concept of a function was modified by P. G. L. Dirichlet (1805–1859), who defined a function of a real variable as a table or correspondence between two sets of numbers. Here is the first hint of a theory of equivalence of point sets. Once more, one observes the breaking away of formal mathematics from theoretical physics. The causes are clear, and the separation is reasonable in view of the frontier problems confronting the specialist in mathematics. From the time of Galileo, physics had acted as a motivation for intuitive discovery in mathematics. At a certain point around 1830 the unsoundness of the logic inherent in intuition became abundantly clear to the great mathematicians. With this perception came the realization that mathematics, if it were to survive as a logical science, must be oriented toward rigor and formalism. This orientation was not without reward, for once mathematics was abstracted, the value of generalization became abundantly clear.

Some analogy may exist between the events of this period and those of an earlier age. One may recall that the Babylonians achieved a number of very useful and valuable algebraic and numerical techniques without any evidence of a logical system of derivation. The succeeding generations of Greek mathematicians demonstrated by logical deduction the consistency of these results using a small number of fundamental axioms or postulates. A disquieting aspect of this observation lies in the fact that, if the analogy holds, the corresponding time intervals for transition and full development may have been shortened considerably; i.e., the lifetime for transition in mathematics became shorter in later periods.

Dedekind, the contemporary of Weierstrass, was primarily an arithmetician and algebraist. The uncertainty of mathematics, which arises from the uncritical application of infinite classes, was perceived and alleviated to some extent in his work. By generalization of the rational integers accompanied by the fundamental theorem of arithmetic, denumerably infinite classes of algebraic integers were introduced in a consistent fashion. These were followed by the creation of the nondenumerable infinite classes. The fundamental theorem of arithmetic states that every natural number other than 1 can be factored into primes in one and only one way, except for the order of the factors. A sketch of the proof of this theorem* can be carried out by assuming the converse and then demonstrating a contradiction. Assume that m is the smallest integer that can be factored into two unique sets of primes, p_j and q_j. Then $m = p_1 p_2 p_3 \cdots p_r$, and $m = q_1 q_2 q_3 \cdots q_s$. The reader will observe that the two sets contain in general different numbers of primes. The first part of the proof requires that we show that the set $p_1 p_2 \cdots p_r$ is entirely different from $q_1 q_2 \cdots q_s$. This is to say that if 7 occurs in one set it will not occur in the second. If, say, p_1 were equal to q_1, then the natural numbers m/p_1 and m/q_1 would have two unique factorizations, contradicting the initial assumption that m is the smallest number with this property.

Once one establishes that the factorizations are unique, $p_j \neq q_j$, and one factor must be smaller than the other. Assume that p_1 is less than q_1 (this is written $p_1 < q_1$). If this is the case, we can create a number n where $n = (q_1 - p_1) q_2 q_3 \cdots q_s$. Carrying out the multiplication of the difference term,

$$n = q_1 q_2 \cdots q_s - p_1 q_2 q_3 \cdots q_s = m - p_1 q_2 q_3 \cdots q_s.$$

Therefore, n is a number less than m. The term $(q_1 - p_1)$, if factored into primes, would not have p_1 as a prime factor because q_1 does not have p_1

* See Ivan Neven, *Numbers Rational and Irrational* (New York: Random House, 1961), p. 118.

as a prime factor. If we now replace m in the expression above by its expansion in p_j, we obtain

$$n = p_1p_2p_3\cdots p_r - p_1q_2q_3\cdots q_s = p_1(p_2p_3\cdots p_r - q_2q_3\cdots q_s).$$

The term $(p_1p_2\cdots p_r - q_2q_3\cdots q_s)$ is not necessarily a prime and thus will have a prime factorization of n which includes p_1. Finally, we have exhibited two different prime factorings of n, although n is smaller than m, violating the initial hypothesis that m is the smallest integer with different prime factorizations. This contradiction completes the proof.

The Galois theory of algebraic equations was acknowledged by Dedekind as the basis for much of his own effort to arithmetize algebra. A basic element of Galois' theory was the domain of rationality or field, and the concept of the field is primary to much of Dedekind's work in algebra—a field is a more generalized system and contains common algebra as a special example. One is immediately struck by the similarities between the definition of a field and a group. Not until 1903 did the structure of a field assume the form shown.

A field F is a system consisting of a set S of elements $a, b, c, \ldots$, and two binary operations, called addition (denoted by $\oplus$) and multiplication (denoted by $\odot$) which may be performed upon any two elements a and b of S, taken in that order, to produce uniquely determined elements $a \oplus b$ and $a \odot b$ of S. Further, five axioms or postulates of the field must be satisfied:*

1. The commutative law for binary operations: $a \oplus b = b \oplus a$, and $a \odot b = b \odot a$.
2. The associative law for binary operations: if a, b, c are any three elements of F, then $(a \oplus b) \oplus c = a \oplus (b \oplus c)$, and $(a \odot b) \odot c = a \odot (b \odot c)$.
3. The field F has two distinct elements 0 and 1, such that if a is any element of F, $a \oplus 0 = a = 0 \oplus a$, and $a \odot 1 = a = 1 \odot a$.
4. For any a of F there exists in F an element x such that $a \oplus x = 0 = x \oplus a$.
5. For any a of F distinct from 0 (i.e., excluding 0), there exists in F an element y such that $a \odot y = 1 = y \odot a$.

The precise meaning of equality was assumed and therefore not stated. Later in the twentieth century, equality was defined as an equivalence relation. If the elements of S are taken as the rational numbers, it is clear that the rational numbers form a field under the binary operations of ordinary addition and multiplication. The first explicit definition of a number field was given by Dedekind in 1879.

Dedekind's theory of algebraic integers was defined by a root of an irreducible equation of the form $\sum_{n=0}^{N} a_n x^n = 0$ of any degree N with rational integer coefficients $a_N, a_{N-1}, \cdots, a_0$. A root of this equation is

* L. E. Dickson, *Algebras and Their Arithmetics* (London: Constable and Co. Ltd., 1923).

called an algebraic number of degree N. If $a_N = 1$, this number is an algebraic integer, and if $a_0 = 1$ or -1, the algebraic integer is unity. When, in a number field, there is a unique decomposition into primes, there are anomalies, and to solve this problem Dedekind reevaluated the divisibility of the rational integers. By this process he was led to the invention of "ideals": an ideal of an algebraic number field F is a subset, say P, of all the integers of F such that if p and q are in P, and a is any integer in F, then $p - q$ and pa are in P. An ideal (subset) P is said to divide the ideal Q if every integer in Q is also in P.

Each field of mathematics was being stripped now of its specialized nature and thereby being reduced to pure form, independent of a particular application. The next step in this direction was that of perturbing a general mathematical structure by suppression of one or more of the basic axioms. This in a sense was done in the theory of groups when the commutative law of combination was eliminated, and was partially accomplished in hyperbolic geometry by replacing an axiom by its converse. Characteristic of this age was the optimistic belief that these abstract movements were the final answer leading to a finished and stable structure. They proved to be quite the opposite; in fact, the generalizations exposed the flimsiness of the structure, and in the twentieth century it was doubted that finite systems free of logical contradiction could even be formed.

Grassmann's generalizations of linear algebra were quickly followed by studies of algebraic invariants conducted by Arthur Cayley (1821–1895) in 1845. Algebraic invariants had been anticipated in the studies of Lagrange and Boole on the discriminants of bilinear forms. Cayley called these invariants "hyperdeterminants"—later he changed the name to "quantics." By 1858 he had invented the matrix algebra, which demonstrated that Hamilton's quaternians could be represented as a two-by-two matrix as a special case. In spite of the fact that the loyal Englishman Tait conspired to attribute the discovery of the matrix to Hamilton's quaternians, Cayley was at pains to explain that his original ideas came from a simple consideration of two linear algebraic equations in two unknowns. By regarding the equations $x' = ax + by$ and $y' = cx + dy$, one can define a symbolic representation of these equations in terms of matrices:*

$$\begin{bmatrix} x' \\ y' \end{bmatrix} = \begin{pmatrix} a & b \\ c & d \end{pmatrix} \begin{bmatrix} x \\ y \end{bmatrix} = \begin{bmatrix} ax + by \\ cx + dy \end{bmatrix}.$$

For convenience we shall employ a slightly more modern notation, where the elements listed vertically in square brackets can be thought of as column vectors. Each position along the vertical then represents a coefficient of the vector or an independent equation of the set. Multiplication is, by necessity, defined as the row of an ordered array on the left

* The notation employed here is relatively modern.

multiplied element by element into a column appearing on its right. The square array of objects is known as a square matrix.

A slightly more appealing form is achieved if the elements of the matrix are assigned two subscripts denoting the row and the column in which they appear in the square array. For instance, our two equations may be written with x and y, replaced by x_1 and x_2; x' and y' may be replaced by x'_1 and x'_2; and a, b, c, and d may be replaced by a_{11}, a_{12}, a_{21}, and a_{22}, respectively:

$$x'_1 = a_{11}x_1 + a_{12}x_2,$$

and

$$x'_2 = a_{21}x_1 + a_{22}x_2;$$

or

$$\begin{bmatrix} x'_1 \\ x'_2 \end{bmatrix} = \begin{pmatrix} a_{11} & a_{12} \\ a_{21} & a_{22} \end{pmatrix} \begin{bmatrix} x_1 \\ x_2 \end{bmatrix} = \begin{bmatrix} a_{11}x_1 + a_{12}x_2 \\ a_{21}x_1 + a_{22}x_2 \end{bmatrix}.$$

In this form the square matrix times a column vector on the right can be written as a sum over the like indices which are adjacent:

$$x'_j = \sum_{k=1}^{2} a_{jk}x_k = a_{j1}x_1 + a_{j2}x_2.$$

By reverting to Grassmann's notation, one can observe that this type of product is the inner product. If, instead of employing the simple notation shown above, one writes all quantities in terms of the bases ϵ_r, the elements of the matrix become $a_{lm}\epsilon_l\epsilon_m$, and the column vector becomes $x_1\epsilon_1 + x_2\epsilon_2$. In this notation, then, the same result occurs if an inner product is taken. For example,

$$(a_{12}\epsilon_1\epsilon_2)\cdot(x_1\epsilon_1 + x_2\epsilon_2) = a_{12}\epsilon_1 x_2 = a_{12}x_2\epsilon_1.$$

The base vectors ϵ_r have been suppressed in the matrix notation and are replaced by the position of the elements in an array, either a square array denoting the presence of a base $\epsilon_r\epsilon_s$ or a column array which implies the presence of a single base with each coefficient. Here the later notation of matrix analysis is being anticipated. By regarding a matrix as a set of column vectors, one can immediately generalize the matrix multiplication to a matrix times a matrix. If we define a new set of equations linking new variables x''_1 and x''_2 with x'_1 and x'_2 by a square array of b_{jk} elements, we obtain

$$x''_1 = b_{11}x'_1 + b_{12}x'_2,$$

and

$$x''_2 = b_{21}x'_1 + b_{22}x'_2;$$

or

$$\begin{bmatrix} x''_1 \\ x''_2 \end{bmatrix} = \begin{pmatrix} b_{11} & b_{12} \\ b_{21} & b_{22} \end{pmatrix} \begin{bmatrix} x'_1 \\ x'_2 \end{bmatrix} = \begin{bmatrix} b_{11}x'_1 + b_{12}x'_2 \\ b_{21}x'_1 + b_{22}x'_2 \end{bmatrix}.$$

Because the column vector of x_1' and x_2' is itself generated by the linear transformation of x_1 and x_2 through the a_{ij}, the double primed array can then be displayed as a single transformation of the x_1, x_2 terms. The original linear equations then determine how the matrix product between $\mathbb{B}$, representing the array of b_{1m} elements, and $\mathbb{A}$, representing the square array of the a_{jk} elements.

$$\begin{bmatrix} x_1'' \\ x_2'' \end{bmatrix} = \begin{pmatrix} b_{11} & b_{12} \\ b_{21} & b_{22} \end{pmatrix} \begin{bmatrix} x_1' \\ x_2' \end{bmatrix} = \begin{pmatrix} b_{11} & b_{12} \\ b_{21} & b_{22} \end{pmatrix} \begin{pmatrix} a_{11} & a_{12} \\ a_{21} & a_{22} \end{pmatrix} \begin{bmatrix} x_1 \\ x_2 \end{bmatrix}$$

$$= \begin{pmatrix} (b_{11}a_{11} + b_{12}a_{21}) & (b_{11}a_{12} + b_{12}a_{22}) \\ (b_{21}a_{11} + b_{22}a_{21}) & (b_{21}a_{12} + b_{22}a_{22}) \end{pmatrix} \begin{bmatrix} x_1 \\ x_2 \end{bmatrix}.$$

The matrix inner product

$$\mathbb{B} \cdot \mathbb{A} = \begin{pmatrix} b_{11} & b_{12} \\ b_{21} & b_{22} \end{pmatrix} \begin{pmatrix} a_{11} & a_{12} \\ a_{21} & a_{22} \end{pmatrix}$$

can be demonstrated by direct substitution of the linear forms in x_1, x_2 for x_1' and x_2'. Such substitution shows that the matrix $\mathbb{B}$ times $\mathbb{A}$ is equivalent to a matrix $\mathbb{C}$, where

$$\begin{pmatrix} c_{11} & c_{12} \\ c_{21} & c_{22} \end{pmatrix} = \begin{pmatrix} b_{11} & b_{12} \\ b_{21} & b_{22} \end{pmatrix} \left(\begin{array}{:c:c:} a_{11} & a_{12} \\ a_{21} & a_{22} \end{array} \right)$$

$$= \begin{pmatrix} (b_{11}a_{11} + b_{12}a_{21}) & (b_{11}a_{12} + b_{12}a_{22}) \\ (b_{21}a_{11} + b_{22}a_{21}) & (b_{21}a_{12} + b_{22}a_{22}) \end{pmatrix}.$$

Here each row of $\mathbb{B}$ operates upon each column of $\mathbb{A}$ separately to produce the elements of a square array.

In terms of our sums, then, if $\mathbb{B} \cdot \mathbb{A} = \mathbb{C}$, we can write

$$c_{lj} = \sum_{k=1}^{2} b_{lk}a_{kj} = b_{l1}a_{1j} + b_{l2}a_{2j}.$$

The fact that there are four different ways of selecting j and l shows that there are four different terms c_{jl}. These examples may be generalized to any $n \times n$ array, where n is an integer.

Algebras could now be constructed as products of different sub-algebras; for instance, the algebra of real quaternions multiplied by the algebra of complex numbers gives a higher algebra of complex quarternions. In general, then, an algebra of order n^2 may be combined with an algebra of order m^2 to give an algebra of order m^2n^2.

Conservation rules and invariances had become a key consideration in the analysis of physical systems. These projections of mathematical security were also present in many of the major theorems of projective geometry. The study of matrices and linear algebras in turn exposed elements of invariance which were to contribute in a remarkable way to mathematical thought and to the quantum physics of the twentieth century. Early recognition by Lagrange and Boole of the invariance of the discriminant of a quadratic form under a linear transformation is basic to the studies of invariance in algebra. Invariances in other areas came to the surface shortly after Cayley: invariances of differential forms, invariances in transformation groups and in geometry, and topological invariances. Actually, these concepts are not as independent of one another as they would appear. As in much of mathematics, the abstract problems of invariance are often treated in analogous ways, with the exception of a specialized notation. In practice, abstracting the steps often illuminates the similarities.

Determinants had been used in one form or another to solve simultaneous algebraic equations since 1100 B.C.E. A major advance came in 1812 when J. P. M. Binet (1786–1856) provided the rule for multiplication. Cayley invented the appealing notation of the square array and, with J. J. Sylvester (1814–1897), carried out major extensions of the techniques. Determinants are an important property of the generalized square arrays of objects known as matrices. By definition the magnitude of a square matrix is the determinant of the matrix. Further, matrices have inverses if their determinants (magnitudes) are not zero.

Consider the transformation $\vec{\mathbf{y}} = \mathbb{A}\cdot\vec{\mathbf{x}}$, or

$$\begin{bmatrix} y_1 \\ y_2 \end{bmatrix} = \begin{pmatrix} a_{11} & a_{12} \\ a_{21} & a_{22} \end{pmatrix} \begin{bmatrix} x_1 \\ x_2 \end{bmatrix}.$$

Since this form represents two linear equations in x_1 and x_2, one can solve for x_1 and x_2 separately as linear combinations of y_1 and y_2, giving, for example,

$$x_1 = \frac{a_{22}}{(a_{11}a_{22} - a_{12}a_{21})} y_1 - \frac{a_{12}}{(a_{11}a_{22} - a_{12}a_{21})} y_2.$$

In general, the elements a'_{jk} of the inverse matrix* (denoted as A^{-1}) are derived from the elements a_{lm} of the original matrix $\mathbb{A}$ by the rule

$$a'_{jk} = \frac{(-1)^{j+k} \text{ minor of } a_{kj}}{\text{determinant of } \mathbb{A}}.$$

Once $\mathbb{A}^{-1}$ is created, we observe that if $\mathbf{y} = \mathbb{A}\cdot\vec{\mathbf{x}}$, and $\vec{\mathbf{x}} = \mathbb{A}^{-1}\cdot\vec{\mathbf{y}}$, then

$$\mathbb{A}^{-1}\cdot\vec{\mathbf{y}} = \mathbb{A}^{-1}\cdot\mathbb{A}\cdot\vec{\mathbf{x}} = \mathbb{I}\cdot\vec{\mathbf{x}} = \vec{\mathbf{x}},$$

so that a matrix times its own inverse (assuming that the determinant is nonzero) produces the identity (or unit) matrix $\mathbb{I}$, or

$$\mathbb{A}^{-1}\cdot\mathbb{A} = \mathbb{A}\cdot\mathbb{A}^{-1} = \mathbb{I} = \begin{pmatrix} 1 & 0 \\ 0 & 1 \end{pmatrix}.$$

Thus, for $n \times n$ square arrays the identity element is a matrix with all diagonal elements having the value 1 and all off-diagonal elements having the value 0. Invariances enter the problem when we ask for transformations

* The prime on a matrix element, such as a'_{jk}, will imply that this is an element of the inverse of $\mathbb{A}$.

which preserve the form of a vector, the form of a matrix, or the magnitude of an associated scalar term. The eigenvalue problem is concerned with transformations which preserve the direction of a vector, such as $\mathbb{A}\cdot\vec{\mathbf{x}} = \lambda\vec{\mathbf{x}}$, or $(\mathbb{A} - \lambda\mathbb{I})\cdot\vec{\mathbf{x}} = 0$, where λ is a scalar quantity. In this equation the magnitude of the matrix $(\mathbb{A} - \lambda\mathbb{I})$ must be 0, giving in the 2×2 case

$$\begin{vmatrix} (a_{11} - \lambda) & a_{12} \\ a_{21} & (a_{22} - \lambda) \end{vmatrix} = (a_{11} - \lambda)(a_{22} - \lambda) - a_{12}a_{21} = 0.$$

Here we have used the property that the sum, or difference, of two matrices involves the sum, or difference, of like elements (or elements occupying the same row-column position). The two-by-two matrix gives a quadratic polynomial, and this polynomial is called the characteristic equation. An $n \times n$ matrix has a characteristic polynomial of degree n. The roots of characteristic polynomials are fundamental properties of the original matrix $\mathbb{A}$. One matrix is said to be congruent to another if both have the same characteristic polynomial.

Properties of the matrix which appear to be important in the polynomial $|\mathbb{A} - \lambda\mathbb{I}| = 0$ are the magnitude (the coefficient of λ^0) and the trace (the coefficient of λ^{n-1}). By examination of the 2×2 example, we observe that $(a_{11} - \lambda)(a_{22} - \lambda) - a_{12}a_{21} = \lambda^2 - (a_{11} + a_{22})\lambda + (a_{11}a_{22} - a_{12}a_{21}) = 0$. The term $(a_{11} + a_{22})$ is called the trace and is the sum of the diagonal elements: $(a_{11}a_{22} - a_{12}a_{21})$ is the negative of Lagrange's discriminant and is the determinant (or magnitude) of $\mathbb{A}$. For polynomials of degree n, there are n coefficients in the polynomial which characterize the matrix.

An interesting investigation then is to find those matrices $\mathbb{M}$ which transform the initial matrix $\mathbb{A}$ to a congruent form. In particular, the most important transformation, when it can be found, is a special case of $\mathbb{M}$, call it $\mathbb{U}$, which transforms some matrices directly to diagonal forms (where the roots of the characteristic polynomial form the diagonal elements of the diagonal form, and the off-diagonal elements are zero). For matrices with complex elements, $\mathbb{U}$ can be found if the matrix is Hermitean,* such that $a_{jk} = a^*_{kj}$, where the symbol * (asterisk) means complex conjugate. When the elements are real, this condition reduces to the requirement of symmetry, such that $a_{jk} = a_{kj}$.

The form of transformations of matrices can be obtained by examining the initial equation plus the transformation of $\vec{\mathbf{x}}$. If $\mathbb{A}\cdot\vec{\mathbf{x}} = \lambda\vec{\mathbf{x}}$, and $\vec{\mathbf{x}} = \mathbb{M}\cdot\vec{\mathbf{y}}$, substitution demonstrates that the equivalent equation in $\vec{\mathbf{y}}$ is

$$(\mathbb{M}^{-1}\cdot\mathbb{A}\cdot\mathbb{M})\cdot\vec{\mathbf{y}} = \lambda\vec{\mathbf{y}},$$

with the result that the transformed $\mathbb{A}$, call it $\mathbb{C}$, is $\mathbb{C} = \mathbb{M}^{-1}\cdot\mathbb{A}\cdot\mathbb{M}$. If the elements of $\mathbb{A}\cdot\mathbb{M}$ are given by

$$(\mathbb{A}\cdot\mathbb{M})_{jl} = \sum_{\text{all } k} a_{jl}m_{kl},$$

then the transformed elements c_{sl} are given by

$$c_{sl} = (\mathbb{M}^{-1}\cdot\mathbb{A}\cdot\mathbb{M})_{sl} = \sum_{\text{all } r}\sum_{\text{all } k} m'_{sr}a_{rk}m_{kl}.$$

A somewhat different transformation arises for bilinear forms. First we must construct a notation for multiplying from the left by a vector. Because our rules involve row times column operations, a vector multiplied from the left must then be a row vector, as opposed to the column vector. This row vector, if it has complex elements, must then be taken from the elements of the column as complex conjugates. Consider the magnitude squared of $\vec{\mathbf{x}} = \begin{bmatrix} x_1 \\ x_2 \end{bmatrix}$ in matrix representation.†

$$\vec{\mathbf{x}}^{\dagger}\cdot\vec{\mathbf{x}} = [x_1^*, x_2^*]\begin{bmatrix} x_1 \\ x_2 \end{bmatrix} = x_1^*x_1 + x_2^*x_2 = |x_1|^2 + |x_2|^2.$$

* Charles Hermite (1822–1901) in 1854 introduced real binary forms of the type $ax^*x + bxy^* + b^*x^*y + cy^*y$, where a and c are real constants and b is a complex number. Hermite initiated the arithmetical theory of bilinear forms.

† This approach corresponds to Hermite's treatment of bilinear forms.

When the elements are real the adjoint vector (or row) denoted as $\vec{\mathbf{x}}^\dagger$ has the same elements as $\vec{\mathbf{x}}$, except that they appear in a row instead of a column.
A bilinear form then is achieved by the expression (a two-dimensional form is used here only for simplicity of display):

$$\vec{\mathbf{x}}^\dagger \cdot \mathbb{A} \cdot \vec{\mathbf{x}} = [x_1^*, x_2^*]\begin{pmatrix} a_{11} & a_{12} \\ a_{21} & a_{22} \end{pmatrix}\begin{bmatrix} x_1 \\ x_2 \end{bmatrix} = K$$

$$= x_1^* a_{11} x_1 + x_1^* a_{12} x_2 + x_2^* a_{21} x_1 + x_2^* a_{22} x_2 = K,$$

where K is a constant. Now if $\vec{\mathbf{x}} = \mathbb{M} \cdot \vec{\mathbf{y}}$ and $\vec{\mathbf{x}}^\dagger = \vec{\mathbf{y}}^\dagger \cdot \mathbb{M}^\dagger$ (notice that because $\vec{\mathbf{y}}^\dagger$ is a row it must appear to the left of $\mathbb{M}^\dagger$), we can transform the initial bilinear equation to $\vec{\mathbf{y}}^\dagger \cdot (\mathbb{M}^\dagger \cdot \mathbb{A} \cdot \mathbb{M}) \cdot \vec{\mathbf{y}} = K$. This transformation of $\mathbb{A}$ has the form $\mathbb{M}^\dagger \cdot \mathbb{A} \cdot \mathbb{M}$. When the transformation matrix $\mathbb{M}$ has the magnitude 1, it turns out that $\mathbb{M}^\dagger$ and $\mathbb{M}^{-1}$ are equal.
Up to this point some interesting properties of matrices have been outlined; however, the reader should be aware that our notation is not that used by Cayley and, further, that to give a rounded picture the description has been carried beyond the work of Cayley. The powerful subscript notation is characteristic of the Italian school of the late nineteenth century. We shall return later to many fascinating properties of the eigenvalue problem ($\mathbb{A} \cdot \vec{\mathbf{x}} = \lambda \vec{\mathbf{x}}$), such as conditions upon $\mathbb{A}$ which give real roots λ_n, and for each λ_n an $\vec{\mathbf{x}}_n$ which is orthogonal to all other solutions, $\vec{\mathbf{x}}_j$ where $j \neq n$.

Although very real and elementary examples have been employed here, the possibilities of this algebra are readily apparent. Two major concerns of Cayley were, first, to determine whether there exists a set of invariants for a given quantic such that any invariant of the quantic is expressible as a polynomial in members of the set, and, second, to determine all independent irreducible algebraic relations among the invariants of any finite set of quantics. A quantic characterized by m and n is a homogeneous polynomial (with arbitrary coefficients) with m independent variables of degree n. Like his predecessors, Cayley was vitally interested in higher spaces, that is, spaces with, say, four coordinate variables instead of three or two. References to higher geometries are also found in the works of Moebius, Cauchy, Sylvester, and Clifford.

In keeping with his intuitive use of the square arrays for a matrix representation, Cayley invented the group multiplication table or, more accurately, the group binary combination table. This mathematical object also takes the form of a square array, demonstrating all of the possible combinations between members of a finite set S taken two at a time.

Before analyzing the group multiplication table in detail, it is instructive to consider a more general class of square tables, defining a set of algebras which are broader than the group algebra. After this it may be demonstrated that Cayley's group multiplication table is a very special case of the more general table.* From ancient times to the eighteenth century, scholars had been fascinated with square arrays of numbers, known as "magic squares." In these arrays the sums of each row, each column, and each diagonal are equal. Euler in 1779 discussed in a paper a

* An entertaining and detailed description of these tables may be found in Sherman K. Stein, *Mathematics, the Man-Made Universe* (San Francisco: W. H. Freeman, 1963), chap. 11.

"new type of magic square." To reduce his original question to a simpler one, we search for a method for arranging 9 officers in a 3×3 square such that in each row and column the three officers would be from different ranks and regiments. Labeling the three ranks a, b, and c and the three regiments A, B, and C, one can solve this problem as the overlay of two tables of the form:

a	b	c
b	c	a
c	a	b

A	C	B
B	A	C
C	B	A

The overlay produces

aA	bC	cB
bB	cA	aC
cC	aB	bA

This problem is an extension of the basic question of tables themselves. One notices that the single tables are formed by taking three objects, say a, b, and c, and then reordering by cycling to form three linear sets. The standard multiplication table is a familiar form of a square array; it differs from the algebraic tables in which we are interested in that in it the binary combination of two elements does not necessarily result in a member of the original set. As an example of the general table which we wish to investigate, consider the combination table of four elements, A, B, C, and D:

	A	B	C	D
A	A	C	D	B
B	D	B	A	C
C	B	D	C	A
D	C	A	B	D

Here rows and columns have been labeled with the ordered set A, B, C, and D. The letter that appears in the box corresponding to row X and column Y is regarded as the binary combination of X and Y:

				Y		Z
X				$X \cdot Y$		$X \cdot Z$

Thus, in the $ABCD$ table shown above, the combination $C \otimes D$ is A. These combinations may represent any number of operations. Because the table is an arrangement of the elements of the set without duplication in any given row or in any given column, in general we can conclude that $X \otimes Y$ in the diagram above is different from $X \otimes Z$. In terms of a column, then, $Y \otimes X$ is different than $Z \otimes X$. Further, each box of the table will contain one of the elements of the guide row (closure requires that the combination of two elements of the set results in a member of the set).

The number of rows (or number of columns) determines the order of the table. It is quite clear that a large number of different algebras can be represented by this method. To impose some order on the systems, one can begin by identifying a few special cases. As an example, if the diagonal elements of a table satisfy the condition that $A \otimes A = A$, $B \otimes B = B$, $C \otimes C = C$, and $D \otimes D = D$, the algebra is known as "idempotent." The $ABCD$ table above is an example of an idempotent algebra.

If the table is symmetric about the diagonal, the algebra is commutative; i.e., if $X \otimes Y = Y \otimes X$. The idempotent $ABCD$ table is not commutative in that $A \otimes B = C$, and $B \otimes A = D$. An example of a commutative table of order four is shown below.

	A	B	C	D
A	A	B	C	D
B	B	A	D	C
C	C	D	A	B
D	D	C	B	A

There are commutative tables of all orders; however, there are *only* idempotent commutative tables for odd orders, and none for even orders. Because of the limit on the elements which can appear, these tables can be analyzed in terms of congruences and modular systems. In other words, by creating congruence rules one can prove the existence of idempotent commutative tables of odd order. Many algebras other than these exist, and among the most important of all are the associative algebras which satisfy the rule $X \otimes (Y \otimes Z) = (X \otimes Y) \otimes Z$, where the symbol $\otimes$ indicates the binary combination rule of the particular table.

An associative table satisfying the inverse and identity axioms is a group and is called the group multiplication table despite the fact that in general it is a table for a binary combination. A commutative group table represents an Abelian group—the table below is an example of an Abelian group of order three:

	I	a	b
I	I	a	b
a	a	b	I
b	b	I	a

A physical example of a commutative group is the set of counterclockwise rotations of an equilateral triangle about an axis through the centroid and perpendicular to the plane which takes the triangle into itself; i.e., rotations of 0°, $\pm 120°, \cdots, \pm(120n)°$, with n an integer.* If we designate a single positive rotation of 120° as r, then the basic set of elements of the group is illustrated in the accompanying diagram and table.

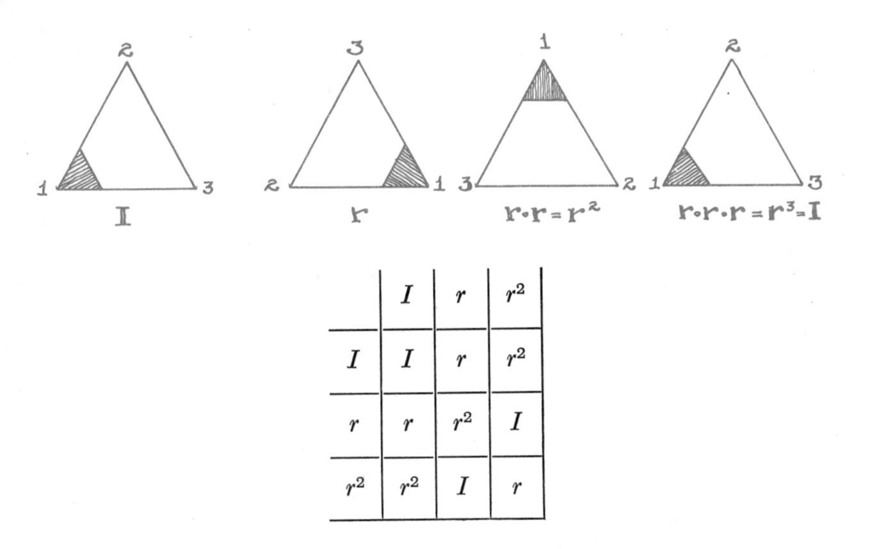

	I	r	r^2
I	I	r	r^2
r	r	r^2	I
r^2	r^2	I	r

* A detailed account of group tables as described here may be found in I. Grossman and W. Magnus, *Groups and Their Graphs* (New York: Random House, 1964), chap. 4.

General rotations of objects in three dimensions are not necessarily commutative; but, because this illustration is a rotation in a plane, the elements do commute under combination. Inverses can be constructed by observing in which row and column the identity element I occurs. When working with group elements, it is often necessary to solve equations of the following type: if a and b are known elements of a group, is there an element x of the group such that $ax = b$? Equations of this type can be cleared because of the existence of inverses. Because of the problem of commutativity, care must be exercised in the ordering of the multiplication. Multiplying the expression $ax = b$ from the left by a^{-1} gives $a^{-1}ax = Ix = x = a^{-1}b$. An expression of the type $ya = b$ may be solved by multiplication from the right by a^{-1}, giving $yaa^{-1} = yI = y = ba^{-1}$.

From these simple cases it is quite clear just how the inverse of an element $d = ab$ is computed. Multiplying in order by b^{-1} and a^{-1} from the right and d^{-1} from the left leaves the result $d^{-1} = b^{-1}a^{-1}$, demonstrating that the order of the combinations is interchanged in the process of taking an inverse. In a table, then, the appearance of identity elements, symmetrically placed about the diagonal, picks up mutually inverse elements. In the example here, r and s are the inverses of each other. Associativity, which is a fundamental property of a group table, is exhibited in the table by a rectangular array of two elements, their product, and the identity element, as shown in the next illustration. Noncommutative groups have at least two elements that do not commute. The existence of the identity element excludes the possibility of a group in which no two elements commute. To illustrate a noncommutative group, the rotation of the equilateral triangle about an axis through the centroid can be extended to include the flipping of the triangle about one of the altitudes—let us say the altitude passing through the vertex labeled 2. Call the flipping operation f.

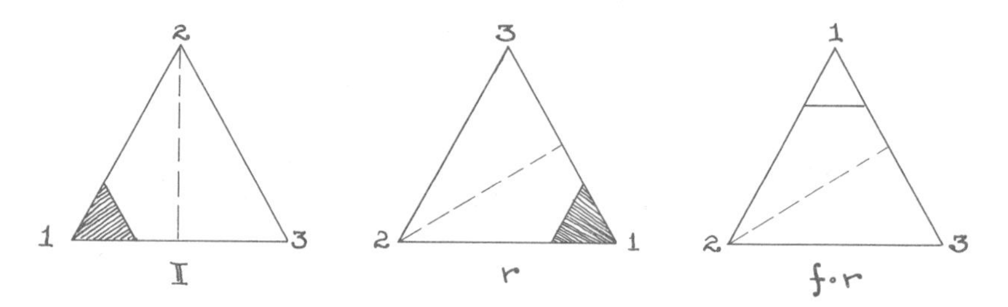

The sixth-order table, representing the complete set of motions composed by the 120° rotations r and the flipping operation f, is shown on the page opposite. Notice that $f \otimes f = I$.

The second table has been constructed to exhibit each entry as one of the six elements of the group. To make this table relations are used of the type: $frfr = I$, $rfr^2 = fr$, $frf = r^2$, $rf = fr^2$, etc. Multiplication from the right or left by r, r^2, and f is sufficient to demonstrate these with, say, $frfr = I$ as a beginning: $ffrfr = f = rfr$, giving $rfr^2 = fr$, etc.

	I	r	r^2	f	fr	fr^2
I	I	r	r^2	f	fr	fr^2
r	r	r^2	I	rf	rfr	rfr^2
r^2	r^2	I	r	r^2f	r^2fr	r^2fr^2
f	f	fr	fr^2	I	r	r^2
fr	fr	fr^2	f	frf	$frfr$	$frfr^2$
fr^2	fr^2	f	fr	fr^2f	fr^2fr	fr^2fr^2

	I	r	r^2	f	fr	fr^2
I	I	r	r^2	f	fr	fr^2
r	r	r^2	I	fr^2	f	fr
r^2	r^2	I	r	fr	fr^2	f
f	f	fr	fr^2	I	r	r^2
fr	fr	fr^2	f	r^2	I	r
fr^2	fr^2	f	fr	r	r^2	I

In this table one observes that in the final reduction each element of the group appears in each row or column. The occurrence of I in each row and column satisfies the axiom of inverses. Some suggestion of the existence of normal subgroups and factor groups is illustrated by the four 3×3 square arrays formed by splitting off the 3×3 square in the upper lefthand corner, representing the original rotations I, r, and r^2.

A fundamental feature of the 3×3 triangular rotation group is that it is a cyclic group of order three. This is apparent when we rewrite the powers of the generators r,

$$I, r, r^2, r^3, r^4, r^5, r^6, r^7, \cdots,$$

as

$$I, r, r^2, I, r, r^2, I, r, \cdots.$$

For many years this form of mathematics was applied to physics in the theory of crystal structure. Since the invention of quantum mechanics, group theory has had more and more impact upon theoretical physics and within the last ten years has achieved striking results in the analysis and description of the elementary particles.

A major step in the creation of higher differential spaces was taken by G. F. B. Riemann (1826–1866). He noted that there are many measurements in physics that are composites of a number of independent magnitudes of the elements of a manifold. A measurement is likened to a line element, and is expressed as the sum of the squares of the variations, or displacements of each coordinate: if ds is the measurement, and the individual coordinates are $x_1, x_2, \cdots, x_n$, with incremental displacements $dx_1, dx_2, \cdots, dx_n$, then

$$(ds)^2 = (dx_1)^2 + (dx_2)^2 + (dx_3)^2 + \cdots + (dx_n)^2,$$

or

$$ds = \left\{\sum_{k=1}^{n} (dx_k)^2\right\}^{1/2}.$$

While Riemann pointed out that this representation of $(ds)^2$ was not unique and could have a variety of other forms, experience has shown that this particular form turns out to have a much wider applicability than many other equations which may be postulated. The analogy with the Euclidean expression for a line element in a three-dimensional space was obvious: the length element in three dimensions appears as

$$(ds)^2 = (dx)^2 + (dy)^2 + (dz)^2,$$

where x, y, and z have been used instead of x_1, x_2, and x_3. Several specialized characteristics of the Pythagorean form in n dimensions are apparent. First, the Pythagorean quadratic differential form has coefficients equal to 1 for each term. Second, because of the explicit assumption that all coordinates are mutually perpendicular, the Pythagorean form contains no quadratic cross terms, such as $(dx_1\, dx_2)$, $(dx_1\, dx_3)$, or $(dx_j\, dx_k)$, in general with $j \neq k$. In the simple case of two dimensions, the appearance of cross terms is associated with a representation of the length element in terms of variables measured along axes which are not mutually perpendicular. Instead of x_1 and x_2 (or x and y), call the variables q_1 and q_2, and let the angle between the coordinate axes be α. In this example the expression for the length element takes the form

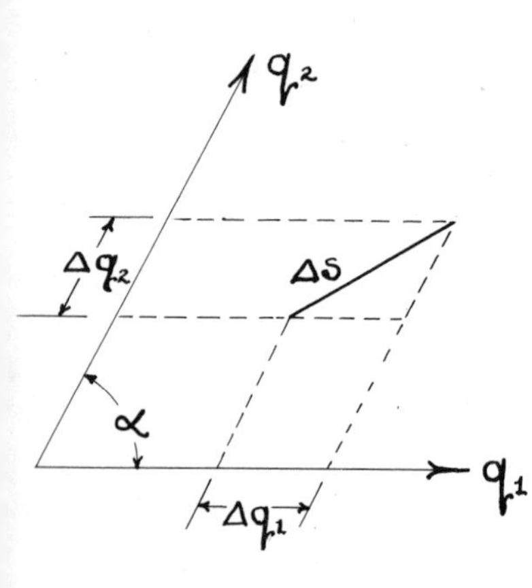

$$(ds)^2 = (dq_1)^2 + (dq_2)^2 - 2(dq_1\, dq_2) \cos \alpha.$$

Therefore, the appearance of cross terms in the quadratic expression for a length element implies that the coordinate axes are not orthogonal (mutually perpendicular). This example is further instructive in that the coefficient of the last term $(-2 \cos \alpha)$ is not necessarily 1.

With profound insight, Riemann passed to the most general quadratic differential form for the square of length element, allowing variable coefficients for each term (we use coordinates q_j in the same manner as the x_k):

$$\begin{aligned}(ds)^2 = g_{11}(dq_1)^2 &+ g_{12}\, dq_1\, dq_2 + g_{13}\, dq_1\, dq_3 + \cdots + g_{1n}\, dq_1\, dq_n \\ &+ g_{21}\, dq_2\, dq_1 + g_{22}(dq_2)^2 + g_{23}\, dq_2\, dq_3 + \cdots + g_{2n}\, dq_2\, dq_n \\ &+ \cdots + g_{n1}\, dq_n\, dq_1 + \cdots + g_{nn}(dq_n)^2,\end{aligned}$$

or

$$(ds)^2 = \sum_{i=1}^{n}\sum_{j=1}^{n} g_{ij}\, dq_i\, dq_j .$$

The coefficients in general are functions of the coordinate variables $q_1 q_2 \cdots q_n$ and define a metric space; the array of elements g_{ij} is called the "metric" of the space in question. From our earlier example we can conclude that when the "off diagonal" g_{ij} (i.e., $i \neq j$) are all zero, the space is orthogonal. Further, if all the (diagonal terms) $g_{ii} = 1$, the space has a Cartesian form.

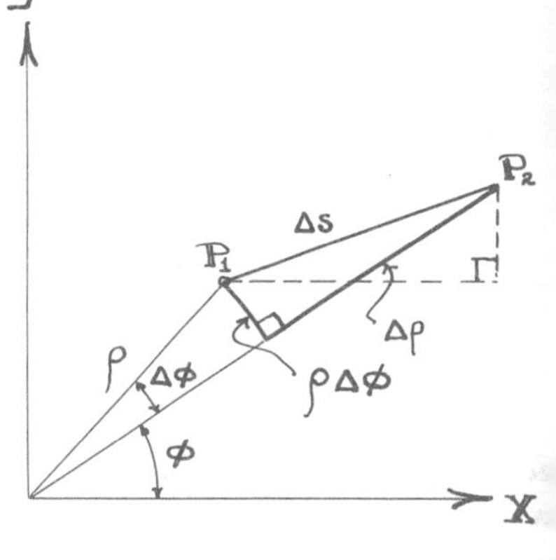

To see the functional forms that the g_{ij} can take, we need only examine the length element in polar coordinates ρ and ϕ. By triangulation, $(ds)^2 = (d\rho)^2 + \rho^2(d\phi)^2$. This simple example sets $q_1 = \rho$ and $q_2 = \phi$. Then $g_{11} = 1$, and $g_{22} = \rho^2$. Observe that there are no terms of the type $g_{12}\, d\rho\, d\phi$, and consequently one may assume that the (ρ, ϕ) coordinates are orthogonal. The curvature of a space also can be related to the metric. By contrast, consider as a special case a two-dimensional surface $F(x, y)$ embedded in a three-dimensional Euclidean space. If the characteristic surfaces (taking a three-dimensional case) are $z = F(x, y)$ [or $x_3 = F(x_1, x_2)$], the corresponding g_{ij} can be written as expansions in the derivatives of F and in terms of the products x_1 and x_2. In particular,

$$g_{ij} \simeq \delta_{ij} + \sum_{k=1}^{n}\sum_{l=1}^{n}\left[\frac{\partial^2 F}{\partial x_i\, \partial x_k}\right]\left[\frac{\partial^2 F}{\partial x_j\, \partial x_l}\right] x_k x_l ,^*$$

where by definition the δ_{ij} are the matrix elements of the identity matrix $\delta_{ij} = 1$, if $i = j$, and $\delta_{ij} = 0$, if $i \neq j$. Regular polar coordinates, such as those of the previous example, form a flat space because the shortest distance between two points is measured along a Euclidean geodesic. Geodesic polar coordinates that reduce to ordinary polar coordinates near the origin have a length element with the form

$$(ds)^2 = (d\rho)^2 + \rho^2[1 - (1/3)k\rho^2](d\phi)^2 .$$

A space defined in this fashion has a curvature. If we designate $g_{22} = \rho^2[1 - (1/3)k\rho^2]$ as a general function g_{22}, then the curvature R at any point is given by

$$R = -\frac{1}{\sqrt{g_{22}}}\frac{\partial^2 \sqrt{g_{22}}}{\partial \rho^2} .$$

Interesting anomalies appear here. The geodesic circle of radius ρ has a circumference $2\pi\rho[1 - (1/6)k\rho^2]$ and an area $\pi\rho^2[1 - (1/12)k\rho^2]$.

Riemann's work covers many of the frontiers of mathematics of his time. He formulated what is known as Dirichlet's principle, which states

* In practice, both subscript and superscript notations are used to denote covariant and contravariant quantities. We have neglected this sophistication here in the interest of simplicity.

that when a mathematical formalism with an associated function is created to conform to a real physical problem, the function must exist. This intuitive position was proved false by Weierstrass in 1870. Supplanting the unfortunate principle artificially assigned to Dirichlet was a reformulation called Dirichlet's problem. This problem is directed toward finding a function $V(x, y, z)$ which, together with its first and second derivatives in all variables, shall be uniform and continuous throughout a region, $\mathscr{R}$, and further, which shall take on preassigned values on the boundary of $\mathscr{R}$.

Both in his geometry and in his approach to the functions of a complex variable, Riemann was primarily an intuitionist, and many of his ideas were first stimulated by physics. In his study of analytic functions, he employed graphic concepts, which not only were appealing as pictures but which turned out to have fundamental importance in topological invariances. He was not unduly concerned with rigor, and it is conceded that Riemann proposed more problems than he solved.* With Cauchy he contributed many of the ideas basic to the theory of functions of a complex variable. Both demonstrated that when a complex function $f(x, y) = u(x, y) + iv(x, y)$ is analytic, the real and imaginary parts of $f(x, y)$ satisfy the Cauchy–Riemann conditions, taking the form $\partial u/\partial x = \partial v/\partial y$, and $\partial u/\partial y = -(\partial v/\partial x)$.

Functions not analytic in a region may have simple properties involving poles or a more complex behavior involving branch points. An elementary function with a first-degree pole at z_0 is $(z - z_0)^{-1}$. Such a function is considered simple because it remains single-valued if a curve of evaluation moves about z_0 for more than one turn. To see this, write $(z - z_0)$ as $\rho e^{i\phi}$. Then a point making one rotation about z_0 on a circle of radius 1 undergoes a phase change of 2π. Thus, for a second revolution about z_0 the point moves with no net phase change relative to the first circuit. Consider, however, a function of the type $(z - z_0)^{1/2}$. From the same point of view, one complete revolution about z_0 in the (x, iy) plane only changes the phase by π. Therefore, on the second traversal about z_0 the function f does not take on the same values that it had in the first traversal, and thus forms of the type $(z - z_0)^{\xi}$, where ξ is not an integer, are in general multivalued functions in the sense illustrated above. The extent of the multiplicity depends on whether ξ is a rational or an irrational number. For our simple example, $(z - z_0)^{1/2}$, the multiplicity is 2. Riemann dealt with this difficulty by creating different (x, iy) (i.e., z) planes for each traversal, known as Riemann sheets. Paths proceed from one sheet to another via a cut in the planes, i.e., a line where the planes are joined. The example of $(z - z_0)^{1/2}$ is pictorially appealing and simple. In the accompanying diagram, the cut and two Riemannian sheets are illustrated for this case.

From this concept, Riemann created the definitions of simply and multiply connected surfaces. The criterion for the multiplicity of the connectedness of a surface lies in the ability to shrink an arbitrary closed

* Bell, *Development of Mathematics*, p. 496.

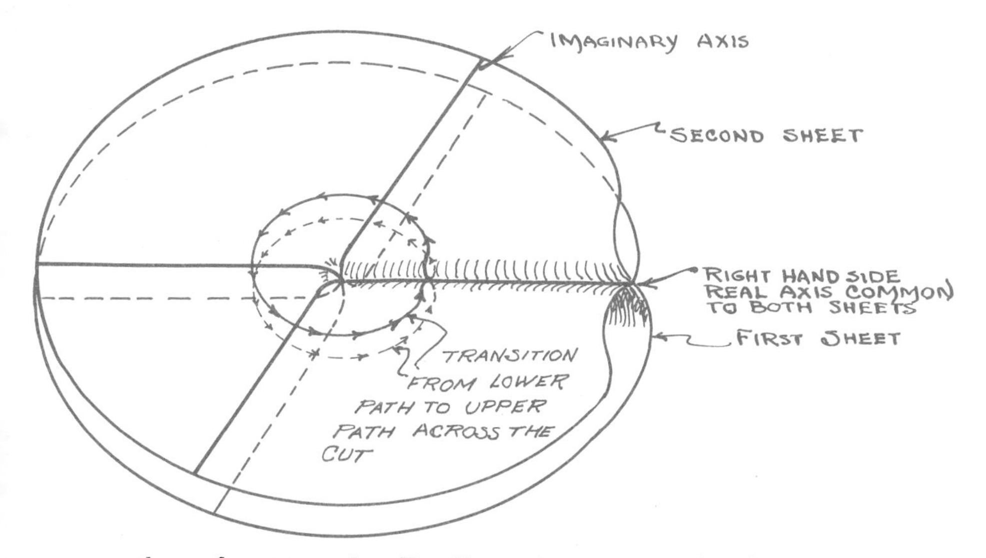

curve on the surface to a point. For illustration, a series of surfaces of different connectivity are shown. A simply connected surface will be divided into two parts by any closed curve within it. An n-connected surface can be transformed into a simply connected surface by $n - 1$ cuts, such as that shown for the doubly connected surface.

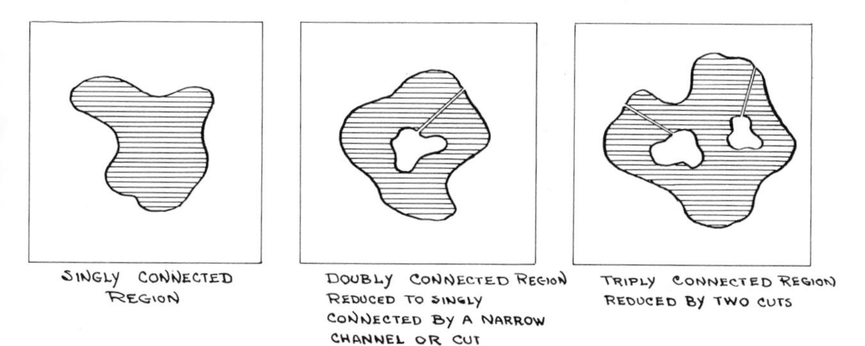

In very complicated diagrams it may be difficult at first to determine whether a point is inside or outside a closed region. The Jordan curve theorem (named for Camille Jordan, 1838–1922) provides a simple technique for doing so. From a point in question, draw a ray P. If the ray crosses regional boundaries an *odd* number of times, the point P is inside; if it crosses them an *even* number of times, the point is outside the region.

Fourier series had, as mentioned, presented analysts with a set of functions exhibiting pathological characteristics in violation of many of the necessary conditions of continuity and limits. These conditions are part of the central theme of the theory of functions of a real variable. In 1823 Cauchy provided a definition for the integral of a function of a real

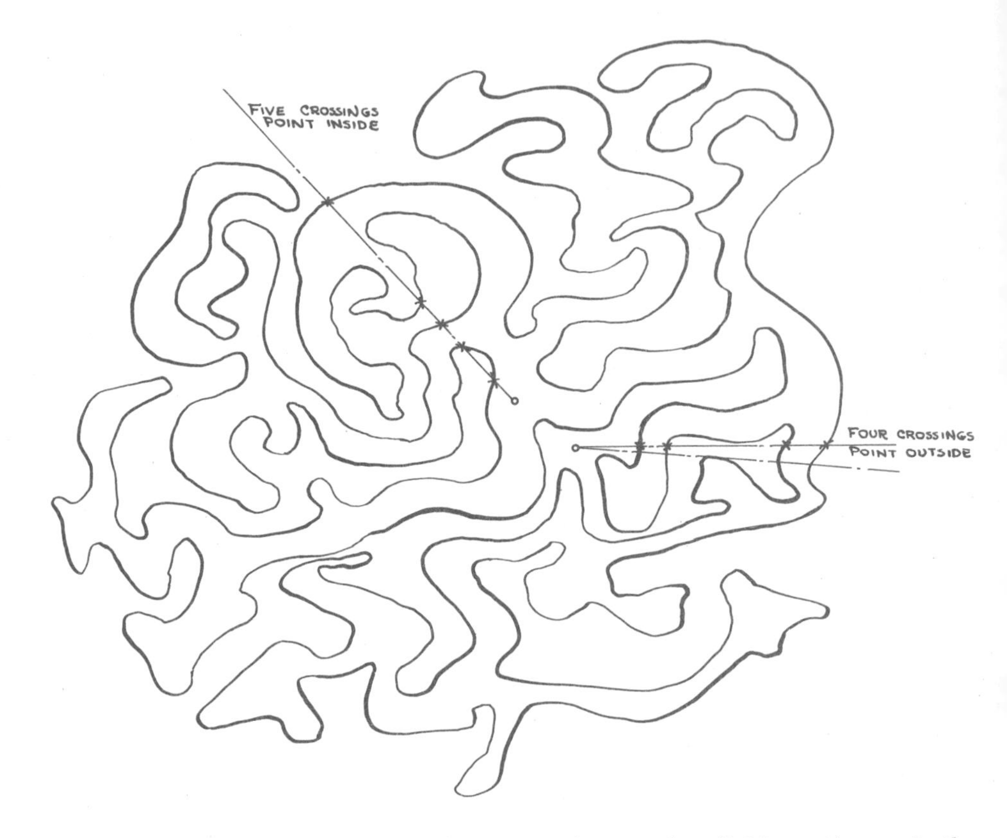

variable—the limit of a finite sum. Assume that $f(x)$ is continuous in the interval $x_0 \rightarrow x_n$; then the integral of $f(x)$ is the limit of the finite sum:

$$\sum_{i=0}^{n-1} (x_{i+1} - x_i) f(x_i),$$

where $x_0 < x_1 < x_2 < \cdots < x_i < \cdots < x_n$. The length of the largest interval $(x_{i+1} - x_i)$ tends to zero as n tends to infinity. According to Riemann's definition, $f(x)$ is bounded in the interval $[x_{i+1}, x_i]$, while U_i and L_i are the upper and lower bounds of $f(x)$ in the interval. Creating two sums,

$$U = \sum_{i=0}^{n-1} (x_{i+1} - x_i) U_i,$$

and

$$L = \sum_{i=0}^{n-1} (x_{i+1} - x_i) L_i,$$

one defines the Riemann integral of $f(x)$ as the common limit of U and L as n becomes arbitrarily large, assuming that a common limit exists; if not, then the integral is undefined. With this modification, the definition of the integral was extended to incorporate functions with a finite number of discontinuities in the interval $[x_0, x_n]$. This definition was modified by the arithmetic work of Cantor in 1884 and Lebesgue in 1902.

Interestingly enough, some of Riemann's early concepts of differential geometry emanated from problems in heat conduction. The study of thermodynamics, with its associated concepts of energy conservation, was of paramount interest at this time. Conservation of thermodynamic energy was implicit in the first law of thermodynamics, while the work of Kelvin and Clausius had expanded the concepts of Carnot into the second law of thermodynamics. In 1847 James Prescott Joule measured the mechanical equivalent of heat, and this work represented an important step in confirming the already established importance of the conservation of energy. During this same year, Hermann von Helmholtz (1821–1894) presented his famous paper on the conservation of energy, in which he dealt with transfers of electrical energy to heat energy. Here and elsewhere he succeeded in correlating the chemical action within a voltaic cell to the electrical energy furnished. This stimulated a reevaluation of the theory of the energy stored in electric and magnetic fields.

The total electrostatic energy U of a system of point charges q_i, each situated at a point in space having an electrostatic potential V_i, was calculated by Helmholtz, who considered increasing the total charge of the system by Δq. The energy U was shown to have a pairing term of 1/2, suggesting the obvious interpretation that when two charges interact to give a total energy $q_1q_2/(4\pi\epsilon_0)r_{12}$, and this energy is shared equally by both of them, the result associated with each charge is an energy $(1/2)q_1q_2/(4\pi\epsilon_0)r_{12}$. The general result he obtained for N charged bodies is

$$U = \frac{1}{2}q_1V_1 + \frac{1}{2}q_2V_2 + \cdots + \frac{1}{2}q_NV_N = \sum_{k=1}^{N} \frac{1}{2}q_kV_k,$$

where V_k is the potential (electrical) set up at the position of the kth body by the *other* $N - 1$ charges.

Continuous distributions of charge $\rho(x, y, z)$, where ρ is the charge density or charge per unit volume at a point (x, y, z), have the same type of relation for the total electrostatic energy, with the modification that the discrete sum transforms to an integral,

$$U = \frac{1}{2}\iiint_{\substack{\text{volume of}\\ \text{charge}}} \rho(x, y, z)V(x, y, z)\, dx\, dy\, dz,$$

where V is the electrostatic potential at the point (x, y, z). The formula given above is interpreted in terms of action at a distance. In Maxwell's formulation of the energy associated with the electric field of a system, this integral will be transformed to an integral over the electric field times the electric displacement field. This, in truth, is nothing more than a field term descriptive of the free singlet charge density ρ. Similarly, the electric field $\vec{\mathscr{E}}$ is directly derived from the potential term V.

These same electrical and thermodynamic concepts were shared by Lord Kelvin. He created an absolute temperature scale based upon cascaded ideal Carnot cycles. To account for the quality of reversibility in a mathematical form, the generalization in terms of the entropy change of the system became the basis for the second law of thermodynamics.

Entropy was the creation of Clausius. If an element of heat dQ is exchanged at an absolute temperature T, the entropy change dS is defined as the ratio dQ/T; the integral (or sum) of dQ/T from one state of the system to another is always the same if the process is reversible. This approach to thermodynamics is analogous to that of the conservative field in dynamics and electricity. A complete cycle of a reversible system has the characteristic that the total entropy change ΔS during the cycle is 0. If $dS = dQ/T$, then the integral (or sum) of the entropy changes about a closed path is 0 for a reversible system:

$$\oint dS = \oint \frac{dQ}{T} = 0.$$

In 1851 Thomson presented an important memoir on the theory of magnetism. In this he distinguished between two magnetic vectors; the primary field, called the magnetic induction field $\vec{\mathbf{B}}$, and a secondary field $\vec{\mathbf{H}}$, which is descriptive of the true electrical currents and all discontinuities in volume distributions of magnetic moments. In early periods a magnetic moment was conceived of as a direct analogue to the electric dipole, namely, as two magnetic monopoles separated by a distance l; however, experimental evidence has never indicated the existence of magnetic monopoles. As discussed earlier, Weber correctly conceived of the magnetic dipole or magnetic moment as a small current loop. A plane loop of area A, carrying a current I about its periphery, has a magnetic moment $\vec{\mathbf{m}}$ having a magnitude equal to AI and a direction oriented perpendicular to the plane of the loop. Further, this directional property was specified by the righthand rule. A volume density of these small current loops, or individual moments, creates a magnetic moment volume density $\vec{\mathbf{M}}_v$. Converting Kelvin's hypothesis to modern units, the quantity $\vec{\mathbf{B}}$ is a linear combination of $\vec{\mathbf{H}}$ and $\vec{\mathbf{M}}_v$, where

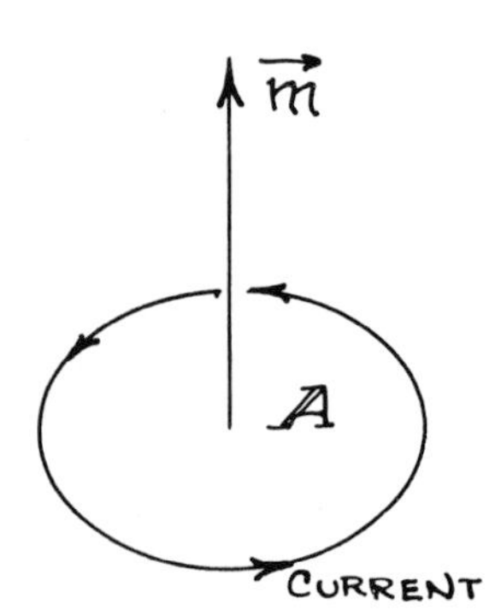

$$\vec{\mathbf{B}} = \mu_0(\vec{\mathbf{H}} + \vec{\mathbf{M}}_v).$$

Here μ_0 is a modern constant giving $\vec{\mathbf{H}}$ in oersteds and $\vec{\mathbf{B}}$ in webers per square meter: 10^4 gauss (the older unit of $\vec{\mathbf{B}}$) equals 1 weber per square meter; μ_0 has the numerical value $4\pi(10)^{-7}$. This relationship in many cases can be expressed as a simpler relation between $\vec{\mathbf{B}}$ and $\vec{\mathbf{H}}$ only. When $\vec{\mathbf{M}}_v$ is proportional to $\vec{\mathbf{B}}$ or $\vec{\mathbf{H}}$ (which is often the case), one can write

$$\vec{\mathbf{B}} = \mu_0(1 + \chi_m)\vec{\mathbf{H}}.$$

χ_m is known as the magnetic susceptibility, and the quantity $\mu_0(1 + \chi_m)$ was called by Kelvin the magnetic permeability. Kelvin in his earliest work gave this result a more general form. The scalar relationship between

$\vec{\mathbf{M}}_v$ and $\vec{\mathbf{H}}$ is quite adequate for isotropic magnetic materials. When anisotropic materials are encountered, the components of the vector $\vec{\mathbf{M}}_v$ become linear combinations of the components of $\vec{\mathbf{H}}$. Thus, to formulate Kelvin's contribution in modern matrix language, the susceptibility χ_m is written in such situations as a 3×3 matrix, and the 1 in $(1 + \chi_m)$ becomes an identity matrix.

Two years later, in 1853, Kelvin formulated a new expression for the energy stored in a magnetic field. In an earlier expression the energy was the volume integral of 1/2 the current density, $\vec{\mathbf{J}}$, times the magnetic vector potential, $\vec{\mathbf{A}}$. As in the electric case, this represents action at a distance. Because $\vec{\mathbf{A}}$ defines $\vec{\mathbf{B}}$, and $\vec{\mathbf{J}}$, including real currents and magnetic moment currents, determines $\vec{\mathbf{H}}$, it is reasonable that Kelvin's form of the total energy U_m, for the system of magnetic fields, became

$$U_m = \frac{1}{2} \iiint_{\substack{\text{all} \\ \text{space}}} \vec{\mathbf{B}} \cdot \vec{\mathbf{H}} \, dx \, dy \, dz .$$

While these important advances in magnetism were taking place, Gustav Kirchhoff (1824–1887) in 1849 generalized Ohm's equations for circuits carrying electrical currents. Further, he solved the problem of the distributions of current in an extended conductor. He removed the uncertainty surrounding the electrical tension of a voltaic cell by identifying this with the electrostatic potential. His investigations of circuits culminated in two statements carrying his name (they are modernized here): first, that the sum of the potential differences across capacitors and resistors equals the algebraic sum of the battery potentials minus the induced voltage drops in inductors; and second (on the conservation of charge), that the algebraic sum of currents entering and leaving any point in a circuit is zero.

A major contribution to the modern formulation of the circuit equations was made by Kelvin in 1853. He noted that, in a closed series circuit (with no batteries) containing only an initially charged condenser C, a resistor R, and an inductance L, the sum of the potential differences and induced voltage across L could be written in terms of the net charge Q flowing in the circuit at any instant. Mathematically, this result appeared as

$$L \frac{d}{dt}\left(\frac{dQ}{dt}\right) + R \frac{dQ}{dt} + \frac{Q}{C} = 0,$$

where the electrical current is given by dQ/dt. Kelvin noticed that when R^2C was greater than $4L$, the discharge did not oscillate; however, when R^2C was less than $4L$, the discharge of the condenser C through R and L produced oscillations having a period $2\pi[(1/LC) - (R^2/4L^2)]^{-1/2}$. Because this differential equation is linear with constant coefficients, solutions appear in the form of exponentials e^{mt}. To observe this, let the equation take the form

$$\frac{d^2Q}{dt^2} + 2a\frac{dQ}{dt} + bQ = 0,$$

where $2a = R/L$ and $b = 1/LC$. By substituting $Q \rightarrow e^{mt}$, we obtain a polynomial in m, $m^2 + 2am + b = 0$, providing two roots, m_1 and m_2 and consequently two independent solutions for Q:

$$m_1 = -a + \sqrt{a^2 - b} = -\frac{R}{2L} + \sqrt{\frac{R^2}{4L^2} - \frac{1}{LC}},$$

and

$$m_2 = -a - \sqrt{a^2 - b} = -\frac{R}{2L} - \sqrt{\frac{R^2}{4L^2} - \frac{1}{LC}}.$$

The most general solution for Q, then, is a linear combination of the two exponential solutions, with m_1 and m_2 as arguments: $Q = Ae^{m_1 t} + Be^{m_2 t}$ (A and B are constants of integration which satisfy a specific problem).

Using the results of Euler for the expansion of the exponential, one observes that when a^2 is less than b, the term in the square root becomes imaginary, giving an imaginary exponential with a resulting sinusoidal variation in time. For the problem posed by Kelvin the coefficients of the linear combination, A and B, are determined by the initial condition that the total charge on the condenser is Q_0 and initially ($t = 0$), the current, dQ/dt, is 0. Then at $t = 0$, $Q_0 = A + B$, and $0 = m_1 A + m_2 B$, giving $A = -(m_2 Q_0)/(m_1 - m_2)$, and $B = (m_1 Q_0)/(m_1 - m_2)$. The reader should notice that by substituting e^{mt} into the differential equation, the problem was reduced to a characteristic polynomial in m.

In the middle of the eighteenth century there was great interest in the propagation of signals on long wires, which culminated in an investigation of the time required for electrical signals to traverse long circuits. This was to result in a theory of telegraphy, and in 1834 Charles Wheatstone demonstrated that the velocities of propagation of signals in extended circuits were close to the velocity of light. The mathematical theory of telegraph signals originated in a correspondence between Kelvin and Sir George Stokes in 1854; their work was limited to a very special case, but Kirchhoff followed it with a remarkable memoir in which he obtained a specific expression for the propagation of signals in a cable formed of coaxial cylindrical conductors. Results of this work suggested that the variation of the electric potential V in space and time along the cable was given by a wave equation similar to that of the vibrating string:

$$\frac{\partial^2 V}{\partial x^2} - \frac{1}{c^2}\frac{\partial^2 V}{\partial t^2} = 0,$$

where x is measured along the symmetry axis of the cylinders, and c is the velocity of propagation. In the derivation of this expression, Kirchhoff showed that c was equal to the equivalent of the modern expression

$1/\sqrt{\epsilon_0\mu_0}$. (Here we have used modern mks units instead of the older cgs units employed by Kirchhoff.) Such was his discovery: the velocity of propagation in this idealized cable was specified from the two constants appearing in the force law of electrostatics and the force law of Ampère for magnetism. Just before the publication of this result, Weber and Kohlrausch experimentally determined the propagation velocity c by measuring the discharge of a condenser. Although it was not apparent at the time, their numerical result of $3.1(10)^{10}$ cm/sec was a relatively accurate measurement of the speed of light. Twenty years later, Oliver Heaviside obtained the general telegraphist equation for real cables which contain a resistive or dissipative term.

During this period of intense speculation about the properties of and connections between electric and magnetic phenomena, James Clerk Maxwell (1831–1879) entered Trinity College, Cambridge, where he became a fellow in 1855. Shortly thereafter he submitted his earliest effort to establish a mechanical model of the electromagnetic field. Under the influence of Kelvin and Faraday, Maxwell, who was a skilled mathematician, seriously examined Faraday's lines of force. The analogy between this concept and the lines of flow of a liquid suggested the mathematical formalism. By analogy with the magnetic induction vector $\vec{\mathbf{B}}$, which is rotational, he created the electric displacement vector $\vec{\mathbf{D}}$, possessing certain similarities. $\vec{\mathbf{D}}$ was defined as a linear combination of the electric force field intensity $\vec{\mathscr{E}}$ and the electric dipole moment per unit volume $\vec{\mathbf{P}}_v$; $\vec{\mathbf{D}} = \epsilon_0\vec{\mathscr{E}} + \vec{\mathbf{P}}_v$.

Once again, the modern units have been employed. This shortcut does not alter the basic ideas presented by Maxwell. In the absence of *free charge*, the divergence of $\vec{\mathbf{D}}$ is zero.* Dielectrics, when subjected to electric fields, exhibit induced bound charges on the surfaces, which lead to reductions of the electric force field in the interior of the dielectric. Because some of the electric field lines must terminate to allow for the reduction, there are sources and sinks of the $\vec{\mathscr{E}}$ field on dielectrics. The electric displacement vector $\vec{\mathbf{D}}$, however, is divergenceless in the presence of dielectrics, and consequently the source properties of $\vec{\mathscr{E}}$ in dielectrics are proportional to the starting and stopping of the $\vec{\mathbf{P}}_v$ vectors.

The only sources of $\vec{\mathbf{D}}$, then, are the free singlet charges in a region; if the free charge density is ρ in a region, the source and sink properties of $\vec{\mathbf{D}}$ are measured by ρ. Mathematically this is written as div $\vec{\mathbf{D}} = \rho$, (the equivalent of Coulomb's Law for a continuum), where the reader must remember that divergence $\vec{\mathbf{D}}$ provides a differential equation of the type:

$$\operatorname{div} \vec{\mathbf{D}} = \frac{\partial D_x}{\partial x} + \frac{\partial D_y}{\partial y} + \frac{\partial D_z}{\partial z} = \rho.$$

* Remember that the earlier definition of the divergence operator states that it provides a measure of the source or sink properties of a field.

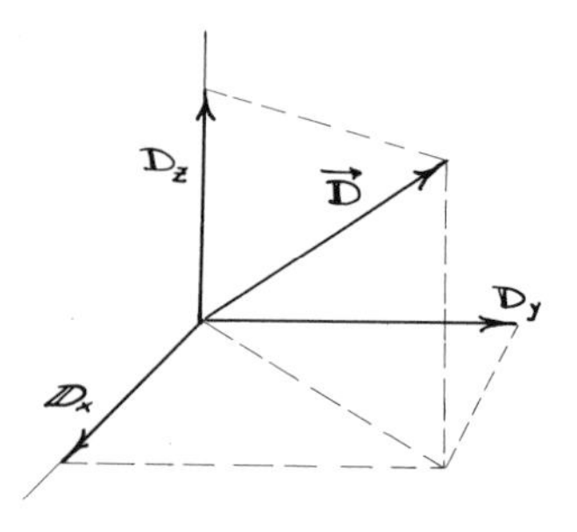

Here D_x, D_y, and D_z are the projections of the vector field $\vec{\mathbf{D}}$ along the x, y, and z axes. This is known as the first Maxwell equation, and is a modified statement of Coulomb's Law (and Poisson's equation) taking into account the effects of dielectric media.

Maxwell's great contribution was to generalize the basic equations of electromagnetic theory. Because $\vec{\mathbf{B}}$ is rotational at all points, with no sources or sinks, the second law of Maxwell is a simple statement of this fact: divergence $\vec{\mathbf{B}} = 0$, or

$$\partial B_x/\partial x + \partial B_y/\partial y + \partial B_z/\partial z = 0.$$

Although the differential operations divergence and curl were outlined earlier in this manuscript (see pp. 158 and 192), they were given their first significance in the writings of Sir George Gabriel Stokes (1819–1903). Stokes made a series of important contributions in hydrodynamics and, with Kelvin, was one of the supporters of the elastic solid model of the aether. He later modified his view, developing the notion of an aether which was dragged by the earth's motion, producing aether turbulence at the surface. Maxwell gave greater physical significance to the operator's divergence* and curl, using them to form the complete set of Maxwell equations. In one of his most significant memoirs† Maxwell attempted to construct a mechanical model of the aether. It was in this work that the first two laws were formulated in terms of the divergence operator. Ampère's law defining the magnetic field of currents and Faraday's law appear in different forms in terms of the curl operator. To Maxwell, Kelvin, and others, this implied a fundamental rotational characteristic of a magnetic field, and this assumption was supported by such experimental evidence as the magnetic rotation of the plane of polarization of light.

Mechanical models of the aether were constructed of vortices with isolating idling wheels between neighboring vortices, and in the late nineteenth century such model building culminated in Kelvin's almost fanatical adherence to a vortex model of the atom. Perturbations in the rotations of microscopic vortices were employed by Maxwell to account for Faraday's law (here we shall refer to this as the fourth Maxwell equation). Faraday's law, when converted from an electromotive force-flux relation to an equation linking the time rate of change of $\vec{\mathbf{B}}$ and a rotational component of $\vec{\mathscr{E}}$, achieves the differential form:

$$\text{curl}\,\vec{\mathscr{E}} = -(\partial\vec{\mathbf{B}}/\partial t).$$

Rhetorically interpreted, this equation states that a time-varying magnetic field induces a solenoidal electric field; i.e., it induces an emf in a closed loop. This particular Maxwell form implies that the total electric field $\vec{\mathscr{E}}$ is determined not only by the negative gradient of a scalar potential function

* Maxwell named the operation "convergence." Because of the confusion with the mathematical convergence of series, the name was later changed to "divergence."

† *Transactions of the Cambridge Philosophical Society* 10 (1864): 27.

V but also by the time variation of the magnetic vector potential $\vec{\mathbf{A}}$: $\vec{\mathscr{E}} = -\text{gradient } V - (\partial\vec{\mathbf{A}}/\partial t)$.

Maxwell's greatest contribution was to occur in his modifications of Ampère's law, which here shall be designated as the third Maxwell equation. Earlier, Faraday and, independently, Mossotti had concerned themselves with the current equivalents produced in dielectrics when the electric dipole charges were in motion to produce a polarization. Across a dielectric condenser which is building up free charge on its plates, the dielectric polarization charge is simultaneously being displaced in such a manner as to constitute a bound current. In order to understand this, regard a conducting sphere initially in a zero field and later in a nonzero field.

Polarization of the sphere implies a flow of positive charge to the upper hemisphere and a flow of negative charge to the lower hemisphere. If in a time interval Δt a net charge ΔQ flows, the current across a plane through the equatorial circle is $\Delta Q/\Delta t$. Rightly then, this current is a displacement current of the bound charge, and further, this displacement current can and will give rise to a magnetic field $\vec{\mathbf{H}}$. Ampère's law in differential form relates the circulation of the magnetic $\vec{\mathbf{H}}$ field to the total current density $\vec{\mathbf{J}}$, and is written simply

$$\text{curl } \vec{\mathbf{H}} = \vec{\mathbf{J}}_{\text{total}}.$$

If the divergence of both sides of this equation is taken, the fact that the divergence of any circulating vector defined by a curl operation is always zero forces the lefthand side of the equation to be zero. This then implies that the divergence of $\vec{\mathbf{J}}_{\text{total}}$ is always zero, and because div $\vec{\mathbf{J}}_{\text{total}}$ is *always* zero, *there can be no sources or sinks for the total current density.* The

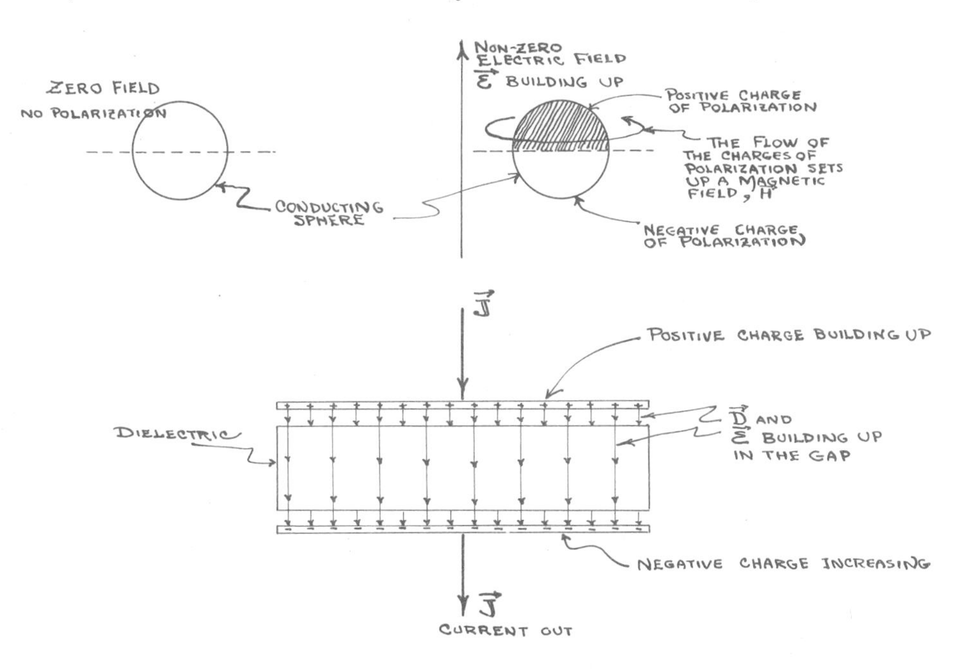

existence of the condenser with a vacuum in the gap, across which there is no real current flow, tends to belie this statement, thereby producing a paradox.

The illustration of the capacitor shows the free positive flow of charge terminating on the upper plate of the condenser, thereby building up a net positive charge in time. Obviously, then, the free charge current has sinks and sources. The Faraday-Mossotti hypothesis, on the other hand, provides a partial escape from the trap when there is a dielectric in the gap. In this case, the displacement current in the dielectric is the time rate of change of the electric dipole moment per unit volume, as was shown with the conducting sphere:

$$\vec{\mathbf{J}}_{\text{displacement}} \rightarrow \frac{\partial \vec{\mathbf{P}}_v}{\partial t}.$$

If the dielectric is simply isotropic, then $\vec{\mathbf{P}}_v$ is proportional to $\vec{\mathscr{E}}$ and $\vec{\mathbf{D}}$:

$$\vec{\mathbf{P}}_v = \epsilon_0 \chi_0 \vec{\mathscr{E}},$$

and because

$$\vec{\mathbf{D}} = \epsilon_0 (1 + \chi_e) \vec{\mathscr{E}},$$

$$\vec{\mathbf{P}}_v = \frac{\chi_e}{1 + \chi_e} \vec{\mathbf{D}}.$$

Algebraically, then, according to Faraday, the displacement current in a dielectric behaved as

$$\vec{\mathbf{J}}_{\text{disp}} \rightarrow \frac{\chi_e}{1 + \chi_e} \frac{\partial \vec{\mathbf{D}}}{\partial t}.$$

Maxwell was convinced that his equations of electromagnetism must possess characteristics similar to those discovered by Kirchhoff in describing the propagation of voltage and current signals in cables. This conviction was the result of the remarkable coincidence that in ideal cables the velocity of propagation was expressed in terms of the fundamental constants of proportionality of electromagnetism. It was further apparent that voltage-current propagation in a cable could be interpreted as an electric field-magnetic field propagation.

With the Faraday-Mossetti hypothesis as an indication of the correct form, Maxwell considered the vacuum as a dielectric of infinite susceptibility;

$$\frac{\chi_e}{1 + \chi_e} = \frac{1}{1 + \chi_e^{-1}}$$

in this case, the term would approach 1 in the limit. This is not the correct way of developing the displacement current, despite the fact that it produces the correct final result, and because of his intuitive approach to this difficulty, Maxwell's hypothesis was severely criticized. The Maxwell hypothesis then stated that the total current at any point was the sum of the true free current $\vec{\mathbf{J}}$ and the displacement current $\partial\vec{\mathbf{D}}/\partial t$. As a result, the third Maxwell equation was postulated, to take the form

$$\text{curl}\,\vec{\mathbf{H}} = \vec{\mathbf{J}} + \partial\vec{\mathbf{D}}/\partial t.$$

This equation adheres to Ampère's original law, with the addition of a redefinition of the total current. More important, this relation provides a symmetry between the last two laws: if a time-varying magnetic field induces a rotational electric field, a time-varying electric field would induce a magnetic field in the same manner.

Although it was not made obvious in the early development of this equation, Ben Franklin's hypothesis of charge conservation is implied by the third equation, for if divergence curl $\vec{\mathbf{H}}$ is always zero, then (applying the divergence operator to the righthand side) divergence $\vec{\mathbf{J}} + \partial/\partial t$ (divergence $\vec{\mathbf{D}}$) $= 0$. Divergence $\vec{\mathbf{D}}$ by the first Maxwell equation is always ρ, the free charge density near a point; thus div $\vec{\mathbf{J}} + \partial\rho/\partial t = 0$. This equation states that if $\vec{\mathbf{J}}$ has a sink at a point, the charge density builds up in time; in the same manner, if $\vec{\mathbf{J}}$ has a source, ρ decreases in time. Modern developments of the third law use the equation of charge conservation as a direct argument to modify Ampère's law.

From the complete set of Maxwell's equations in free space, i.e., $\rho = 0$ and $\vec{\mathbf{J}} = 0$, a set of wave equations for $\vec{\mathscr{E}}$ and $\vec{\mathbf{H}}$ can be derived; the result depends upon taking the curl of the last two equations.* As an example,

$$\text{curl}\,(\text{curl}\,\vec{\mathbf{H}}) = \frac{\partial}{\partial t}\,(\text{curl}\,\epsilon_0\vec{\mathscr{E}}) = \frac{\partial}{\partial t}\left(-\epsilon_0\mu_0\,\frac{\partial\vec{\mathbf{H}}}{\partial t}\right),$$

giving

$$\nabla^2\vec{\mathbf{H}} = \epsilon_0\mu_0\,\frac{\partial^2\vec{\mathbf{H}}}{\partial t^2} = \frac{1}{c^2}\frac{\partial^2\vec{\mathbf{H}}}{\partial t^2},$$

where ∇^2 is the Laplacian operator div grad. The fact that a vector quantity (here $\vec{\mathbf{H}}$) appears behind the scalar Laplacian operator ∇^2 implies three equations, one for each component of $\vec{\mathbf{H}}$.

The wave equation has a velocity of propagation c, and from the Maxwell equations one finds that c for electromagnetic waves in free space is given by $1/\sqrt{\epsilon_0\mu_0}$, the inverse of the square root of the electric permitivity times the magnetic permeability. The initial aim was justified in

* An expansion of the curl curl operator shows that it is equal to gradient of the divergence minus the divergence of the gradient. The last term in this expansion (i.e., div grad) is the Laplacian,

$$\nabla^2 = \frac{\partial^2}{\partial x^2} + \frac{\partial^2}{\partial y^2} + \frac{\partial^2}{\partial z^2}.$$

The first term of the expansion always vanishes in free space because div $\vec{\mathbf{H}}$ and div $\vec{\mathscr{E}}$ are zero. In free space div $\vec{\mathscr{E}}$ is zero because ρ and $\mathbf{P}_v$ are zero.

that the analogue of waves on a cable had been produced for unconstrained electromagnetic radiation.

Maxwell also provided a field form of the energy storage integral U_e for the electric field. The expression for the energy density $(1/2)\rho V$ transformed directly to $(1/2)\vec{\mathscr{E}}\cdot\vec{\mathbf{D}}$, giving the total electromagnetic energy U in terms of the four field quantities:

$$U = U_e + U_m = \frac{1}{2}\iiint_{\substack{\text{all}\\ \text{space}}} \{\vec{\mathscr{E}}\cdot\vec{\mathbf{D}} + \vec{\mathbf{B}}\cdot\vec{\mathbf{H}}\}\, dx\, dy\, dz.$$

By this time the velocity of light had been measured by methods other than those utilizing astronomical aberrations. In 1849 A. H. L. Fizeau determined c by a method involving the reflection of light through the teeth of a rapidly rotating toothed wheel. The result of $3.15(10)^{10}$ cm/sec was so close to the electromagnetic result that, like Kirchhoff, Maxwell postulated light as indeed being electromagnetic radiation.

The anomalous dispersion of light was discovered in 1862 by F. P. Leroux. Refractive indexes for light had been believed to be slowly varying functions of the frequency, behaving as an even power series in the frequency. This dispersion relation was a result of Cauchy's elastic solid model of the aether in 1830. Leroux found that a prism filled with iodine vapor gave the inverse behavior, refracting red light to a greater extent than blue. Maxwell dealt with this effect by considering matter to be built up of atoms, each composed of a single massive particle supported symmetrically by springs from the inner surface of a massless spherical shell. Assuming a fine-grained medium consisting of large numbers of these mechanical atoms, he developed the response of this theoretical medium to the propagation of waves and showed that the ratio of the square of the velocity of light of radial frequency ω in a vacuum to the square of the velocity in the medium is frequency-dependent, giving the square of the refractive index, n:

$$n^2 = \frac{c^2}{v^2} = 1 - \frac{K\omega_0^2}{\omega_0^2 - \omega^2}.$$

Here ω_0 was the characteristic frequency of the mechanical system and K was a constant. Quite clearly, then, by expanding this result in a power series in ω^2/ω_0^2 for $\omega > \omega_0$, Cauchy's formula was obtained as an approximation. This was a significant step, for the later electron theories of matter were to produce a similar type of result.

Although Maxwell was able to apply his theory to the propagation of light in isotropic dielectrics and in metals, he was unable to apply his equations to reflection and refraction at a boundary. The missing element was a clear understanding of the boundary conditions which had to be imposed upon the four vector fields $\vec{\mathscr{E}}$, $\vec{\mathbf{D}}$, $\vec{\mathbf{B}}$, and $\vec{\mathbf{H}}$ at a boundary surface. He did, however, create the concept of electromagnetic stress and strain, which led him to postulate that light incident upon a reflecting surface would exert light pressure. Lord Kelvin never accepted the predictions of the Maxwell equations, and, in particular, he rejected the thesis that the index of refraction varied as the square root of the dielectric constant together with the postulate of light pressure.

Several years after Maxwell's first description of an electromagnetic theory of light, Ludwig Lorenz (1829–1891), following in the tradition of Gauss and Riemann, created a system equivalent to that of Maxwell but based upon potential functions instead of vector fields. Lorenz invented quantities known today as the retarded potentials of the electromagnetic field. The concept of retardation was essential because of the finite speed of light: a disturbance in the charge density and/or current density $\vec{\mathbf{J}}$ near the origin at a time t' can only be detected at some point x, y, and z at a time t, where $t = t' + r/c$ with $r \simeq \sqrt{x^2 + y^2 + z^2}$. This approach was quite suggestive; for the first time four source coordinates, x', y', z', and t', were distinguished from four field coordinates, x, y, z, and t—the two sets being connected by a transformation involving a frame of reference moving with a velocity c. By treating time on an equal footing with the space coordinates, electromagnetic theory played a primary role in the invention of a four-dimensional theory of relativity. The retarded potentials of Lorenz were presented in integral form and led to a set of defining equations for the scalar and vector potentials, V and $\vec{\mathbf{A}}$.

Much of the deterministic philosophy of the nineteenth century centered about efforts to create a purely mechanical picture of thermodynamics—especially the thermodynamics of gases. The view that matter was to be regarded as an aggregation of hard indivisible particles had been held by Lucretius, Gassendi, and Hooke. Daniel Bernoulli was able to deduce Boyle's law assuming a gas to be formed of finite-sized molecules, spherical and absolutely hard. Clausius used such a kinetic theory of gases to develop the ideal gas law together with values for the specific heats at constant volume and at constant pressure. In 1859 Maxwell became interested in the kinetic theory of gases, and derived the law of the distribution of velocities, known today as the Maxwell-Boltzmann distribution. By considering large numbers of colliding spheres and calculating the probabilities of transitions into and out of elements of velocity space, Maxwell arrived at the conclusion that at equilibrium the probability, f (or fraction), of molecules having velocity components along the three Cartesian coordinate axes with values lying between u and $u + du$, v and $v + dv$, and w and $w + dw$ is $f = Ae^{-K(u^2+v^2+w^2)}$, where A and K are constants. The constant K was to be shown to be $m/2kT$, where m is the mass of a molecule, k is the Boltzmann constant, and T is the absolute temperature. Maxwell's original proof is unsatisfactory, but was historically important.

From this result, assuming that a gas consisting of N molecules has $3N$ degrees of freedom, Maxwell deduced the relation between the average kinetic energy of a molecule and the temperature of the gas as

$$[(1/2)mv^2]_{\text{ave}} = (3/2)kT,$$

where m is the mass of a molecule and v is the magnitude of its velocity.

Classical statistical analysis emerged largely from the efforts of Josiah Willard Gibbs (1839–1903) and Ludwig Boltzmann (1844–1906). Statistical mechanics has the special advantage of providing reasonable methods for treating the behavior of mechanical systems even when we know less of the system than the maximum theoretically possible. Since our encounters with the physical world are such that one never has maximal knowledge of systems, the methods either of idealization or of statistical analysis are essential. Both Gibbs and Boltzmann considered the many-body gas system as a continuous function of the $6N$ degrees of freedom corresponding to $3N$ spatial coordinates plus $3N$ momentum coordinates: this $6N$-dimensional space was designated as the phase space of the system. The instantaneous state of an ensemble can then be viewed as a point in phase space: for an ensemble a density function is defined which specifies the state of each individual system within an ensemble. Observable quantities are obtained by taking averages over phase space. For example, the average value of the density function must have the value unity, giving the density function an interpretation as the probability per unit volume of finding a phase point in phase space.

Boltzmann proceeded by constructing a distribution function which provided a measure of the extent to which a system deviates from equilibrium. Gibbs took a probabilistic point of view, utilizing the number of ways that N things can be partitioned into n_j subgroups. Both approaches gave the same equilibrium distribution for a gas found by Maxwell. Using this distribution function for an ideal gas made up of molecules of mass m, the pressure can be evaluated, demonstrating that Maxwell's constant K was of the form $m/2kT$. If the energy of a state is E, the Maxwell-Boltzmann distribution for the number of particles dN with energy lying between E and $E + dE$ is $dN = N_0 e^{-E/kT}\, dE$. This broad and powerful approach to the analysis of physical systems served as a guide for the later development of the quantum theory of matter.

Gibbs was one of the first famous theoretical physicists in the United States. His understanding of mathematics was thorough. By the turn of the century he had established a world-wide reputation in statistical mechanics; he had invented the modern vector analysis, and his other contributions ranged from the Fourier series to studies of the propagation of electromagnetic radiation. His *Elements of Vector Analysis* was privately printed in New Haven in 1884: in this work concepts equivalent to those of the multiple algebra of Grassmann, the matrices of Cayley, and the quaternions of Hamilton were prevalent. The success of this small manuscript annoyed the followers of Hamilton, led by Tait, and the ensuing controversy was to muddy the waters of vector notation for some sixty years. In his later clear and impartial account of the history of multiple algebra,* Gibbs conceded that the algebra of Grassmann had been the most important influence upon his vector analysis.

* J. W. Gibbs, *Proceedings of the American Association for the Advancement of Science*, 35 (1886): 37.

Throughout this work we have been implicitly employing vectors without explicitly writing down their form in terms of a set of bases. A set of unit bases created by Gibbs consisted of the parameters $\vec{\mathbf{i}}$, $\vec{\mathbf{j}}$, and $\vec{\mathbf{k}}$, aligned along the x, y, and z axes, respectively. These were defined together with a set of rules for multiplication. Multiplication was composed of two varieties; an inner product, or dot product, producing a scalar, and an outer product, or vector product, producing a pseudovector. A directed quantity (or vector) $\vec{\mathbf{F}}$ in three dimensions was represented in terms of its projections on the three axes. Thus, $\vec{\mathbf{F}}$ would have components $\mathbf{F}_x$, $\mathbf{F}_y$, and $\mathbf{F}_z$, obeying the Pythagorean condition that the magnitude of $\vec{\mathbf{F}}$ (designated as $|\vec{\mathbf{F}}|$) was the square root of the sum of the squares of the components: $\vec{\mathbf{F}} = F_x\vec{\mathbf{i}} + F_y\vec{\mathbf{j}} + F_z\vec{\mathbf{k}}$, and $|\vec{\mathbf{F}}| = \sqrt{F_x^2 + F_y^2 + F_z^2}$.

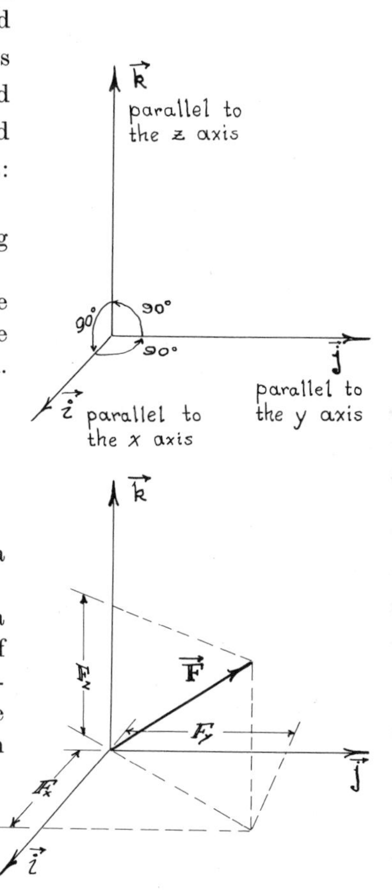

Rules for multiplication were contained in the rules for multiplying bases.

1. *The dot or inner product*: here the product is the product of the magnitudes of the two vectors times the cosine of the angle between the two, and is specified by a bold dot between vectors which are multiplied. Thus:

$$\begin{array}{lll} \vec{\mathbf{i}}\cdot\vec{\mathbf{i}} = 1 & \vec{\mathbf{j}}\cdot\vec{\mathbf{i}} = 0 & \vec{\mathbf{k}}\cdot\vec{\mathbf{i}} = 0 \\ \vec{\mathbf{i}}\cdot\vec{\mathbf{j}} = 0 & \vec{\mathbf{j}}\cdot\vec{\mathbf{j}} = 1 & \vec{\mathbf{k}}\cdot\vec{\mathbf{j}} = 0 \\ \vec{\mathbf{i}}\cdot\vec{\mathbf{k}} = 0 & \vec{\mathbf{j}}\cdot\vec{\mathbf{k}} = 0 & \vec{\mathbf{k}}\cdot\vec{\mathbf{k}} = 1. \end{array}$$

These rules are just the inner product rules of Grassmann, limited to a space of three dimensions.

2. *The vector, cross, or outer product*: this product, specified by a cross, is the product of the magnitudes of two vectors times the sine of the angle between the two. The direction of the resulting vector is perpendicular to the plane of the two vectors in the product, with a sense given by the righthand rule when the first vector is rotated in the direction of the second:

$$\begin{array}{lll} \vec{\mathbf{i}}\times\vec{\mathbf{i}} = 0 & \vec{\mathbf{j}}\times\vec{\mathbf{i}} = -\vec{\mathbf{k}} & \vec{\mathbf{k}}\times\vec{\mathbf{i}} = \vec{\mathbf{j}} \\ \vec{\mathbf{i}}\times\vec{\mathbf{j}} = \vec{\mathbf{k}} & \vec{\mathbf{j}}\times\vec{\mathbf{j}} = 0 & \vec{\mathbf{k}}\times\vec{\mathbf{j}} = -\vec{\mathbf{i}} \\ \vec{\mathbf{i}}\times\vec{\mathbf{k}} = -\vec{\mathbf{j}} & \vec{\mathbf{j}}\times\vec{\mathbf{k}} = \vec{\mathbf{i}} & \vec{\mathbf{k}}\times\vec{\mathbf{k}} = 0. \end{array}$$

With these rules the dot product of a vector $\vec{\mathbf{A}}$ and a vector $\vec{\mathbf{B}}$ is:

$$\vec{\mathbf{A}}\cdot\vec{\mathbf{B}} = |\vec{\mathbf{A}}|\,|\vec{\mathbf{B}}| \cos \measuredangle_A^B = A_xB_x + A_yB_y + A_zB_z.$$

The connection with part of the quaternion product and with Grassmann's inner product is obvious. The cross product (or outer, or vector) between these vectors is:

$$\begin{aligned} \vec{\mathbf{A}}\times\vec{\mathbf{B}} &= |\vec{\mathbf{A}}|\,|\vec{\mathbf{B}}| \sin \measuredangle_A^B \,\vec{\mathbf{n}}_{AB} \\ &= (A_yB_z - A_zB_y)\vec{\mathbf{i}} + (A_zB_x - A_xB_z)\vec{\mathbf{j}} + (A_xB_y - A_yB_x)\vec{\mathbf{k}}. \end{aligned}$$

In the expression above, $\vec{\mathbf{n}}_{AB}$ is a unit vector directed perpendicular to the plane formed by $\vec{\mathbf{A}}$ and $\vec{\mathbf{B}}$. Gibbs provided a complete rendition of all the

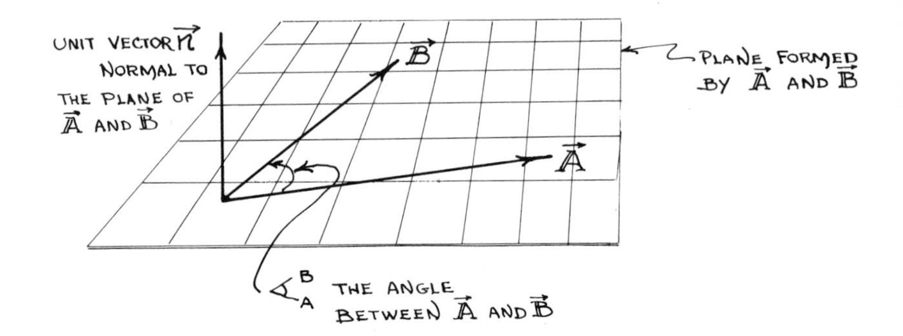

possibilities of his system, including the calculus of vectors, the analysis of linear vector functions including the use of diadics, and the analysis with complex vectors. The operators gradient, divergence, and curl were easily represented in Gibbs's Cartesian system by the symbol ∇, defined as:

$$\nabla = \vec{\mathbf{i}}\frac{\partial}{\partial x} + \vec{\mathbf{j}}\frac{\partial}{\partial y} + \vec{\mathbf{k}}\frac{\partial}{\partial z}.$$

With this definition the standard vector operations can be readily expressed as:

$$\text{gradient } V = \nabla V = \vec{\mathbf{i}}\frac{\partial V}{\partial x} + \vec{\mathbf{j}}\frac{\partial V}{\partial y} + \vec{\mathbf{k}}\frac{\partial V}{\partial z},$$

$$\text{divergence } \vec{\mathbf{F}} = \nabla \cdot \vec{\mathbf{F}} = \frac{\partial F_x}{\partial x} + \frac{\partial F_y}{\partial y} + \frac{\partial F_z}{\partial z},$$

and

$$\text{curl } \vec{\mathbf{F}} = \nabla \times \vec{\mathbf{F}} = \left(\frac{\partial F_z}{\partial y} - \frac{\partial F_y}{\partial z}\right)\vec{\mathbf{i}} + \left(\frac{\partial F_x}{\partial z} - \frac{\partial F_z}{\partial x}\right)\vec{\mathbf{j}} + \left(\frac{\partial F_y}{\partial x} - \frac{\partial F_x}{\partial y}\right)\vec{\mathbf{k}}.$$

As an example of the vector analysis of integrals, consider a line element $d\vec{\mathbf{r}} = dx\vec{\mathbf{i}} + dy\vec{\mathbf{j}} + dz\vec{\mathbf{k}}$ and a force $\vec{\mathbf{F}}$; the element of work dW done by the force $\vec{\mathbf{F}}$ acting through $d\vec{\mathbf{r}}$ is the inner product:

$$dW = \vec{\mathbf{F}} \cdot d\vec{\mathbf{r}} = F_x\,dx + F_y\,dy + F_z\,dz;$$

the total work along a path P between points a and b is then

$$W_{ab} = \int_a^b \vec{\mathbf{F}} \cdot d\vec{\mathbf{r}}.$$

The paper on multiple algebra outlined the manner in which matrix operators could be applied as transformations of vectors in a linear vector space. In spite of the arguments over priority, Gibbs's summary of the vector analysis stands as an excellent example of a major advance brought about by the creation of a superior notation.

Very basic correspondences exist between the theory of linear transformations and the theory of finite groups. For example, a specific set of rotational transformations is linear and may also form a group. With continuous groups the situation is quite different. The theory of groups strips the whole of mathematics of its matter and reduces it to pure form. Marius Sophus Lie (1842–1899), who, oddly enough, showed little if any interest in mathematics before the age of twenty-six, was able to accomplish in weeks what normally requires years. Invariance (that which remains unchanged) was the dominant concept in Lie's work. The secret of his approach was to reduce a nonlinear problem, in which the transformations are finite, to a linear problem utilizing infinitesimal transformations. In mathematical physics Lie's method is often called that of "contact transformation" because when two curves or surfaces touch before transformation, they also are required to touch after transformation.

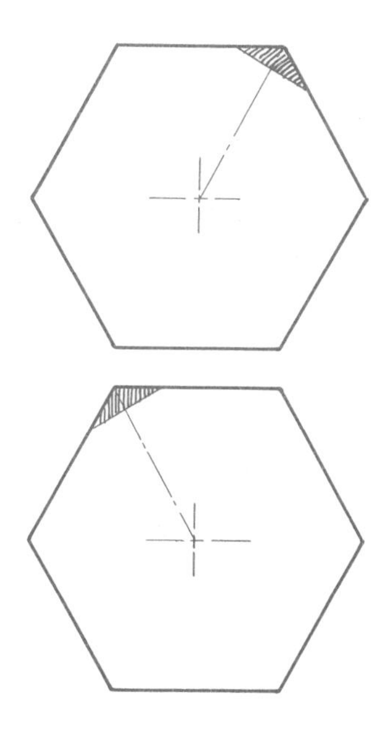

As an illustration, consider a plane hexagon which can be rotated through 60°, 120°, 180°, 240°, or 300°; each of these special rotations leaves the figure in coincidence with the original figure.

These rotations form a cyclical group of order six, and each of the six rotations forms a class by itself; hence there are six classes and six representations. A representation of this group is that in which a rotation through an angle $\Theta_0 = \pi/3$ corresponds to the complex number $e^{i\Theta_0}$. Other representations are $e^{2i\Theta_0}$, $e^{3i\Theta_0}$, $e^{4i\Theta_0}$, and $e^{5i\Theta_0}$. Because $e^{6i\Theta_0} = 1$, the sixth representation gives the identity element common to every group.

Similar representations occur for all regular n-sided figures with $e^{i\Theta_0}, e^{2i\Theta_0}, \cdots, e^{(n-1)i\Theta_0}$, and $e^{ni\Theta_0} = 1$ representing the rotations. Proceeding to the limit as n approaches infinity, the group of rotations of a circle are obtained. The elements of this group form a continuous group, and rotations from zero to an angle Θ can occur by a sequence of infinitesimal displacements $\delta\Theta$. Matrix methods, for instance, can be applied with decided success to this infinite continuous group. The rotations in a plane form a commutative group of degree one, whereas rotations in three or more dimensions, as well as the group of all nonsingular linear transformations on n variables, although continuous groups, may not be commutative.

In general, transformation groups are comprised of a set of n functions $f_1 \cdots f_n$, such that the n equations,

$$x'_j = f_j(x_1, x_2, \cdots, x_n; a_1, a_2, \cdots, a_r),$$

are solvable for the n independent variables x_j. The a_l represent r parameters with which the variables may vary continuously.

The n equations in f_j define transformations of the x_k to new coordinates x'_j with the specific set of parameters $a_1 \cdots a_r$. By the application of a second set of parameters, $b_1 \cdots b_r$, it may happen that the x'_k variables are transformed to x''_m, such that

$$x''_m = f'_m(x'_1, \cdots, x'_n; b_1, \cdots, b_r).$$

By eliminating the primed variables x'_k in the equation above, one may find that the x''_m are related to the x_l through the same function f_m:

$$x''_m = f_m(x_1, \cdots, x_n; c_1, \cdots, c_r).$$

This is not the usual case but a very special one. If this does occur, the f_j's satisfy one of the cardinal properties of a group, namely, that the set of all such transformations is closed. The theory of transformation groups is concerned solely with sets having the property of closure. Because the f_l are continuous in the x_j and a_s, the group is said to be a finite continuous r-parameter group.

As an example, consider the one-parameter group of transformations $x_1' = f_1(x_1, x_2; a)$, and $x_2' = f_2(x_1, x_2; a)$. It is assumed that there is an identity corresponding to a value of the parameter a given by a_0. Then $x_1 = f_1(x_1, x_2; a_0)$, and $x_2 = f_2(x_1, x_2; a_0)$. An infinitesimal displacement Δa from a_0 gives $x_l' = f_l(x_1, x_2; a_0 + \Delta a)$.

By expanding the functions f_m in a Taylor's series about a_0, a "linear infinitesimal transformation" can be constructed by neglecting all terms of order $(\Delta a)^2$ and higher; then

$$x_l' \simeq f_l(x_1, x_2; a_0) + \left[\frac{\partial f_l}{\partial a}\right]_{a=a_0} \Delta a,$$

and the displacement in x_l is

$$\Delta x_l = x_l' - x_l = \left[\frac{\partial f_l}{\partial a}\right]_{a=a_0} \Delta a = g_l(x_1, x_2; a_0)\, \Delta a,$$

where g_l is the first derivative of f_l evaluated at a_0. Unless g_l is 0 or infinite, this expression defines the linear infinitesimal contact transformation. Lie also showed the converse theorem that every infinitesimal contact transformation defines a one-parameter group. Lie's theory of transformation groups had immediate application to algebraic invariants. Further, it incorporated the broad concepts of groups into the theory of ordinary and partial differential equations. In physics these techniques had an important impact upon analytic mechanics, particularly in the realm of the Hamilton-Jacobi theory and including the powerful methods of action-angle variables. When represented in a Euclidean space, a rigid body retains its shape and size under transformations, and all possible motions generate a continuous group. Lie determined all of the subgroups of Euclidean and non-Euclidean geometry.

Beginning with Weierstrass and Dedekind, arithmetic and the theory of numbers reached a high point in the last half of the nineteenth century. Georg Cantor (1845–1918) may be regarded as one of the founders of the modern approach to the mathematically infinite. As early as 1638 Galileo had observed that there is a one-to-one correspondence between the set of positive integers and the set of the squares of the integers, while the distinction between denumerable and nondenumerable classes was first recognized by N. Bolzano (1781–1848) in 1840 and by Cantor in 1878. In the earlier work of Bode, sets of elements, as opposed to the individual elements, were shown to be of interest, and with the work of Cantor the theory of classes and sets was to become of fundamental importance. In Cantor's theory there is a strict distinction between cardinal and ordinal numbers. For instance, finite classes have the same cardinal number if and only if they are similar. The number 2, 5, or N, denoting the cardinal number of a class, is a mere tag characteristic of the class without reference to the order in which its elements are arranged. The cardinal number of a class finite or infinite can itself be a class. For cardinal numbers Cantor proved that the class of algebraic numbers is denumerable, and he gave a procedure for constructing an infinite nondenumerable class of real numbers. As a simple example, one can show that the rational numbers correspond in a bi-unique manner with the rational points on a line and consequently are denumerable. To demonstrate this,* the rational numbers may be displayed in a plane array, giving the integers n along the first horizontal row and the rational fractions $1/n$ in the appropriate row. The rational fractions can then be connected by the broken line, as shown: i.e., all are represented on a line. In this sequence all numbers a/b are canceled for which a and b have a common factor, so that each rational number r appears but once.

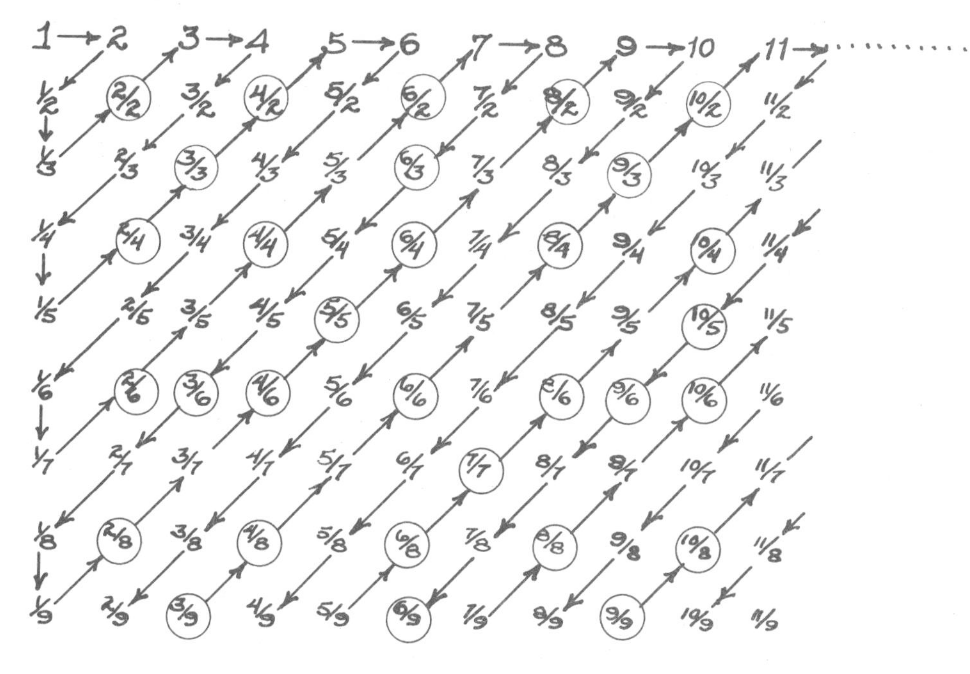

Cantor's proof, that the set of real numbers is uncountable, relies upon the assumption that if the set is countable we can list the numbers in a given base representation. Suppose that the set of real numbers between 0 and 1 is countable, say in the form $n_1, n_2, n_3, n_4, \cdots$; then these numbers can be written in decimal form, avoiding terminating decimals by employing an infinite periodicity of 9s. For example, 0.5000 is written as the periodic decimal 0.4999. . .; then the list of numbers can be tabulated in terms of the coefficient a_{ij}, where a_{ij} refers to the ith number and is the coefficient of 10^{-j}. Therefore

* See Courant and Robbins, *What Is Mathematics?*, p. 80.

$$
\begin{aligned}
n_1 &= 0.a_{11}a_{12}a_{13}a_{14}\cdots.\\
n_2 &= 0.a_{21}a_{22}a_{23}a_{24}\cdots.\\
n_3 &= 0.a_{31}a_{32}a_{33}a_{34}\cdots.\\
&\vdots\\
n_k &= 0.a_{k1}a_{k2}a_{k3}a_{k4}\cdots.^*
\end{aligned}
$$

Because of the base, the coefficients a_{lm} lie between 0 and 9. If this is a complete set of the real numbers, there should be no number excluded from the set. The proof of nondenumerability is simple if we can construct at least one number outside the set.

Let β_1 be any digit between 1 and 9, except that it must be different from a_{11}. Similarly, let β_2 be any nonzero digit other than a_{22}. In general, we take β_k as any nonzero digit other than a_{kk}. We now construct the number β where $\beta = 0_0\beta_1\beta_2\beta_3\beta_4\cdots$. Under the conditions above, β is different from every number n_k because they differ at least in the diagonal position. Obviously, β is a real number, providing a contradiction to the statement that we can list or count all of the real numbers.

This proof of nondenumerability is useful in showing that the set of transcendental numbers is uncountable. To accomplish this, we note that the set of algebraic numbers is countable. The proof of countability of the algebraic numbers relies on the fact that the set of all algebraic equations with integral coefficients is countable, and that with an equation the roots are countable. Granting this, because the set of real numbers is composed of the two independent subsets of algebraic and transcendental numbers, the fact that the subtraction of a countable set (the algebraic) from an uncountable set leaves an uncountable set, the transcendentals. Other interesting aspects of countability can be demonstrated by showing that there are the same number of points on a line of unit length as on a line of infinite length, reminiscent of Galileo's cone and bowl. This is demonstrated by projecting one line onto the other, with rays extending through a point inside the unit line when it is bent into a curve.

A more intuitive demonstration of nondenumerability can be achieved by assuming that all of the points on a unit line $(0 \rightarrow 1)$ can be arranged in a countable sequence $a_1, a_2, a_3, \cdots$. Now, enclose the coordinate a_1 in an interval of length $1/10$, a_2 in an interval $(1/10)^2$, and a_n in an interval of length $(1/10)^n$. If all points on the unit line lie in the sequence a_k, then the

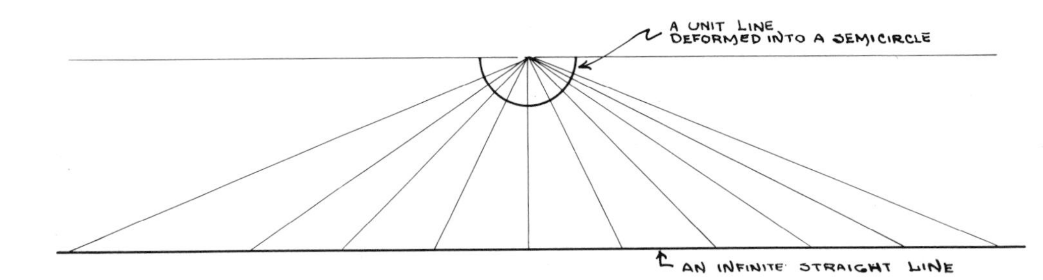

* Neven, *Numbers Rational and Irrational*, p. 127.

interval should be spanned or exceeded by the overlapping subintervals $1/10$, $(1/10)^2, \cdots, (1/10)^n$, etc. On the other hand, the sum of these lengths over a countable set may not be* 1, demonstrating a contradiction:

$$\sum_{n=1}^{\infty} \left(\frac{1}{10}\right)^n = \frac{1}{10} + \frac{1}{100} + \cdots + \left(\frac{1}{10}\right)^k + \cdots = \frac{1/10}{1 - 1/10} = \frac{1}{9}.$$

In the equation above we have used the truncated geometric series;

$$\sum_{n=0}^{\infty} x^n = \frac{1}{1-x}, \qquad (x < 1);$$

and therefore

$$\sum_{n=1}^{\infty} x^n = \frac{x}{1-x}.$$

With Boole, Cantor is considered to be one of the major influences in the establishment of set theory. The concept of a set or a class of objects is one of the most basic in modern mathematics. A set is defined by any property or attribute which each object considered must or must not possess. Those possessing the property form a set. Of the integers, if we consider the property of being prime, the corresponding set A is the set of all primes, $2, 3, 5, 7, \cdots$, etc. The algebra of sets comprises a study of the operations by which two or more sets are combined to form other sets. The algebra of numbers has many formal similarities but also many differences from the algebra of sets. In recent times, set theory has been found basic to measure theory and probability theory and is useful in all branches in reducing concepts to a logical base. To state as briefly as possible some of the most elementary ideas, consider the fixed set of objects I of any nature, called the universal set or set of discourse. Within I include I itself and the empty set $\bigcirc$, which contains no elements. The symbol for containment is $\subset$ or $\supset$. Thus, $A \subset B$ means A is a subset of B, or B contains A. If $A \subset B$, and $B \subset A$, the sets A and B are said to be equal, and $A = B$. Two operations are then defined, $+$ and $\cdot$. By the "union" or "logical sum" of A and B one means the set which contains objects in A or in B, or in both A and B. This written $A + B$. The intersection or logical product $A \cdot B$ refers to the set of objects which are common to both A and B.

The compliment of a set A in I consists of all objects in I which are not in A. This complimentary set is designated as A'.

The 26 basic laws of set theory are, for the most part, apparent. They are:

1. $A \subset A$
2. If $A \subset B$ and $B \subset A$, then $A = B$
3. If $A \subset B$ and $B \subset C$, then $A \subset C$
4. $\bigcirc \subset A$ for any set A

* Notice that if the basic interval is greater than or equal to 1/2 the unit line is spanned. This, however, does not contradict the argument presented because if the denumerable infinite set of points spans the line, a denumerable infinite set of non-zero segments must also span the line.

5. $A \subset I$
6. $A + B = B + A$ (commutative law for the union of two sets)
7. $A \cdot B = B \cdot A$ (commutative law for the intersection of two sets)
8. $A + (B + C) = (A + B) + C$ (associative law for the union of three or more sets)
9. $A \cdot (B \cdot C) = (A \cdot B) \cdot C$ (associative law for the union of three or more sets)
10. $A + A = A$
11. $A \cdot A = A$
12. $A \cdot (B + C) = A \cdot B + A \cdot C$ (distributive law)
13. $A + (B \cdot C) = (A + B) \cdot (A + C)$
14. $A + \bigcirc = A$
15. $A \cdot I = A$
16. $A + I = I$
17. $A \cdot \bigcirc = \bigcirc$
18. $A \subset B$ is equivalent to $A + B = B$ or $A \cdot B = A$
19. $A + A' = I$
20. $A \cdot A' = \bigcirc$
21. $\bigcirc' = I$
22. $I' = \bigcirc$
23. $A'' = A$
24. $A \subset B$ is equivalent to $B' \subset A'$
25. $(A + B)' = A' \cdot B'$
26. $(A \cdot B)' = A' + B'$

A remarkable symmetry is apparent in that the interchange of $\subset$ and $\supset$, $\bigcirc$ and I, and $+$ and $\cdot$ in any of the laws again results in one of the twenty-six. To see this, consider law 14, $A + \bigcirc = A$ By changing $\bigcirc$ to I and $+$ to $\cdot$, we obtain $A \cdot I = A$, which is law 15.

The algebra of sets can be translated to the Boolean algebra of logic by the following rhetorical substitutions:*

$A + B \rightarrow$ either A or B
$A \cdot B \rightarrow$ both A and B
$A' \rightarrow$ not A
$(A + B)'$ and $A' \cdot B' \rightarrow$ neither A nor B
$(A \cdot B)'$ and $A' + B' \rightarrow$ not both A and B
$A \subset B \rightarrow$ "all A are B," or "if A then B," or "A implies B"
$A \cdot B \neq \bigcirc \rightarrow$ some A are B
$A \cdot B = \bigcirc \rightarrow$ no A are B
$A \cdot B' \neq \bigcirc \rightarrow$ some A are not B
$A = \bigcirc \rightarrow$ there are no A

Much of the abstract symbolism of this aspect of mathematical thought is applicable to other areas. Cantor's theory of point sets had a marked influence upon the developments in topology which followed. Group theory in many of its applications achieves a combinational aspect. G. Frobenius (1848–1912) invented the algorithm of group characters and the algebra of groups in the nineteenth century. The work of Frobenius

* Courant and Robbins, *What Is Mathematics?*, p. 113.

covers differential equations, hypercomplex number systems, and the theory of group algebras; in these he showed a detailed technical interest in problem-solving, an approach which has been lost in the philosophical abstractions of modern mathematics.

In physics the quiet complacency of a determined world, wherein all things were understood in principle and the future state of systems was thought to depend entirely upon a knowledge of the kinematic parameters at some earlier time was soon to be confronted with revelations of microscopic particles which would attract and confound the concentrated efforts of generations to come. Following Maxwell, the world of physics for the most part refined the equations of electromagnetic theory, influenced to a great extent by the conviction that electromagnetic radiations required a supporting aether to convey radiation from one point to another. In 1889 Oliver Heaviside demonstrated that the force on a point charge q moving with a velocity $\vec{\mathbf{v}}$ in a magnetic field $\vec{\mathbf{B}}$ was given by the vector equation $\vec{\mathbf{F}} = q\vec{\mathbf{v}} \times \vec{\mathbf{B}}$. Detailed examination of this equation reveals that it is nothing more than a form of Ampère's original force equation.

Henry Augustus Rowland (1848–1901) had demonstrated in 1876 that the moving static charge was equivalent to a current. With this, in practice one can convert Ampère's law for the force between current elements to a law for force between a moving charge and the magnetic field of a current element. The skin effect characteristic of the propagation of electromagnetic radiation in conductors was derived by Heaviside in 1885. This discovery was virtually a statement that a perfect conductor is impenetrable to magnetic lines of force.

Maxwell's postulate of radiation pressure was further strengthened by the construction of a general theorem on the transfer of energy in an electromagnetic field, discovered in 1884 by J. H. Poynting and independently by Heaviside. This theorem is known as the power balance equation, and it relates the field terms to the power developed in the sources of the field. Regard a volume V containing all currents $\vec{\mathbf{J}}$, surrounded by a closed surface S; the total power developed by generators or batteries delivering a generator electric field $\vec{\mathscr{E}}_B$ is

$$\underbrace{\iiint_V \vec{\mathscr{E}}_B \cdot \vec{\mathbf{J}}\, dv}_{\text{power generated in } V} = \underbrace{\frac{1}{\sigma}\iiint_V \vec{\mathbf{J}}^2\, dv}_{\text{resistive losses in } V} + \underbrace{\iiint_V \frac{\partial u}{\partial t}\, dv}_{\text{power stored in } V} + \underbrace{\iiint_V \operatorname{div}(\vec{\mathscr{E}} \times \vec{\mathbf{H}})\, dv.}_{\text{power radiated across the the surface of } V}$$

In the expression above, the generator power loss is equal to the resistive losses $(1/\sigma)\mathbf{J}^2$ (σ is the conductivity), the power stored in the field ($u = (1/2)(\vec{\mathscr{E}} \cdot \vec{\mathbf{D}} + \vec{\mathbf{B}} \cdot \vec{\mathbf{H}})$ = the energy density), plus the radiative escape across the surface S of the volume V. The quantity $\vec{\mathscr{E}} \times \vec{\mathbf{H}}$ is associated with the momentum of the electromagnetic radiation field and is called the Poynting vector. Actually, the existence of $\vec{\mathscr{E}} \times \vec{\mathbf{H}}$, is not sufficient to account for radiation losses; the quantity, divergence $(\vec{\mathscr{E}} \times \vec{\mathbf{H}})$, must be nonvanishing if a radiation field is present; this qualification rules out quasistatic artifices such as the superposition of two independent field vectors.

Heinrich Hertz (1856–1894) attracted the attention of Helmholtz in 1880 in Berlin, and, after serving as an assistant to the now famous Helmholtz, he began researches in electromagnetic theory. These culminated in his famous experiments which demonstrated the existence of radiation from macroscopic electric circuits. Hertz's theoretical investigations centered about attempts to unify electric and magnetic fields through a unique choice of potentials. His results were ingeniously wrought; however, they did nothing more than produce the equivalent of Maxwell's equations. After attempting to justify theoretically the basic equations of electromagnetism, Hertz turned his attention to the possibility of verifying them experimentally. He began with an intuitive belief that electrical oscillations would produce radiation and in 1886 achieved an effect. Ultimately he was able to produce a spark in an open gap in a secondary coil, stimulated by a primary excitation sufficiently far removed to disallow ordinary induction. By 1888 he had shown that the electromagnetic propagation proceeded at a finite velocity, and that these waves could be reflected to produce standing waves and interference.

Not all attempts to account for the theory of the electromagnetic field by the motion of ponderable charged bodies had been altogether successful. Various theories to describe the aberration of light had been based on assumptions concerning the relative velocity of the earth's surface and the velocity of the aether at the surface. Stokes in 1845 presumed that the aether near the earth is at rest relative to the surface of the earth. Such a configuration would imply that a star would appear to be displaced toward the direction in which the surface of the earth is moving.

H. A. Lorentz (1853–1928) indicated that Stokes's hypothesis would imply a nonphysical fluid motion for the aether. Since the time of Newton, the world of physics had been constantly preoccupied with the paradoxes presented by the assumption of an aether, and both Maxwell and Lord Kelvin hoped to establish a reliable mechanical model of the medium. Vortex models proved a very attractive representation of microscopic matter to Helmholtz in 1858 and to Maxwell in 1861. Helmholtz was able to show that the vortex rings of a perfect fluid possess permanent individuality, showing mechanical attraction and repulsion. Kelvin constructed a vortex-atom model of ponderable matter in 1867, and this charming but untenable theory held his attention until the end of his life.

Like Weber, Clausius, and Riemann, Lorentz adopted an electron theory of matter; this is to say that all electromagnetic phenomena were ascribed to moving charged particles, and that the interaction of electromagnetic waves with these particles was believed to be the mechanism by which the waves interacted with ponderable matter. The atomic properties of matter were by this time apparent from studies of conduction in liquids and in gases via the discharge tube. Hertz's experiments strongly indicated a source of the field, which consisted of vibrating and consequently accelerating charged particles—the size of these atoms or particles could be estimated from Avogadro's number—the number of atoms in a gram atomic weight of a substance.

Assuming the existence of an aether, Lorentz employed a description which was a combination of Clausius' theory of electricity and Maxwell's concept of the aether, using four aethereal equations together with the equation for the force on a charged particle. This is known today as the Lorentz force, but it is, in fact, the culmination of the efforts of Weber, Riemann, Clausius, and Heaviside.

The total force on a charged particle q moving with velocity $\vec{\mathbf{v}}$ in a region containing both an electric field $\vec{\mathscr{E}}$ and a magnetic field $\vec{\mathbf{B}}$ is $\vec{\mathbf{F}} = q(\vec{\mathscr{E}} + \vec{\mathbf{v}} \times \vec{\mathbf{B}})$, where the product between $\vec{\mathbf{v}}$ and $\vec{\mathbf{B}}$ is the vector product, as defined by Gibbs. By utilizing the picture of matter in which charged particles determine the electrical characteristics of dielectrics and magnetic media Lorentz derived an expression for the dielectric constant K under the action of an electromagnetic excitation having a radial frequency of ω ($\omega = 2\pi$ times the frequency), as

$$K = 1 + \frac{e^2N/m\epsilon_0}{(\omega_0^2 - \omega^2)} \cdot$$

This form assumes a linear restoring force in the atom, acting upon the charged particles e which are being accelerated simultaneously by the external field. If damping (a force proportional to the velocity) is also included, a very good approximation to anomalous dispersion is obtained, exhibiting absorption of the light as well as refraction. Lorentz's equation incorporates the number of atoms per unit volume N and the characteristic response frequency of the material, ω_0. Although the detailed picture of the atom was only achieved early in the twentieth century, it is apparent that many of the suggestive concepts had been worked out in the second half of the nineteenth century.

The aether view of the vacuum was found to be seriously in error by Albert Abraham Michelson (1852–1931), who began these experiments in

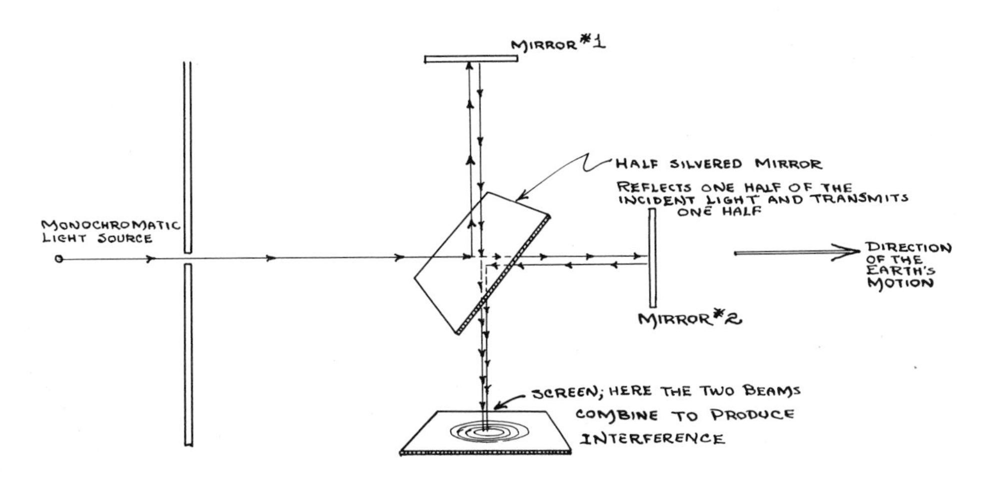

1881. It was believed that the motion of the aether at the earth's surface should be manifested by a measurable difference in the time of flight of rays describing equal paths parallel and perpendicular to the earth's surface. Michelson performed this experiment with the interferometer named after him. A sketch of the basic plan is shown above.

Let the distance between the half-silvered reflector and the mirror perpendicular to the motion of the earth's surface be d_1. The distance between the reflector and the mirror in the direction of the earth's motion is d_2. Light coming from the monochromatic (one frequency ω_0) source S is split at the reflector, part being reflected along d_1 and the remainder transmitted along d_2. The wave trains initially are in phase, coming from a single source.

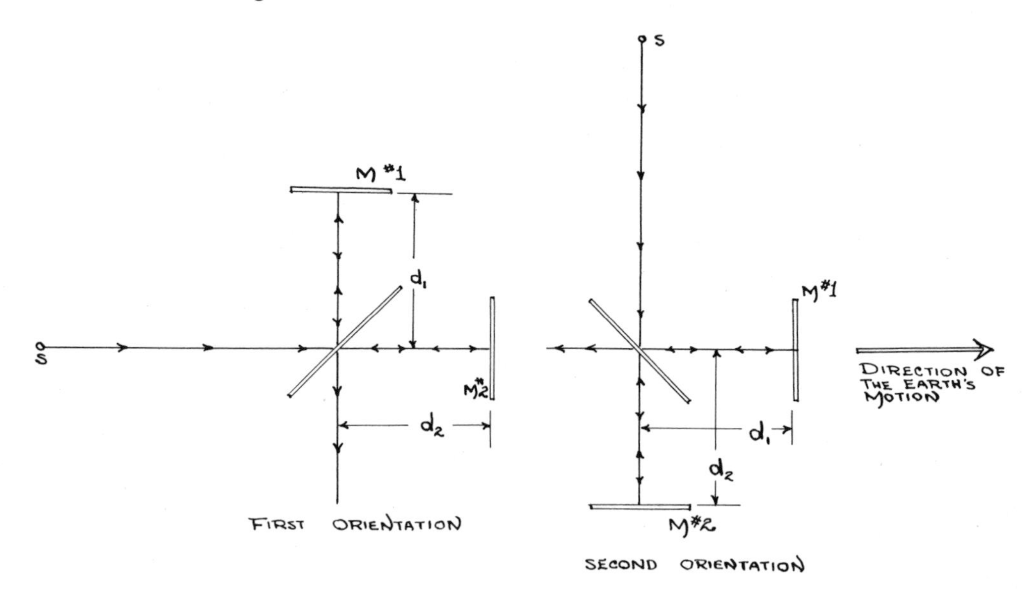

Because mirror 1 is traveling relative to the aether, in a time t_1 the total distance which wave 1 travels through the aether is given by the two legs of the isoceles triangle of base Vt_1 and of altitude d_1 [i.e., the distance traveled by wave 1 is $2\sqrt{d_1^2 + (Vt_1)^2}$].* In the stationary aether the velocity of light is c; thus the total time for wave 1 to move to mirror 1 and to return to the observation plane is

$$t_1 = \frac{2}{c}\sqrt{d_1^2 + (Vt_1)^2},$$

or

$$t_1 = \frac{2d_1}{c\sqrt{1 - V^2/c^2}}.$$

* This experiment can also be analyzed from the point of view of Galilean relativity. If the vector velocity of the light is compounded from the velocity c and the velocity of the earth V, the resultant phase shift after reflection and recombination is also proportional to V^2/c^2. The net phase shift between orientations is just twice the amount calculated in the following presentation.

Wave 2 which is emitted in the direction of motion (at the same time as wave 1) moves at a fixed velocity c in the direction of the earth's motion. Because the observation point (or origin) and mirror are moving first parallel and then antiparallel to wave 2, the wave takes $(1/2)t_2 + \Delta$ to reach mirror 2 and $(1/2)t_2 - \Delta$ to return to the origin, giving a total time of transit t_2. The first distance of travel for wave 2 is

$$c[(1/2)t_2 + \Delta] = d_2 + V[(1/2)t_2 + \Delta],$$

and the return distance is

$$c[(1/2)t_2 - \Delta] = c[(1/2)t_2 + \Delta] - Vt_2.$$

Solving for Δ in the last equation and substituting into the preceding equation, one finds that

$$t_2 = \frac{2d_2}{c(1 - V^2/c^2)}.$$

This last result is the same as that obtained by assuming that the light travels to a stationary mirror with a velocity $c + V$ and then upon reflection with a velocity $c - V$.

The time difference, $(t_2 - t_1)$, between the two waves initially in phase arriving at the origin introduces a phase shift between the recombined wave trains of

$$\phi = \omega_0(t_2 - t_1) = \frac{2\omega_0}{c}\left(\frac{d_2}{1 - \beta^2} - \frac{d_1}{\sqrt{1 - \beta^2}}\right),$$

where $\beta = V/c$.

By rotating the apparatus through 90° so that d_1 is now parallel to the earth's motion and d_2 is perpendicular, a second time difference $(t_2' - t_1')$ and phase shift ϕ' are produced:

$$t_1' = \frac{2d_1}{c(1 - \beta^2)},$$

$$t_2' = \frac{2d_2}{c\sqrt{1 - \beta^2}},$$

and

$$\phi' = \omega_0(t_2' - t_1') = \frac{2\omega_0}{c}\left[\frac{d_2}{\sqrt{1 - \beta^2}} - \frac{d_1}{(1 - \beta^2)}\right].$$

The measurement produced by the apparatus is the net phase shift $\Delta\phi$ between the first orientation and the second:

$$\Delta\phi = \phi - \phi' = \frac{2\omega_0(d_1 + d_2)}{c(1 - \beta^2)}\{1 - \sqrt{1 - \beta^2}\}.$$

If β^2 is much less than unity, the term $\sqrt{1 - \beta^2}$ can be expanded to first order as $1 - (1/2)\beta^2$, giving

$$\Delta\phi \simeq \frac{\omega_0(d_1 + d_2)}{c(1 - \beta^2)}\beta^2.$$

This numerical result suggests that there will be a second-order phase shift between the two orientations of the device proportional to V^2/c^2, where V is the relative velocity between the earth and the aether. The shift in phase shows up as a shift in the interference pattern at the exit aperture of the instrument. Michelson's apparatus was sufficiently sensitive to detect such an effect; the ratio was expected to be of the order of 10^{-4}.

Michelson's experiment showed that there was no phase shift between the two wave trains of the order predicted. The concept of space contraction, which was to be an interesting paradox of the special theory of relativity, was suggested by G. F. Fitzgerald in 1892; Fitzgerald's hypothesis assumed that the dimensions of material bodies are slightly altered when they are in motion relative to the aether. Using our notation, then, his postulate claims that for motion anti-parallel to V, in the first orientation d_2 actually becomes d_2', a distance smaller than d_2. Thus, t_2 becomes sufficiently small to equal t_1 in second order. Using a contraction of d_2 for only one-half the path gives a contraction $d_2(1 - V^2/2c^2)$. As we shall observe in the theory of relativity, this is merely the first-order expansion of $d_2\sqrt{1 - V^2/c^2}$.

When Lorentz set about to develop a theory of electromagnetism in moving media, he was presented with the basic failure of classical transformations. The transformation of variables from one frame of reference to another, moving at constant relative velocity, was ordinarily assumed to be Galilean; that is, if a particle has a position vector $\vec{\mathbf{r}}$ in frame $\bigcirc$, and a position vector $\vec{\mathbf{r}}'$ in frame $\bigcirc'$, the relation between the time derivatives of $\vec{\mathbf{r}}$ and $\vec{\mathbf{r}}'$ were linear and contained the relative velocity $\vec{\mathbf{V}} = d\vec{\mathbf{R}}/dt$ between the two frames $\bigcirc$ and $\bigcirc'$. In the diagram $\vec{\mathbf{R}}$ is the position vector locating $\bigcirc'$ relative to $\bigcirc$, and with these definitions the linear vector relationship holds with $\vec{\mathbf{r}}' = \vec{\mathbf{r}} - \vec{\mathbf{R}}$.

Upon differentiation with respect to time the linear velocity transformation is produced:

$$\vec{\mathbf{v}}' = \frac{d\vec{\mathbf{r}}'}{dt} = \frac{d\vec{\mathbf{r}}}{dt} - \frac{d\vec{\mathbf{R}}}{dt} = \vec{\mathbf{v}} - \vec{\mathbf{V}},$$

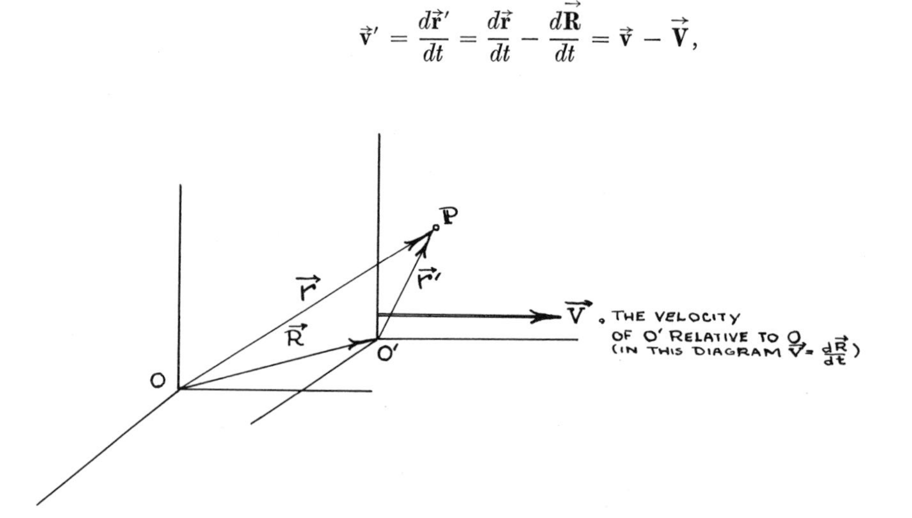

or $\vec{v}' = \vec{v} - \vec{V}$. This Galilean transformation does not affect Newton's laws because $\vec{V}$ is a constant vector. Differentiating again to relate accelerations, $\vec{a} = d\vec{v}/dt$ with $\vec{a}' = d\vec{v}'/dt$ shows that $\vec{a}' = \vec{a}$ when $\vec{V}$ is constant. Because the Maxwell equations contain second derivatives in the space coordinates, these equations do not transform (under the Galilean transformation) covariantly. This means that the equations in the new coordinates x', y', and z' do not have the same form as the equations in the original coordinates. If the motion of ○′ relative to ○ is along the z axis, the four Maxwell equations plus the expression for the Lorentz force do not retain their form: we relate x', y', and z' by $x' = x$, and $y' = y$, and $z' = z - V_z t$. Lorentz was able to achieve covariance below second order by treating time as a fourth variable. His earliest attempt involved the introduction of a local time, t', for the coordinates of ○′, where $t' = t - (zV_z/c^2)$. Such an assumption gave the five equations expressed in x', y', z', and t' in the same form as the initial set of equations. The results were only in agreement, however, when all terms of the order V^2/c^2, $(dz/dt)(V/c^2)$, or higher were neglected. From this approximate success, it became obvious that a more general transformation must be sought which would provide covariance to all orders in V/c. A further clue to the form which the new transformation would take was apparent; it was essential to view the time variable as a fourth dimension, on an equal basis with the three space coordinates. In 1903 Lorentz obtained the appropriate transformation. When only spatial coordinates are transformed, one implies the invariance condition on an element of length ds such that

$$(ds)^2 = (dx)^2 + (dy)^2 + (dz)^2$$

after transformation becomes

$$(ds)^2 = (dx')^2 + (dy')^2 + (dz')^2.$$

If the time variable were treated equally with the spatial coordinates, it would be necessary to impose an invariance condition on the transformation of the displacement between events (events are points in space

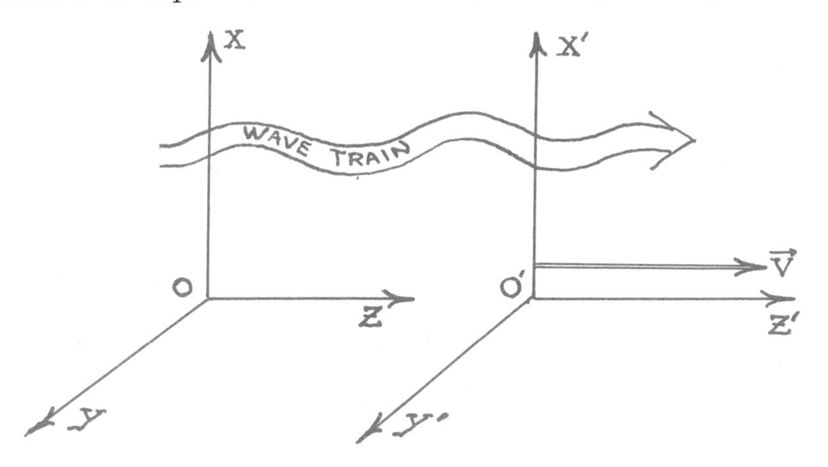

and time). By analogy with a point in space, an event is a point in space time with four coordinates (x, y, z, t) or (x', y', z', t'); therefore the invariance condition requires that the four-dimensional displacement ds transform invariantly, with the condition that

$$(ds)^2 = (dx)^2 + (dy)^2 + (dz)^2 - c^2(dt)^2$$

be equal to

$$(dx')^2 + (dy')^2 + (dz')^2 - c^2(dt')^2.$$

Modern notation is similar to the Grassmann notation. Let

$$\begin{aligned} x &= x_1 \qquad & x' &= x_1' \\ y &= x_2 \qquad & y' &= x_2' \\ z &= x_3 \qquad & z' &= x_3', \end{aligned}$$

and

$$ict = x_0 \qquad ict' = x_0'.$$

Then our expression for the invariance may be written in a Pythagorean form:

$$\begin{aligned} (ds)^2 &= (dx_0)^2 + (dx_1)^2 + (dx_2)^2 + (dx_3)^2 \\ &= (dx_0')^2 + (dx_1')^2 + (dx_2')^2 + (dx_3')^2. \end{aligned}$$

The solution for the appropriate expressions linking the primed and the unprimed variables (the transformations) was first found by Woldemar Voigt in 1887. To maintain this Pythagorean invariance, Voigt chose his transformations in the form:

$$\begin{aligned} x' &= \gamma(x - V_x t), \\ y' &= y, \\ z' &= z, \end{aligned}$$

and

$$t' = \gamma\left(t - \frac{xV_x}{c^2}\right);$$

here the relative motion is assumed to be parallel to the x and x' axes. Taking the differentials, and substituting for the dx_n' in the expression for ds, one can evaluate the parameter, γ, which turns to be a constant given by:

$$\gamma = \frac{1}{\sqrt{1 - V_x^2/c^2}} = \frac{1}{\sqrt{1 - \beta^2}}.$$

This result can be achieved much more cleverly with matrices by searching for the 4×4 unitary (magnitude equal to plus 1) matrix which transforms invariantly the four-dimensional vector in x_n to the four-dimensional vector in x_j'. Such an approach is more elegant and illustrates the underlying structure of the transformation; however, the results are the same as those obtained by straightforward algebraic substitution. One must immediately be impressed by the appearance of $(ds)^2$. Riemann's foresight is more clearly apparent in the expression being transformed—it is a four-dimensional manifold of fundamental importance in physics.

Lorentz arrived at this transformation, which is named for him, in 1909; in his manuscript he acknowledged Voigt's paper of 1887 and remarked that he, in fact, had not known about Voigt's work in deriving the same result. Henri Poincaré (1854–1912) gave this set of transformations the name "Lorentz transformations" in 1905. Although this work makes it quite obvious that the laws of physics must include these new transformations, it was not immediately apparent how they could be incorporated into mechanics. Albert Einstein (1879–1955) integrated the concepts of relativity with our intuitive understanding of dynamics.

Poincaré was one of the last of the great physicist-mathematicians. His interests ranged from number theory and groups through the theories of electromagnetism and relativity. Through his many contributions in mathematics he gave an air of confidence to that field. In 1900 he made the claim that obscurity had been dispelled from continuum analysis by the contributions of number theory, together with the concepts of infinite classes.

The modern theory of difference equations began with a memoir of Poincaré in 1885; here he considered the general linear difference operator

$$\sum_{j=0}^{N} a_j(x)u(x+j) = b(x),$$

where $a_j(x)$ and $b(x)$ were given functions and $u(x)$ was to be found. This general form contained most of the earlier problems in difference equations as special cases. Poincaré's primary objective became the qualitative or topological study of all solutions of a system and the relations between them.

Topology, which had been latent in the earlier mathematical discoveries, is the study of those properties of space which are invariant under homeomorphic transformations (mapping). While formal topology is a creation of the last hundred years, the early discoveries of the invariances of polyhedra by Descartes and Euler were fundamental. Gauss, Moebius, Tait, Jordan, and Cayley all contributed to the subject, mainly in terms of combinatorial analysis. In 1895 Poincaré published his *Analysis situs* and with it brought forth modern topology. Along with these contributions he worked with integral equations and with the general analysis of abstract spaces.

During the middle of the nineteenth century the spirit of determinism reached a climax. It was generally believed that if the positions and velocities of a specific isolated system were completely determined at any time, the future behavior of that system would be absolutely fixed. In the field of mathematics it was speculated that, given the basic axioms of arithmetic, the resulting logical structure would be completely consistent and free of contradiction. By the turn of the century, this complacent view was crumbling, and the discoveries of the microscopic world of physics presented the scientists of the twentieth century with probabilities in the place of certainties.

IX

Gulliver's Travels

Early experiments on the conduction of electricity in solutions led to inquiry into the conduction of gases. While working with discharge tubes containing rarefied gases (gases at low pressure), Faraday (1838) noticed that the glow of the discharge stopped short of the cathode (the negative electrode). The Geissler mercury pump invented in 1855 permitted the construction of superior discharge tubes, and in 1858 Julius Plucker, using a point cathode, discovered that the arc in a discharge tube was deflected by the presence of a magnetic field. These were then called cathode rays. Sir William Crookes (1832–1919) regarded these rays as a molecular torrent, a description which seemed to be borne out by the fact that the Faraday dark space lengthened as the gas pressure was reduced. This, in turn, suggested that a basic length was involved, such as a mean free path for collisions. Cutting holes in the anode permitted the rays to escape into the region behind, allowing further experimentation. Using this technique, Sir Joseph John Thomson (1856–1940) in 1894 measured the velocity of cathode rays as the order of 10^7 cm/sec. Before cathode rays were identified, another radiation from a highly evacuated discharge tube was discovered.

On November 8, 1895, Wilhelm K. Roentgen (1845–1923) noticed that when a current was passed through a discharge tube, a nearby paper painted with barium platinocyanide exhibited fluorescence and that photographic plates in the same proximity became exposed or darkened. He deduced that the cathode rays striking the glass envelope produced a radiation and called them X-rays. Attempts at magnetic deflection and absorption of the rays soon convinced Sir Arthur Schuster and others that X-rays were composed of very high frequency electromagnetic radiations. A clue to the electron was uncovered by Thomson, who found that X-rays while passing through a gas render it conductive; from this he deduced that the X-radiation somehow split up the molecules of the gas or, in the language of electrolysis, "ionized" it.

In 1897 Thomson advanced a new hypothesis to reconcile the molecular torrent concept with the presence in a discharge tube of the long dark space (or mean free path) and the high velocity of cathode rays. The last two notions were untenable for molecules, and Thomson suggested that cathode rays were, in fact, charged particles much smaller than atoms or molecules. He found that these cathode rays appeared to have properties independent of the nature of the gas in the discharge tube. He then subjected isolated cathode rays (i.e., rays emanating from the cathode) to electric and magnetic fields. A charged particle moving through appropriately crossed (mutually perpendicular) fields in a direction at right angles to both can traverse the field region without deflection if the ratio of the magnitudes of the two fields are taken in such a manner as to set the Lorentz force to zero. According to the second law of mechanics, a particle of mass m and charge $-e$ moves according to

$$m \frac{d^2\vec{\mathbf{r}}}{dt^2} = m\vec{\mathbf{a}} = -e(\vec{\mathscr{E}} + \vec{\mathbf{v}} \times \vec{\mathbf{B}}),$$

when $vB = \mathscr{E}$, and when the directions are such that $\vec{\mathscr{E}}$, $\vec{\mathbf{B}}$, and $\vec{\mathbf{v}}$ form a righthanded system in that order, the total force is zero. Moreover, the relative values of the magnitudes of the fields determines the magnitude of the velocity v: $v = |\vec{\mathscr{E}}|/|\vec{\mathbf{B}}|$ for the geometry shown.

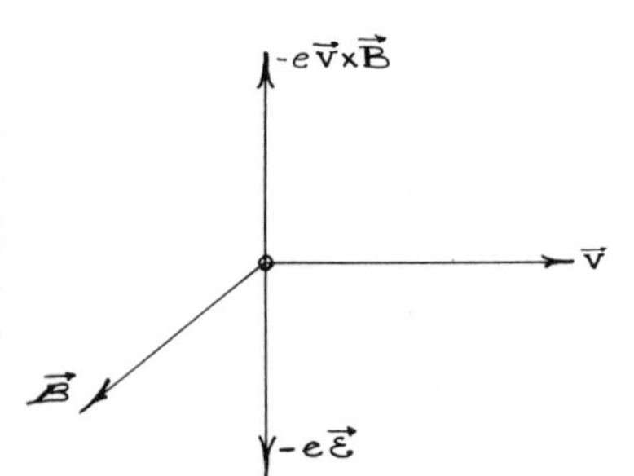

With this convenient method for determining the velocity of a charged particle, a simple magnetic or electric deflection enables the determination of the ratio of charge to mass of a moving particle. Deflection in a magnetic field alone occurs from the presence of the magnetic force $\vec{\mathbf{F}} = -e\vec{\mathbf{v}} \times \vec{\mathbf{B}}$. By setting $\vec{\mathbf{B}}$ at right angles to $\vec{\mathbf{v}}$, after determining $|\vec{\mathbf{v}}|$ from the crossed fields, Thomson found the value of m/e to be 1/1000 of the value of M/e for the hydrogen atom in electrolysis. At the British Physical Association meeting in 1899 Thomson read a paper entitled "On the Existence of Masses Smaller Than the Atoms." This focus on the mass of the electron was brought about by Thomson's conviction, supported by studies of the charges of heavy ions, that the charge of the negative particles was equal to that of the ions. At first, Thomson's ideas were met with some skepticism; however, within a few years the existence of the electron (named by Stoney) was a confirmed and integral part of physics.

Few of these discoveries can be classified as mathematical or theoretical; the X-ray and the electron were simply identified. As a theoretical object, the X-ray was soon adequately described as a form of high-frequency electromagnetic radiation. The electron has remained a fundamental building block of matter; however, to this day one cannot claim that it is understood in terms of some subelectronic structure. There are expressions, such as the Dirac equation in quantum mechanics, which describe the behavior of electrons when subjected to various external stimuli, but the role that the electron plays with respect to the other elementary particles is not apparent. This is to say that all formalisms which have been created to contain the heavier elementary particles, such as the proton and neutron, in no way suggest the existence of the electron. To complicate the picture of microscopic matter further, radioactivity was discovered in 1896 by Antoine Henri Becquerel (1852–1908). While studying the phosphorescence of the double sulfate of uranium and potassium, he found that a photoplate was darkened when placed nearby. The ensuing investigations had far-reaching effects on prevailing concepts of the atom—they led to the postulate of new forces, the weak interactions and nuclear forces, and, in part, validated the theory of special relativity, in particular, the equivalence of mass and energy.

With the discoveries of the elementary particles and radioactivity, the laws of thermal radiation simultaneously led to early hints of quantization. At the beginning of the nineteenth century, Pierre Provost had noticed that a red-hot body cools in a vacuum by emitting thermal radiation. He created the concept of radiative equilibrium, wherein a number of hot bodies are placed such that they intercept the whole of each other's radiation, thereby setting up a state of fixed temperature wherein the heat lost equals the heat gained by absorption. By 1859, Kirchhoff and Balfour Stewart had proposed the rule that the radiating power of every substance is equal to its absorbing power for a fixed frequency of radiant heat. Kirchhoff formalized this to a rule that the ratio of the coefficient of emissivity to the coefficient of absorption is the same for all bodies, and is equal to the emissive power of a black body at that temperature. If one imagines an evacuated, closed cavity with walls at a uniform temperature T, the cavity will be filled with electromagnetic radiation, ranging in wave length from zero (infinite frequency) to a wave length of infinity (zero frequency). The energy of the radiation per unit volume then will be associated with a distribution function $F(\lambda, T)$ which predicts the fraction of the total energy of the radiation at an absolute temperature T with wave lengths lying in the interval between λ and $\lambda + d\lambda$.

With this description, the total energy per unit volume for all wave lengths in the cavity is just the area under the curve of $F(\lambda, T)$ versus λ. Experimental observations were made by drilling a small hole in the wall and examining the wave length distribution of the radiation which leaked out. Thus, the basic problem confronting physicists of that time was to account for the observed function $F(\lambda, T)$. Either a thermodynamic or a statistical analysis was indicated, and in 1884, through thermodynamic arguments, the problem of the total radiated energy per unit volume by a black body at a temperature T (call it u) was solved by Boltzmann.

Combining the second law of thermodynamics with that for radiation pressure, he showed that the radiation energy density, u, was proportional to the fourth power of the absolute temperature, T: $u = aT^4$, where a is a constant, giving the emissivity of the radiating body. This law has been called the Stefan-Boltzmann law, named in part after Josef Stefan, who had obtained the relation empirically.

Wilhelm Wien (1864–1928) in 1893 studied the problem of a perfectly reflecting spherical cavity which contracts in time. The total radiant energy in the sphere was found to vary in inverse proportion to the radius. He supposed that the sphere was initially filled with radiation of all wave lengths, the energy being distributed in such a manner that the radiation is in equilibrium with a black body at a temperature T. The second law of thermodynamics then predicts that as the cavity contracts the partition of the energy spectrum remains the same. Obviously, if this were not the case the shift in the spectrum would correspond to a change in the temperature associated with the spectrum. A nonequilibrium distribution would permit the operation of a heat engine which could employ bodies capable of absorbing the wave lengths corresponding to the higher temperatures, thus extracting mechanical work. It follows, then, that the function $F(\lambda, T)$ is of the form $T\lambda^{-4}\phi(\lambda T) = c\lambda^{-5}e^{-b/\lambda r}$, where ϕ is a function of the product λT only. From this form Wien discovered that the product of the wave length at the maximum of F and the corresponding T is always a constant. Wien's distribution function was found to have a limited application, in that it only predicted the high frequency or small wave length portion of the observed function.

From the kinetic theory of gases, the theorem of equipartition of energy points out that in a state of statistical equilibrium at an absolute temperature T the total energy is partitioned in such a way that on the

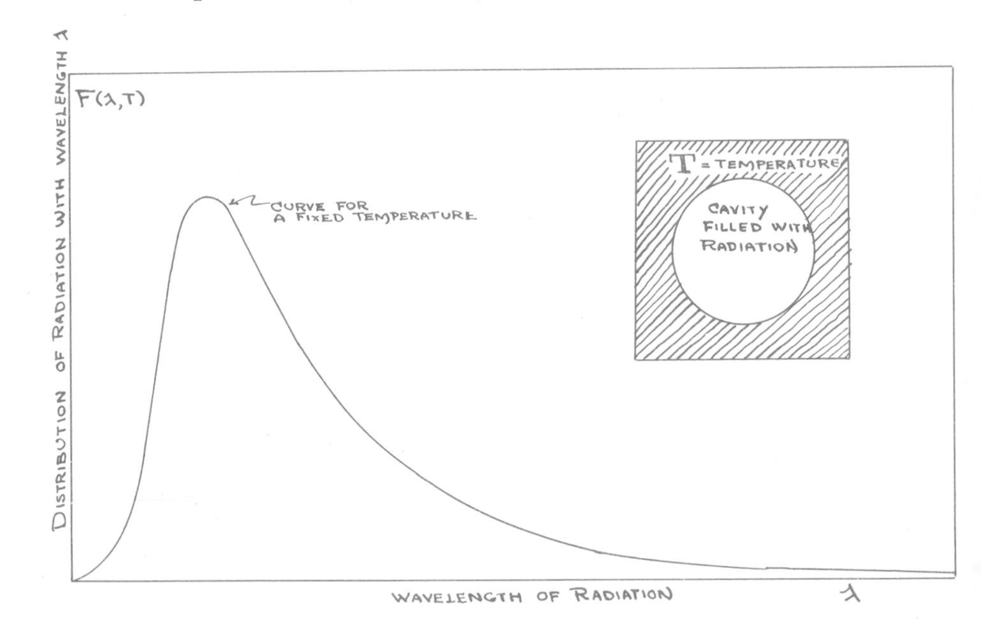

average the same kinetic energy is attributed to every degree of freedom, namely, $1/2(kT)$, where k is the Boltzmann constant (k is the ideal gas constant divided by Avogadro's number). Lord Rayleigh (1842–1919) in 1900 employed the theory of equipartition of energy to derive the energy distribution of black body radiation. He assumed that radiation in a closed cubical enclosure of dimension L could be Fourier-analyzed into standing waves of the type

$$\cos.\left(\frac{n\pi x}{L}\right) \cos\left(\frac{m\pi y}{L}\right) \cos\left(\frac{l\pi z}{L}\right),$$

where sines or cosines could be used, and where n, m, and l are integers. The arguments of the trigonometric functions, such as $n\pi x/L$, make up the components of a three-dimensional wave vector, k, which is related to the angular frequency $\omega = 2\pi\nu$ (ν is the frequency) by $kc = \omega$. This expression is the same as that for wavelength λ and frequency ν; i.e., $\lambda\nu = c$. Thus $k = 2\pi\nu/c = 2\pi/\lambda$. The magnitude of the vector k is proportional to the square root of the sum of the squares of the three components of the arguments; thus,

$$k = \frac{\pi}{L}\sqrt{n^2 + m^2 + l^2},$$

and

$$\nu = \frac{c}{\lambda} = \frac{kc}{2\pi} = \frac{c}{2L}\sqrt{n^2 + m^2 + l^2}.$$

Because n, m, and l are integers, the allowed points form a cubic array, each cell having a volume $(c/2L)^3$, and represent the allowed values of ν. The number of waves in the frequency interval $d\nu$ between ν and $\nu + d\nu$ is equal to the number of points in the first octant of a spherical shell of volume $(1/2)\pi\nu^2\,d\nu$. This differential number then turns out to be

$$dN = \frac{1}{2}\pi\left(\frac{2L}{c}\right)^3 \nu^2\,d\nu,$$

or, in terms of λ,

$$dN = 4\pi L^3\lambda^{-4}\,d\lambda.$$

Here the relation that $\nu = c/\lambda$ implies that $d\nu = -(c/\lambda^2)\,d\lambda$. Doubling this number because of the two transverse modes of electromagnetic vibration and multiplying by the average energy per degree of freedom $[1/2(kT)]$ gives $F(\lambda, T) = 8\pi kT\lambda^{-4}$. This result was known as the Rayleigh-Jeans law; it was found to be valid only for the long wave length, low frequency portion of the spectrum. The high frequencies are favored improperly because the density of states becomes a continuum for very large frequencies.

Here then was a dilemma: two different approaches to the black body radiation distribution function were only partially in agreement. Looking back, it is apparent that the failure of the Rayleigh-Jeans analysis lay in the assumption of a continuum of states in the high frequency region, and that a mechanism which reduced the density of states in the high frequency region would approach the observed distribution.

By 1900 some attempts had been made to fit the function $F(\lambda, T)$ empirically. On October 19, 1900, Max Planck (1858–1947) read a paper in which the correct function was derived by a rather obscure thermodynamic argument which, in a sense, was adjusted to conform to both the Wien and the Rayleigh-Jeans formulas. Later in the same year Planck reformulated his derivation and considered a system consisting of a large number of electromagnetic oscillators in a hollow chamber with reflecting walls. Most important in his derivation was the assumption that the energy of the vibrators was made up of small discrete elements of energy in the amount ϵ, and that the emission and absorption of radiation occurred in energy jumps rather than continuously. Assuming N vibrators at the frequency ν with P elements excited to provide the total energy of this set, Planck concluded that the number of distinguishable configurations was a binomial-type relation of the form $(N + P - 1)!$, the total number of configurations, divided by the number of undistinguishable arrangements, $(N - 1)!\,P!$. In other words, if $(N + P - 1)$ things are taken P at a time, the total number of complexions is given by the binomial or random walk coefficient,

$$\frac{(N + P - 1)!}{(N - 1)!\,P!}.$$

By finding the extreme value of this function, subject to the constraints that the total number of energy quanta ϵ and the total energy of the system be constant, the average energy of an oscillator was derived in the form,

$$\frac{\epsilon}{(e^{\epsilon/kT} - 1)}.$$

Because earlier derivations committed this expression to a form which was linear in the frequency, Planck assumed that the energy quanta, ϵ, must be given by a universal constant h times the frequency ν: $\epsilon = h\nu$. With this assumption, the function F for black body radiation was evaluated in terms of ν instead of λ as

$$F(\nu, T) = \frac{8\pi h}{c^3} \frac{\nu^3\, d\nu}{(e^{h\nu/kT} - 1)}.$$

Thus modern quantum theory was born. Accurate determination of the Wien displacement law and derivative measurements of the distribution curve gave relatively accurate values for the new constant h and the Boltzmann constant k. Using the more accurate value of k and the ideal gas constant, Planck was able to provide a much better value of Avogadro's number. By this time this number was known to be related to the Faraday in electrolysis through the charge on the electron. Two years earlier J. J. Thomson had evaluated e and obtained a number too large by 30 per cent. Planck's value of e was good to 1 part in 60.

These immediate rewards, however, gave no inkling of the profundity of Planck's discovery. Not only had the concept of discrete quantized levels of the atom been indicated, but it was also apparent that the Maxwell-Boltzmann statistics which held for a gas would not suffice to describe the statistical nature of the radiation field. The path was open for a very young man, Albert Einstein (1879–1955), to climax these discoveries with the theory of the photon and a quantum theory of specific heats.

Classical analysis was in the throes of critical reexamination. The modern giant of mathematical physics and critical mathematics David Hilbert (1862–1943), reintroduced the postulational basis of geometry in 1899. His attempts to solidify the foundations of geometry led him to a similar effort upon the structure of common arithmetic. In 1900 he observed that the outstanding challenge of mathematics was to find proof that, proceeding from the postulates of arithmetic, it is impossible to reach a contradictory result by means of a finite number of logical deductions. This attitude placed the traditional Pythagorean program of deductive reasoning on trial. Like his predecessors, Hilbert realized that one of the fundamental questions of mathematics was the significance of the concept of the infinite and its clarification. In 1890 Hilbert proposed his famous basis theorem, founded upon Leopold Kronecker's concept of modular systems. Such a system is a set M of all polynomials in the n variables $x_1, x_2, x_3, \cdots, x_n$ defined by the property that if the polynomials P_k belong to the system, then so do the linear combinations of the P_k and all polynomials QP_r, where Q is any polynomial in $x_1, x_2, \cdots, x_n$. A basis of a modular system M is any set of polynomials $B_1, B_2, \cdots$ of M such that every polynomial of M can be represented as an expansion in the B_j of the form

$$A_1B_1 + A_2B_2 + \cdots + A_jB_j + \cdots;$$

here the coefficients A_j are constants or polynomials in the n variables. Hilbert's basis theorem states that every modular system has a basis consisting of a finite set of bases. This work was severely criticized by P. Gordan (1837–1912) for its use of a nonconstructive existence proof. One can perceive immediately the extension of this concept to Hilbert's function spaces of infinite dimensions, which he laid down in 1906. The Fourier expansion is a representation in terms of an infinite set of infinite series. Modern terminology regards a Hilbert space as a set M (finite or infinite) consisting of a complete array of linearly independent base functions. This concept has interesting analogies with the base vectors of a finite vector space. The simplest examples of this are observed in the vector algebras of Gibbs. Here, as in a Fourier series, a mathematical object is expanded in terms of its projections along the basis of the system. In this respect, then, the coefficients of a Fourier series are mathematically similar to the coefficients of an ordinary vector.

Along with his notable contributions to postulational geometry, algebra, arithmetic, and number theory, Hilbert had a most decided influence upon the course of modern theoretical physics. Dirichlet's problem of demonstrating the existence of a function harmonic within a closed region and having preassigned continuous values on the boundary seemed intuitively obvious to the physicists. With suitable restrictions, Hilbert solved this problem in 1901, making it adequate for most applications. In 1904 he showed that in many significant cases a single integral equation is equivalent analytically to a differential equation together with its boundary conditions.

The history of integral equations began about 1826, with Abel's generalization of the general problem of the tautochrone. An integral equation is a form in which the dependent variable or function $u(x)$ appears at least once under an integral sign. Abel's equation is now known as a Volterra equation of the first kind. The linear integral equations are the best known and most widely used, appearing in the form

$$f(x) = g(x)u(x) + \lambda \int_a^b K(x, y)u(y)\, dy.$$

$K(x, y)$ is the kernel of the equation and is a known function of x and y, and $g(x)$ is also a known function; $f(x)$ is either a known function or, in some instances, the unknown function $u(x)$. This last alternative can be obtained also by allowing $f(x)$ to be zero and by redefining $g(x)$. The scientific significance of these forms was noticed by Liouville but exhibited more completely by Hilbert.

Appealing as these equations appear, the methods of solving them are often difficult to achieve. Although Hilbert and others created many ingenious solutions for special problems, the method of solution of the general linear equation leaves much to be desired, even today. One of the more decisive advances was made by 1903 by I. Fredholm (1866–1927), who found solutions in terms of the ratio of two-power series in the parameter λ.

Returning to our central theme, the application of mathematical methods to physics, the publication in 1924 of *Methoden der Mathematischen Physik* by Richard Courant and Hilbert proved to be an invaluable contribution to physics. Presented there were the modern methods of linear transformations, quadratic forms, expansions of arbitrary functions in a complete basis set, integral equations, partial differential equations, and the calculus of variations. This work appeared just two years before the wave mechanics was invented, and since that time more than three generations of physicists have been involved in one way or another with the ramifications of this single work.

As the leader of the formalist school of mathematics, Hilbert in his later years was deeply engaged in debate with the supporters of the newer concepts of intuitionism put forward by L. E. Brouwer (1882–). Hilbert attempted to separate the symbolic game of mathematics from its interpretation and to reformulate the rules to preclude contradictions. Brouwer's intuitionism asserted that mathematics per se is prior to logic—that there are certain questions unanswerable within the present framework. These spirited polemics continue to the present, although the discoveries of K. Gödel and P. J. Cohen have exposed the rigidity of Hilbert's position.

Problems of logic and formalism were of the least importance in the frontiers of physics between 1890 and 1920. The discovery of the electron and the X-ray were but a part of the breakdown of the older beliefs in determinism and classicism. Around 1896 Becquerel discovered radioactivity in uranium phosphate, and within two years Marja Skłodowska Curie (1867–1934) and Pierre Curie (1859–1906), searching for other substances with such properties, discovered that the compounds of thorium emitted similar radiations. More important, the Curies demonstrated that these radiations were atomic in nature, since they were unaffected by chemical changes. Within three months after they had carried out a painstaking systematic analysis of pitchblende the Curies discovered a new element, polonium, and in the same year discovered radium, which turned out to be many millions of times more radioactive than uranium. They spent the next few years obtaining the atomic weights of the new elements by a series of successive fractionations.

The nature of the rays given off by the radioactive substances was investigated by many groups. Giesel, Becquerel, and others found that one component was easily deflected in a magnetic field, while a second was not; Rutherford separated the two components by absorption; and the Curies found that the readily deflected rays carried negative charges and had a charge-to-mass ratio characteristic of the electrons of Thomson.

In 1903 Ernest Rutherford (1871–1937) succeeded in magnetically deviating two types of rays and found still a third component undeviated. The negatively charged particles were called beta rays and the positively charged heavy particles alpha rays, while the undeviated rays discovered in 1900 by Villard were later identified as electromagnetic radiations,

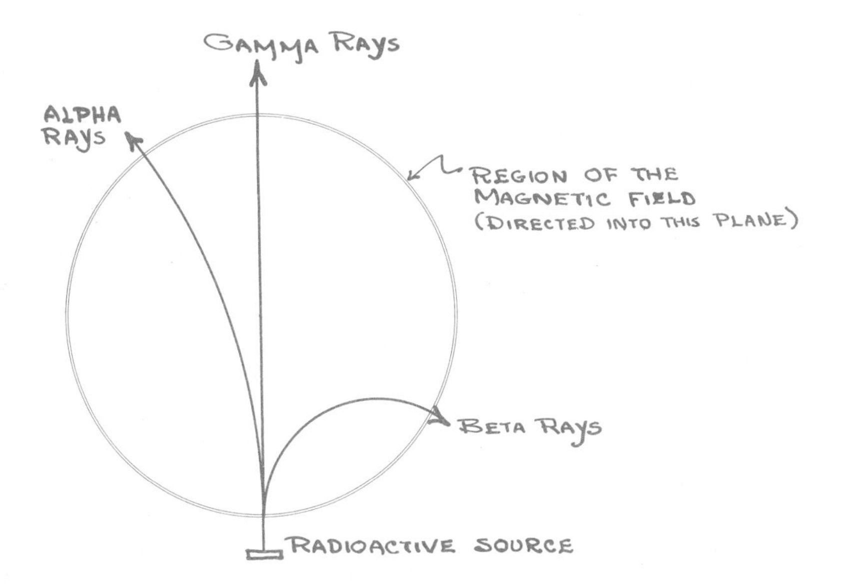

similar to X-rays but often of higher frequency: these were called gamma rays.

Earlier, Rutherford had discovered that the intensity of the radiations from thorium fell off exponentially with time. Such an effect was not surprising: it merely implied that the rate of disintegration was proportional to the number of radioactive atoms (nuclei) present. If N is the number at any time t, then $dN/dt = -\lambda N$, giving $N = N_0 e^{-\lambda t}$, where λ, the constant of proportionality, is the inverse of the mean life and N_0 is the total number of radioactive parents present at $t = 0$. As early as 1900 Sir William Crookes had discovered that the decay of uranium led to a new substance, also radioactive but with a different activity. All of the new active elements exhibited the same behavior, and Rutherford and Frederick Soddy attempted a general theory of radioactivity in 1903. They asserted that there is a continuous production of new kinds of matter; that radioactivity is a spontaneous transformation, and that the rate of disintegration was proportional to the number of atoms available. They further correctly asserted that when there are several changes in activity, these do not occur simultaneously but rather in a chain of successive radioactive types. Because they detected helium in the radioactive emanations, Rutherford, Ramsey, and Soddy initially considered that helium might be the end of the chain of successive radioactive decays. Using a double-walled tube which permitted alpha ray penetration from the inner tube containing the activity, in 1909 Rutherford finally unambiguously identified the alpha particle as a doubly ionized helium atom. Investigations on the three natural radioactive chains began in 1903, and some systemization was achieved by 1913, when A. S. Russel and Kasimir Fajans announced the displacement laws: an alpha disintegration causes a two-place displacement lower in the elements of the periodic table and represents a decrease of 4 in atomic number, and a beta decay results in a single displacement higher on the periodic table.

In 1869 Dmitri Ivanovich Mendeleev (1834–1907) first proposed that the known chemical elements could be arranged into a periodic table. From this postulate he inferred that there were elements in nature not yet discovered (scandium, gallium, and germanium). As early as 1886 Crookes suggested the concept of the isotope, namely, that for a given element there would exist a number of atoms of different atomic weights, all grouped about a central weight corresponding to that which occurs in the highest proportion in nature. All of this was highly speculative because the nuclear constituents and the model of the atom were not as yet known, but they doubtless had a profound effect on the later conceptions of atoms and nuclei. Although J. J. Thomson succeeded in separating two isotopes of neon, Ne^{20} and Ne^{22}, accurate mass spectrometry was not accomplished until 1919, when F. W. Aston performed experiments on a wide range of elements, taking oxygen 16 as a base isotope. The radioactive chains, starting with thorium 232, uranium 238, and uranium 235, were beginning to be understood in terms of successive alpha and beta transitions which carried the radioactive chains through the periodic table.

One very complex effect, called branching, was discovered independently by Rutherford and Fajans. Both B. B. Boltwood and Rutherford had suggested that lead was the end point of the radium series. As it turned out, lead 208, lead 207, and lead 206 were the residuals of the three series. The difference in branched decays occurs because in a successive set of alpha and beta decays the intermediate species may be quite different, depending upon whether the total decay begins with an alpha or a beta ray. A diagram of the three chains is shown. Here the modern atomic weights and numbers are employed.

The invention of the Geiger counter by Rutherford and Hans Geiger in 1907 represented a major advance in nuclear technology, as it allowed the detection of individual particles of disintegration. Combining this count of individual decays with measurements of the amounts of helium

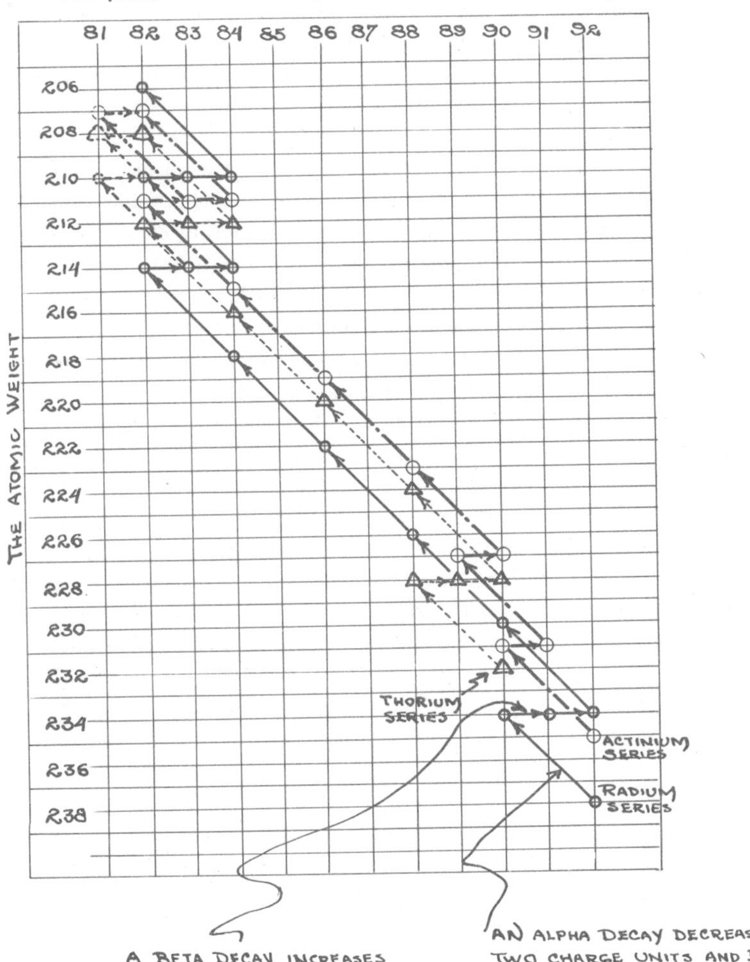

liberated by a known amount of radium, it is possible to evaluate the number of atoms in a known weight of material, thus providing an additional evaluation of Avogadro's number.

Avogadro's number, then, can be obtained in various ways. Planck obtained a value by evaluating the Boltzmann constant. An original, powerful method for obtaining the number was proposed by Albert Einstein in a paper published in 1905. It was a sequel to two earlier papers dealing with a statistical theory of heat, and it derived the basic equations representing Brownian motion. When very small particles are in liquid suspension they exhibit a random motion. In a given time t, these small random jumps of a particle culminate in a net displacement from its original position. For spherical particles, Einstein showed that the mean square displacement $(\Delta s)^2$ is $RTt/3\pi a\mu N$, where R is the ideal gas constant, T is the absolute temperature, a is the radius of the spherical particle, μ is the coefficient of viscosity, and N is Avogardo's number. This result was confirmed experimentally by a number of physicists, and the value of N which was obtained was quite accurate for that period.

The year 1905 was a singular one for theoretical physics. Within a period of six months Einstein published three major works, each of which profoundly influenced the future of physics. In December of 1906, using the quantum statistics of Planck, he proposed a quantum model for the theory of specific heats. In terms of spontaneity alone, this was a remarkable demonstration of creative thought. Even more significant is the range of topics to which Einstein applied his powers. Influential as it was, the specific heat paper of 1906 had but limited application, being based upon the assumption that all of the atoms of a crystal have the same frequency of vibration. Einstein's approach was generalized in 1912 by Peter Debye (1884–), who included all of the normal modes of vibration of a crystal. Debye's theory was, and still is, in relatively good agreement with the experimental data for many substances.

Three of these major works reveal the influence of statistical mechanics upon Einstein's thought at this time. J. J. Thomson and P. Lenard showed independently in 1894 that when the surfaces of metals were irradiated with ultraviolet light, negative electrons were emitted. By 1902 Lenard had demonstrated that the number of electrons was proportional to the intensity of the light, while the maximum energy of the liberated electrons was dependent only upon its frequency. These experimental results led Einstein to propose the photon (G. N. Lewis introduced the name "photon" in 1926), or light quantum. He initially considered monochromatic radiation of low density confined in a cavity with perfectly reflecting walls of volume V_0. Combining Planck's formula and the kinetic probability theorem, Einstein calculated the probability of finding all of the radiation in a smaller volume V; this result was proportional to $(V/V_0)^{E/h\nu}$, where E was the total energy of radiation and ν was the frequency of the radiation. Ordinary kinetic theory of gases predicts that the probability of finding all the gas molecules in a container of volume V_0 confined in a volume V is $(V/V_0)^N$, where N is the number of molecules.

Therefore, by analogy Einstein inferred that the light behaved as though it were composed of $E/h\nu_0$ quanta of energy. This result can almost be developed from the concept that the probability of finding one particle in a volume V when the container has a volume V_0 is $(V/V_0)^1$. Because the probability for each particle is the same, the probability for N is simply the product of the individual probabilities leading to the result $(V/V_0)^N$.

Considering electron emission from metals in the light of the concept of the quanta of energy, each of energy $h\nu$, Einstein proposed that the interaction of a light parcel of energy, or photon, with the electrons in the metal could result in the liberation of an electron having a kinetic energy equal to the energy of the photon minus some constant energy, $e\phi$, characteristic of the metal. If the maximum kinetic energy of a liberated electron is $(1/2)mv^2$ (where m is the electron mass and v is the maximum velocity), the appropriate relation is $(1/2)mv^2 = h\nu - e\phi$, where $e\phi$ is the energy lost by the electron in its escape from the surface of the metal. This effect will also occur when X-rays fall upon atoms, liberating some of the more tightly bound electrons. In such a case, the emitted electron energy is equal to the energy of the photon minus the initial binding energy of the electron in the atom.

In his other 1905 paper, Einstein leaped into the active arena of relativity physics. One should recall that the essential quality of the Lorentz transformation had been recognized for some time, although there was little physical understanding of the equations involved. The work of Woldemar Voigt, Lorentz, G. F. Fitzgerald, Sir Joseph Larmor, and Poincaré had established the mathematics of the transformation. Einstein focused upon the major feature of relativity, asserting that the velocity of light in vacuo is the same for all systems of reference moving relative to each other at constant velocity. This was proposed in conjunction with a more general statement—now known as the principle of relativity—which holds that it is impossible by physical experiment to label any one coordinate system as intrinsically stationary or absolutely moving with constant velocity.

Earlier calculations of moving charged bodies suggested that these bodies moved as though the mass were altered by an amount proportional to the energy of the electrostatic field, and to climax these indications of mass variations, Einstein asserted that the mass of a body is a measure of its energy content. Relativistic dynamics were studied in great detail in 1906 by Planck, who derived expressions giving Newton's second law in modified form along with expressions for the kinetic and total energy of a particle. By 1909 G. N. Lewis and R. C. Tolman had re-derived his results, adding the laws for the resolution of velocity components in a relativistic system. Because the transformation relations (named the Lorentz transformation in 1905 by Poincaré) had been known for some time, there was a reluctance to credit Einstein with the discovery of relativity. Lorentz and others had actually regarded the time t', calculated for the second reference frame, as an artifice, which they called "proper time." Einstein grasped the full physical significance of these transformations and generalized their application to all physical systems.

To appreciate some of the consequences of the concept of relativity, one can imagine two coordinate systems moving relative to one another with constant velocity V in such a manner that the instant that the origins of the two systems coincide in passing constitutes a synchronization of the clocks in the two systems. For convenience we assume that clock t in O and clock t' in O′ are set to zero at the instant when O and O′ coincide. Basically, there are two methods by which clocks can be synchronized in the same reference frame, by a light flash from a third system or another, by the slow transport of telescopes (or by a television screen—this method is actually the first).

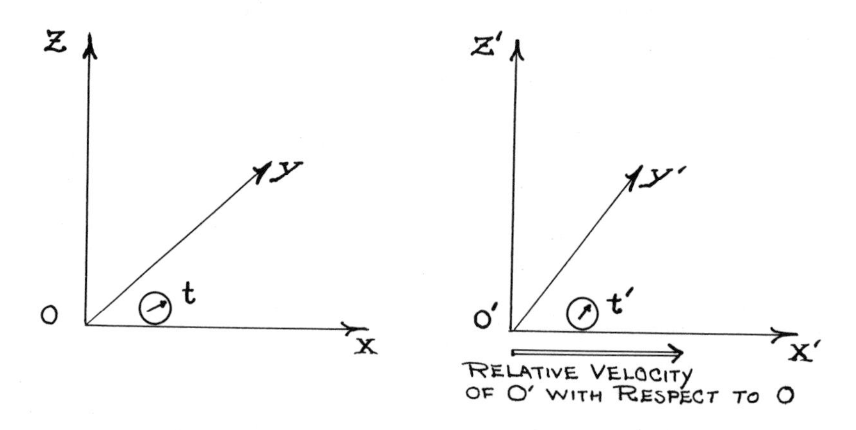

If we assume that the relative motion is along the x and x' axes, one can represent the space-time motion in terms of events plotted in (x, t) space and in (x', t') space. The Pythagorean forms employed by Voigt and Lorentz imply that the coordinate axes, x and t, should be taken perpendicular to one another. One must recognize that a point in a space-time diagram is an event in that it has a specific location only at a specific time: events moving along the x axis with the speed of light form a straight line in (x, t) space. To further simplify the diagram one ordinarily uses as coordinates x and ct. In this case the light ray or light cone intersects the (x, ct) space in a line, bisecting the angle between x and ct. Consider two events, A and B, in the (x, ct) diagram.

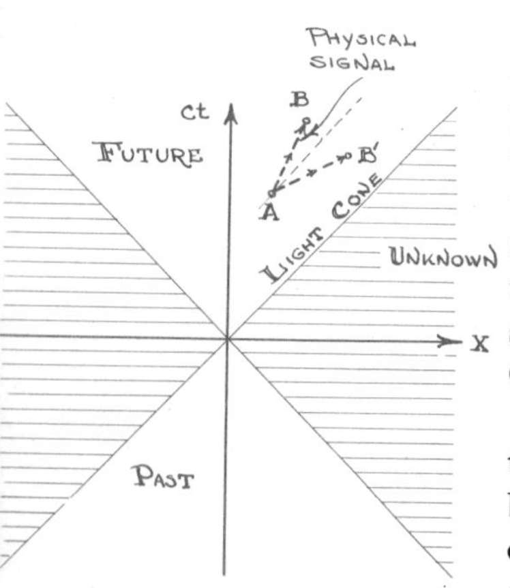

If the line connecting A and B makes an angle with the x axis of less than 45°, it is impossible to connect these two events with a physical signal because the transmission would require a velocity greater than c. A line connecting A and B making an angle greater than or equal to 45° with the x axis represents a situation in which the two events can be connected by a physical signal.

Hermann Minkowski (1864–1909) in 1906 reformulated the physical problems of relativity using four-dimensional manifolds and introduced some of the appropriate tensors which were to dominate the language of relativity. He provided a very simple diagram for graphic analysis, known as the Minkowski diagram. In these diagrams the loci of $(x', 0)$ and $(0, ct')$ are plotted to provide a picture of the events relative to both frames, O and O′. By observing the motion of O′ in time, one obtains the ct' axis which makes an angle α with the ct axis. The tangent of this angle is V/c. The axis $ct' = 0$ (or x') appears as a straight line, making an angle α with respect to the x axis. From the Minkowski diagram one observes that the symmetry of the light cone is retained with respect to x' and ct'.

Some care must be exercised because world points or events must be projected onto the x', ct' axes by drawing lines through the world point parallel to x' and ct'. The intersections of these parallels with ct' and x' give the prime coordinates of a world point.

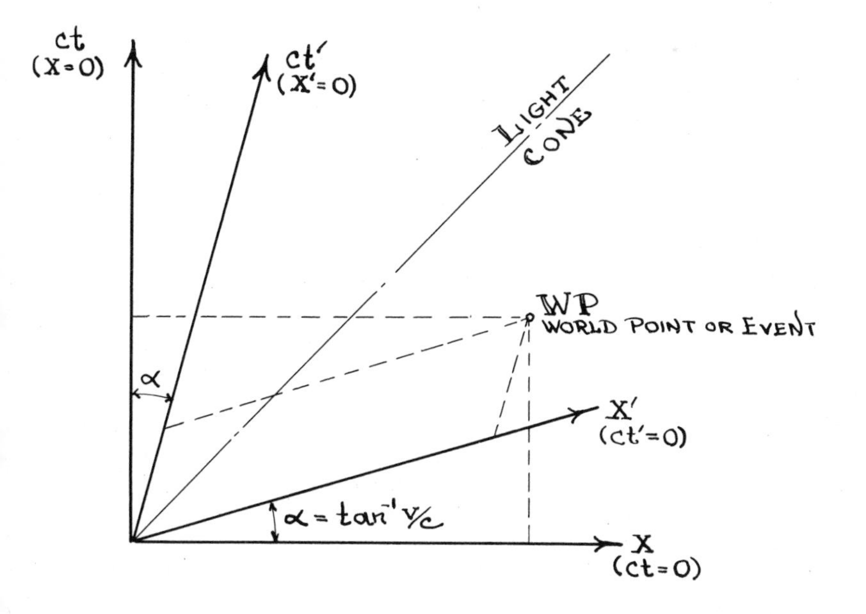

A further precaution must be taken; not only is the (x', ct') space non-orthogonal in this graphing, but the scale of x' and ct' is also different from the scale of x and ct. Thus, to make this plot consistent, a relative stretching of the axes occurs. This phenomenon is equivalent to the Fitzgerald contraction of lengths and time dilation, and the change in scale can be calculated readily by employing the invariant Pythagorean forms which connect the two spaces. We define a scale term G as

$$G = x^2 - c^2t^2 = x'^2 - c^2t'^2.$$

Such an expression is that of hyperbola in (x, t) space and also in (x', ct') space:

$$G = (x + ct)(x - ct) = (x' + ct')(x' - ct').$$

Under the condition that G is a constant, the corresponding values of x' and t' can be determined; for instance, if $x = 0$, $-c^2t^2 = G$; if

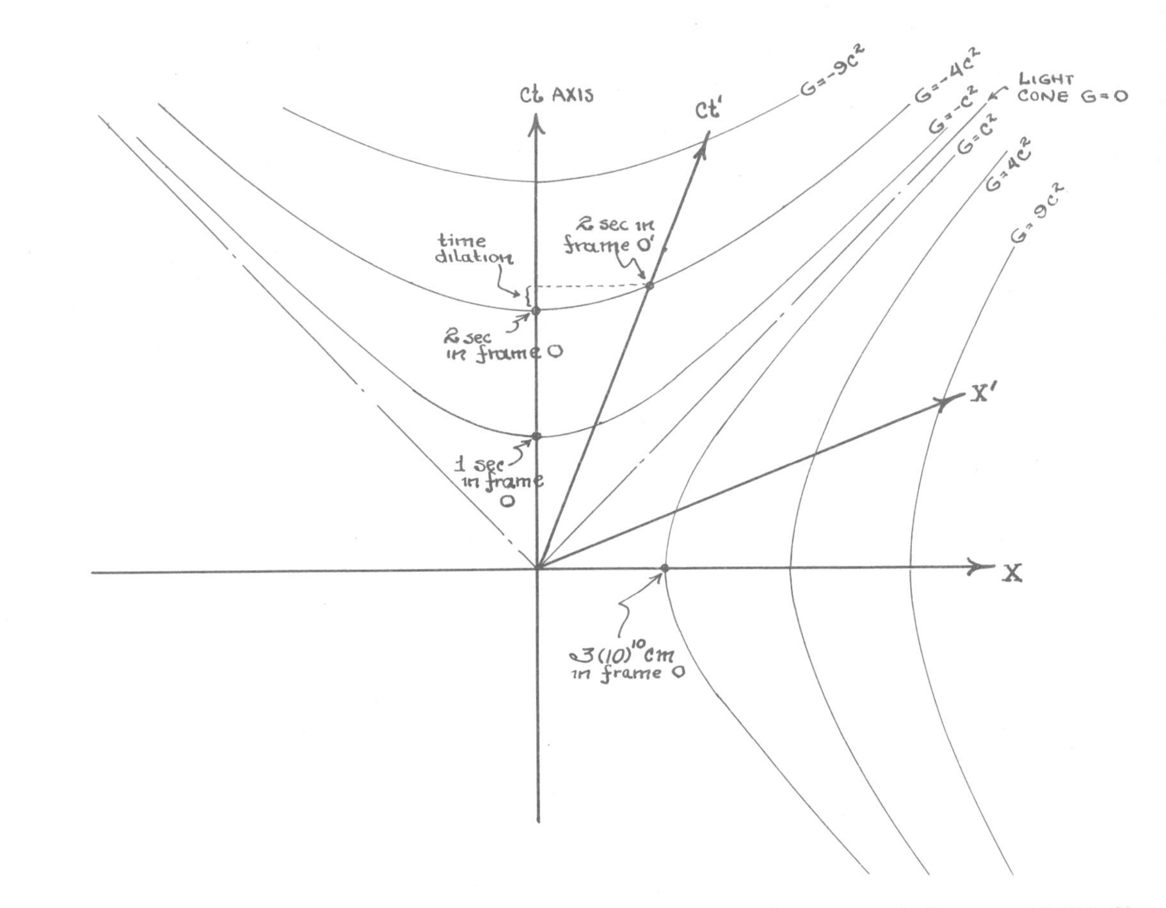

$x' = 0$, $G = -c^2t'^2$. As an example, let $G = -1$ cm^2; then $t = 1/9(10)^{-20}$ seconds and $t' = 1/9(10)^{-20}$ seconds. On the diagram, however, the marked lengths $t = T$ seconds and $t' = T$ seconds are different. Because we are plotting all events relative to ◯, the observer in ◯ believes that the clocks in ◯′ are running more slowly than those in ◯. Taking $G = +1$ cm^2 allows one to obtain the appropriate scale of x' in the (x, ct) space.

This diagram can be drawn with (x', ct') as the initial axes. Then the roles are reversed, since all measurements would be referred to the ◯′ instruments. From this we deduce that, relatively, the longer time interval is always read by the proper clock in the proper system. In other words, regardless of the system to which one refers all measurement, the referral system will always measure the longest time for corresponding events.

As an example, if a decaying system, such as a mu-meson, moves in upon the earth at high velocity, the lifetime of the moving system will always appear longer to us than it would to an observer on the moving system. This effect is called time dilation, and the relationships between the clock intervals can be obtained from the Lorentz transformations. The postulate that the speed of light in vacuo is the same in all frames moving relatively at constant velocity can be employed in a simple thought experiment illustrating time dilation. Consider two frames O and O′ moving parallel to their x axes. Assume that the frame O′ carries a mirror M at a distance d along the y' axis. Assume that a light pulse is emitted from the origins at $t = 0$ and $t' = 0$ when O and O′ coincide; the reflection from the mirror M is different when viewed from the two frames. The diagram shows this process as a series of still pictures.

The distance traveled in O is $2r = 2\sqrt{d^2 + 1/4(V\Delta t)^2}$, and Δt the time interval in O is $2r$ divided by c. Therefore,

$$\Delta t = \frac{2r}{c} = \frac{2}{c}\sqrt{d^2 + \frac{1}{4}V^2(\Delta t)^2},$$

giving Δt as

$$\Delta t = \frac{2d}{c\sqrt{1 - V^2/c^2}}.$$

Because the path of the light ray in O′ is up and back along the y' axis, the time interval in O′, $\Delta t'$ is given by $\Delta t' = 2d/c$. Finally, the relation between Δt and $\Delta t'$ is then

$$\Delta t = \frac{\Delta t'}{\sqrt{1 - V^2/c^2}}.$$

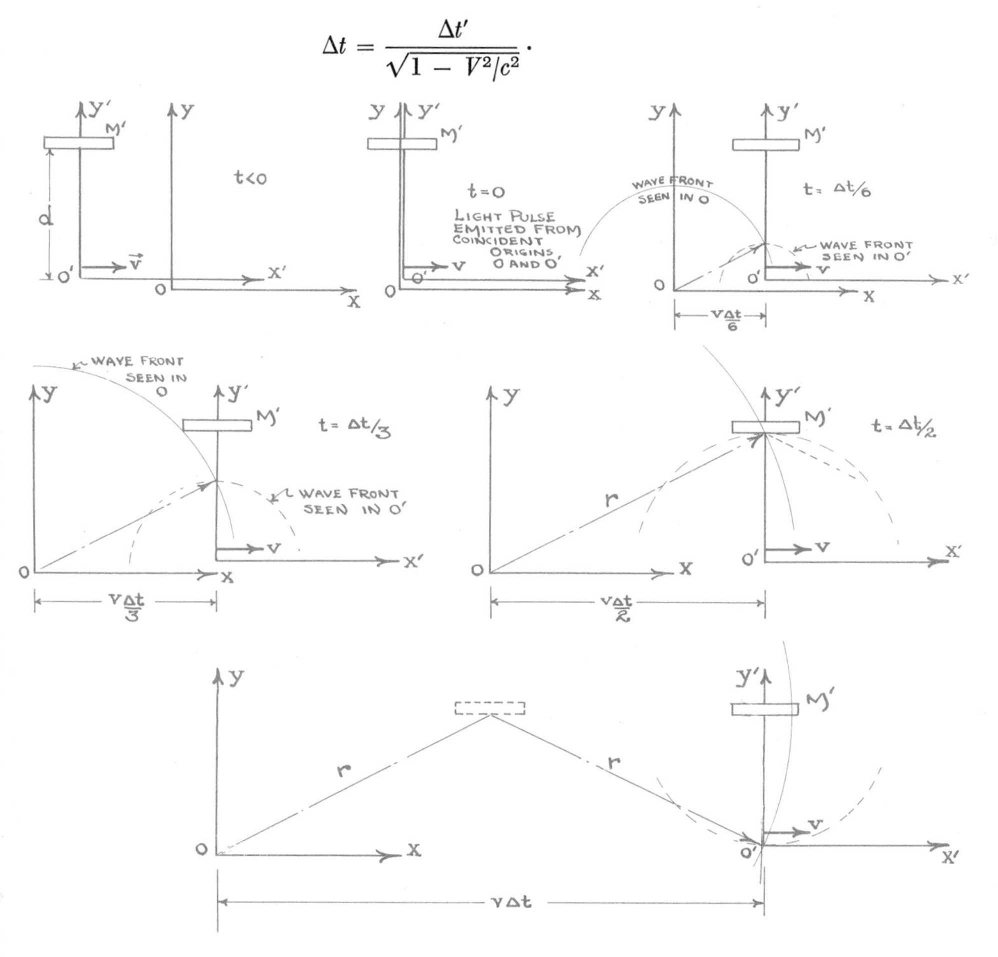

Since the events are being viewed in O, the clock in O′ seems to be running more slowly than that in O. This calculation has a great deal of bearing on the twin paradox. Assume that one of a pair of twins takes off in a rocket ship, leaving the first twin on earth, and travels at high velocity for, say, N years. The ship then turns about and returns to earth at the same speed, taking N years for the second leg of the journey. The paradox arises because time dilation allows one to predict that the biological clock in the rocket ship runs more slowly than the clocks on earth, with the result that the twin who traveled in the rocket is younger at the conclusion of the journey than the twin who remained on earth. This same paradox could be illustrated by using two identical radioactive sources instead of humans. If one radioactive source is sent off in a rocket and returned, the argument of relativity predicts that the half-life of the source in the rocket will appear to have been larger than its natural half-life during the trip because of time dilation.

These arguments do not violate the symmetry implied in relativity because accelerations relative to the center of mass of the universe are

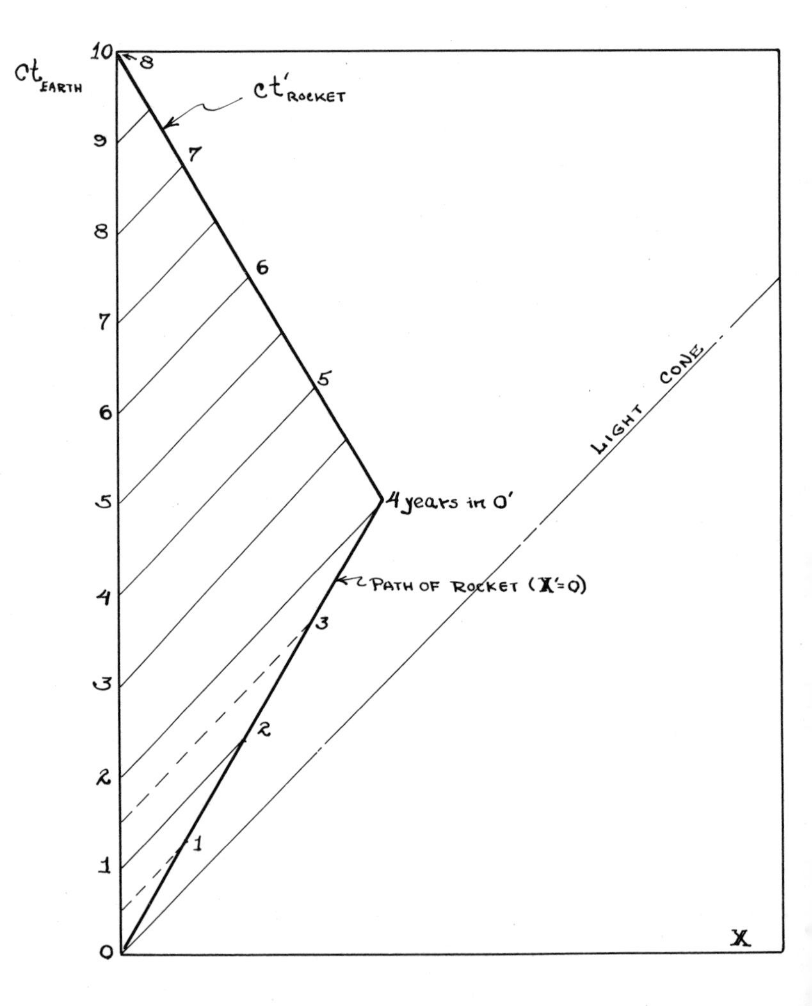

involved. In the rocket paradox the frames are distinguishable; thus the observer in the rocket must conclude at the termination of the trip that his clock ran at a slower rate than the earth clock. In 1940 this effect was checked experimentally through observations of the decay of the mu-meson. A stationary mu-meson has a half-life of approximately $2(10)^{-6}$ seconds. If a mu is formed 6,000 meters above the earth traveling at a speed close to $c = 3(10)^8$ meters/sec, it requires $2(10)^{-5}$ seconds to reach the earth. Thus, if decay progressed with the natural half-life, very few mu-mesons would reach the earth's surface. Experiments have shown that most do reach the earth's surface, and this is explicable in terms of time dilation. Assume that the velocity of the mu-meson is such that $(1 - V^2/c^2)^{-1/2}$ is of the order of 100. Relative to the earth, the time dilation of the moving mu gives an apparent half-life of $2(10)^{-4}$ sec instead of $2(10)^{-6}$ seconds. Under these conditions, in 10^{-5} seconds most of the mus would reach the earth's surface without decaying.

The Minkowski diagram of the twin in the rocket shows quite clearly the time scales on earth and on the rocket. Possible light signals sent from the earth to the rocket are also indicated.

Remarkable results were exhibited in the reformulation of mechanics in terms of relativity. Velocity addition was modified because of the limitations imposed by the velocity of light. By considering two electrons moving antiparallel to one another relative to the laboratory, both with speeds close to c, one concludes that velocity addition can no longer be represented by the vector sum of the two. The equations of the Lorentz transformation indicate that an appropriate calculation of the ratios of space intervals to time intervals will obviously provide the appropriate velocity transformations.

The Lorentz transformations from ○ to ○′ are:

$$x' = \gamma(x - Vt), \qquad z' = z,$$
$$y' = y, \qquad t' = \gamma(t - Vx/c^2),$$

where

$$\gamma = \frac{1}{\sqrt{1 - V^2/c^2}}.$$

Using these forms one can compute the relationships between velocity components of a moving point as observed in ○ and ○′:

$$\frac{dx'}{dt'} = \frac{dx/dt - V}{1 - (V/c^2)(dx/dt)}, \qquad \text{or} \qquad v'_x = \frac{v_x - V}{1 - v_x V/c^2};$$

$$\frac{dy'}{dt'} = \frac{dy/dt}{\gamma[1 - (V/c^2)(dx/dt)]}, \qquad \text{or} \qquad v'_y = \frac{v_y}{\gamma(1 - v_x V/c^2)};$$

$$\frac{dz'}{dt'} = \frac{dz/dt}{\gamma[1 - (V/c^2)(dx/dt)]}, \qquad \text{or} \qquad v'_z = \frac{v_z}{\gamma(1 - v_x V/c^2)}.$$

Here the relative motion of the frames is parallel to the x and x' axes. As an example, regard two electrons moving away from each other along the x axis; one with a velocity of $v_{Ax} = 0.9c$ relative to the lab, and the other with a velocity $v_{Bx} = -0.9c$. The velocity of A relative to B, instead of being $0.9c + 0.9c$, is, by the equations above,

$$v_{AB} = \frac{v_{Bx} - v_{Ax}}{1 - v_{Bx}v_{Ax}/c^2} = -\frac{1.80}{1.81}\,c\,.$$

Special relativity gave rise to a modification in the definition of the mass and momentum of a point particle. By regarding the collision of two particles having the same mass in a reference frame, where the two velocities are equal and opposite, relativistic analysis shows that the two particles have different masses in a reference frame, moving with velocity V relative to the first frame. Conservation of momentum in both frames requires that a moving mass must be defined as $m_0/\sqrt{1 - V^2/c^2}$, where m_0 is the rest mass (the mass at velocities which approach zero). This result is consistent with the definition of the remaining dynamical quantities:

momentum of a point particle: $\vec{\mathbf{p}} = \gamma m_0 \vec{\mathbf{v}}$

Newton's second law: $\frac{d}{dt}(\gamma m_0 \vec{\mathbf{v}}) = \vec{\mathbf{F}}$

kinetic energy: $T = m_0c^2(\gamma - 1) = (m - m_0)c^2$

total energy: $E = \sqrt{c^2p^2 + (m_0c^2)^2} + V(r)$

$$\gamma = \frac{1}{\sqrt{1 - v^2/c^2}}\,.$$

In the expression for E, V is the potential energy.

In keeping with the notion of four coordinate dimensions, Planck reformulated the momentum vector as a four-component momentum consisting of the three components of the relativistic momentum plus a fourth component which behaved as the energy. The impact of this theory was far-reaching: in addition to more immediate gains, interest in the tensor analysis was renewed and ultimately Einstein and others presented the general theory of relativity—a question which has entertained the attention of many scientists since its inception and which in large part remains unsolved.

The Newtonian theory of gravitation contained some marked discrepancies—departures from elliptic motion, periodic inequalities which corrected and cycled again, and secular inequalities, such as the net changes in the mean angular velocities of Jupiter and Saturn. The mean angular velocity of Jupiter increases continually, while that of Saturn is decreasing. The problem was considered by Euler and Lagrange with no success. In 1773 Laplace discovered that within the composite system of Jupiter and Saturn the net effect canceled. This inequality turned out to be periodic, with a period of 929 years. Laplace also made a partial accounting for the secular acceleration of the mean motion of the moon. His solution suggested that in part this perturbation was also periodic and was associated with the approach of the earth's orbit to a circular form.

One of the major anomalies of planetary motion was (and still is) the precession of the perihelion of Mercury. The semi-major axis of the elliptical orbit of Mercury precessed by about 38″ more per century than could be accounted for by the perturbations introduced by the presence of the other planets. This problem was investigated by Weber, Tisserand, and others through the introduction of artificial perturbing forces into Newton's law of gravitation.

Speculation about the gravitational law was renewed by the theory of relativity. Earlier, in 1900, Lorentz had looked for electromagnetic connections; after 1905 the Newtonian law of gravitation was regarded in light of relativity. A fundamental question about the role of mass was revealed in 1907 when Planck, investigating the experiments of József Eötvös (which indicated the equality of gravitational and inertial mass), suggested that all energy having inertial properties must gravitate. During the same year, Einstein introduced this concept as the principle of equivalence, which, in brief, concludes that all objects appearing to be in a field of force could equally be viewed as objects in a closed chamber which is accelerating. In accordance with this principle, Einstein demonstrated that photons which are emitted from the same type of atom and which correspond to the same transitions within the atom will exhibit a longer wave length when emitted from a high gravitational potential and observed in a lower gravitational potential than when the emission and observation are at the same potential. As an example, he took an atom on the sun's surface and an observation on the earth. If Ω is a measure of the potential on the sun's surface, the energy lost by an escaping photon, $h\nu_0$, is Ω times the mass of the photon, or $\Omega h\nu_0/c^2$. As a result,

$$h\nu_{\text{observed}} = h\nu_0(1 - \Omega/c^2),$$

and

$$\lambda_{\text{observed}} = \frac{\lambda_0}{1 - \Omega/c^2} \xrightarrow[\Omega \ll c^2]{} \lambda_0(1 + \Omega/c^2),$$

where λ_0 and ν_0 are the natural wave length and frequency, respectively, of the photon. The overall conclusion of this argument was that, because light gravitates, a light ray passing near a massive body (such as the sun) must be curved.

Generalizations of algebra by Grassmann and Gibbs and the invention of n-dimensional manifolds by Riemann led to the creation of a general tensor calculus by M. M. G. Ricci (1853–1925). Together with Tullio Levi-Civita (1873–1942), Ricci developed tensor analysis and its application to mathematical physics in 1901. Quadratic differential forms and their transformations constituted the basis of this work, and although it received very little attention when published, it became of fundamental importance in the development of the general theory of relativity. The efforts of H. Bateman (1882–1946) in 1909 suggested that the customary differential form for the light ray,

$$(dx^{(0)})^2 + (dx^{(1)})^2 + (dx^{(2)})^2 + (dx^{(3)})^2 = 0,$$

should be replaced by the more general equation

$$\sum_{m=1}^{4} \sum_{n=1}^{4} g_{mn}\, dx^{(m)}\, dx^{(n)} = 0.*$$

Here a light ray progresses from the point $(x^{(0)}, x^{(1)}, x^{(2)}, x^{(3)})$ to the point $(x^{(0)} + dx^{(0)}, x^{(1)} + dx^{(1)}, x^{(2)} + dx^{(2)}, x^{(3)} + dx^{(3)})$ in the presence of a field of force. Such broad views of the geometry of physics led, quite naturally, to a surge of interest in generalized field theories, wherein all of the laws of physics would be encompassed by a few compact axioms. These efforts at unification essentially have failed.

In 1913 Einstein and Marcel Grossmann completely abandoned the concept of the gravitational field as described by a single scalar potential function. Basic to Einstein's new theory was his demonstration that the single scalar potential led to some invalid inferences; in particular, he constructed a problem which involved the raising of a fixed volume of black body radiation in two cavities with perfectly reflecting walls. He showed that if one cavity was contained in a movable box, while the other was formed from a vertical tube with movable pistons a fixed distance apart, the work required in raising the two was quite different. The starting point of his theory involved the variational integral $\delta\{\int ds\} = 0$, where

$$(ds)^2 = \sum_{m=1}^{4} \sum_{n=1}^{4} g_{mn}\, dx^{(m)}\, dx^{(n)} = g_{mn}\, dx^{(m)}\, dx^{(n)},\dagger$$

and ds is the path of a particle in the presence of a force field which must be explicitly described by the ten independent quantities g_{mn}. All in all, there are sixteen metric coefficients g_{mn}; however, the symmetry of the array $(g_{mn} = g_{nm})$ reduces this number to ten independent quantities.

As in the special theory, the velocity of light in any reference frame has the value c: some contradictions appear, however, when the phenomena of anomalous dispersion is considered. Louis Brillouin and Arnold Sommerfeld removed this difficulty by demonstrating that the limitation on the velocity of light is in practice a limitation on the velocity with which energy (or information) can be transported. Therefore, the limitation is applied to the group velocity of a wave and not the phase velocity.

When a pulse is formed and moves through a medium, the mathematical representation involves the linear superposition of a large number of monochromatic waves (Fourier components). The independent components, each with radial frequency $\omega = 2\pi\nu$ and wave number $k = 2\pi/\lambda$, move with a phase velocity which can be larger than c. The group velocity,

* The superscripts on the xs, such as $x^{(m)}$, are not powers but indices, in the same sense as subscripts, such as x_m. This notation is employed in the theory of relativity and is therefore appropriate here.

† To reduce the complexity of the notation, a sum rule was invented in which the summation symbol $\sum$ was implied whenever an index was repeated. Thus, $g_{mn}\, dx^{(m)}$ means $\sum_{m=1}^{4} g_{mn}\, dx^{(m)}$. In actual practice the superscript is not enclosed in parentheses, and the reader is required to differentiate between powers and superscripts.

or velocity of the centroid of the pulse, moves with a velocity less than or equal to c. A group velocity, v_G, is defined in terms of the angular frequency ω and the wave number k as $v_G = \partial\omega/\partial k$. This expression then suggests that the phase velocity $v_p = \omega/k$ may be quite different from v_G in a dispersive medium (where ω is a function of k) and that in a sense it is not measurable, but rather is a quantity generated in a consistent way by the representation.

Before 1915 Einstein gave a formal representation of the equations of electromagnetic theory and dynamics in a space for which the gravitational field is specified. These were derived from the principle of equivalence, assuming that the systems behave as free systems when referred to a coordinate system accelerating in such a manner as to compensate for gravitation. The formulation of these equations employed the full power of the tensor analysis, and the basic equations for the gravitational field, developed in 1915, put forward the basic relations for the curvatures of his four-dimensional space.

Ernst Mach (1838–1916) had introduced a fundamental postulate stating that the inertia of a body must be reducible to the interaction between that body and the universe. Adopting this idea, Einstein generalized it into an axiom, which he called Mach's principle. From Einstein's point of view this postulate assumed that the ten independent potentials g_{mn} must be determined only by the masses of bodies or energies as expressed by the Minkowski energy tensor, and from this principle he produced ten equations in ten unknowns of the form:

$$K_{pq} - (1/2)g_{pq}K = -\kappa T_{pq}$$

where T_{pq} is the Minkowski energy tensor, K_{pq} is the Ricci tensor, and the g_{pq} are the unknowns with

$$K = \kappa \sum_{m=1}^{4} \sum_{n=1}^{4} g^{mn} T_{mn}.$$

Adoption of Mach's principle led to the conclusion that the curvature of space is determined by the presence of physical phenomena and that therefore if there were but one particle in space it would have no inertia. David Hilbert in the same year postulated a scalar world function, H, which involves the g_{mn}, their derivatives, and an electrodynamic four potential ϕ. He showed that the Einstein gravitational equations and electromagnetic equations of general relativity could be obtained from the null variation of the integral of H over the volume element dx^0, dx^1, dx^2, dx^3 in four space.

Changes occurred rapidly in the concepts of physics and in those of mathematics. An abrupt transition in analysis took place shortly after the reconstruction of the real number system. The difficulties in analysis were, as in the previous period, difficulties of rigor, and a reconsideration of the theory of integration was initiated by T. J. Stieltjes (1856–1894) and by Poincaré. Efforts during the late 1890's and at the beginning of the twentieth century culminated in a new interest, which has been classified as the theory of real variables. Henri Lebesgue (1875–1941) produced a generalization of the Riemann integral in 1902 which directed attention to an ancient field. Early definitions regarded integration as the inverse of differentiation (Newton) or as the limit of a sum (Leibniz). The Riemann definition favored Leibniz's view: basically the difficulties of all approaches lay in the assumption of intuitive notions of smoothly varying velocities and lengths. Particularly when the limit of an infinite sum was taken with the intervals associated going to zero, there were intrinsic misconceptions in the type of infinities to be employed. As demonstrated earlier, the length or measure of a straight line segment of unit length is not spanned by a denumerable set of points, even when intervals of finite width are associated with each point. Although the inclusion of the nondenumerable set of points corresponding to all of the real numbers seems deceptively clear, the convergence of such a sum is by no means guaranteed. Foremost, then, was a need for a usable definition of measure for a point set and for a consistent technique in the use of diverging series. This was developed and contained in Lebesgue's theory of integration. The sense of his contribution was to divide the domain of integration into measurable sets and to define the limit of a particular sum for all sets as their number is indefinitely increased.

Discoveries of the electron and radioactivity led to considerations of the microscopic constitution of matter; these reached their climax in 1910 and 1911. Shortly after the discovery of the electron, J. J. Thomson, in order to preserve a neutral atom of radius approximately 10^{-8} centimeters, assumed that an amount of positive charge equal to the total charge of the electrons of the atom must be distributed uniformly throughout a sphere having the radius of the atom. This and other models were shown to be inadequate by Rutherford in 1910. As a result of experiments designed to study the intensity of alpha particles scattered from atoms, Rutherford found that many more alphas were scattered through large angles than could be accounted for by the uniform spherical atomic models proposed. In 1911 he published the results of calculations which accounted for the observed alpha scattering; these were based upon the assumption of a planetary atom which had a positively charged nucleus of radius approximately $3(10)^{-12}$ cm surrounded by negative electrons contained in orbits and distributed throughout the remaining volume of the atom.

To achieve a neutral atom, the charge on the nucleus was determined solely by the number of electrons. If there were Z electrons in an atom, the nuclear charge would then be $+Ze$, where e is the magnitude of the charge on the electron.

To a large extent, theorists in the nineteenth century assumed that the natural lines in the spectrum of light from an atom came about from the vibrations of normal modes of the electronic structure of the atom. A study of line spectra initiated by Fraunhofer culminated in the work of J. J. Balmer (1825–1898) and J. R. Rydberg (1854–1910), who gave general

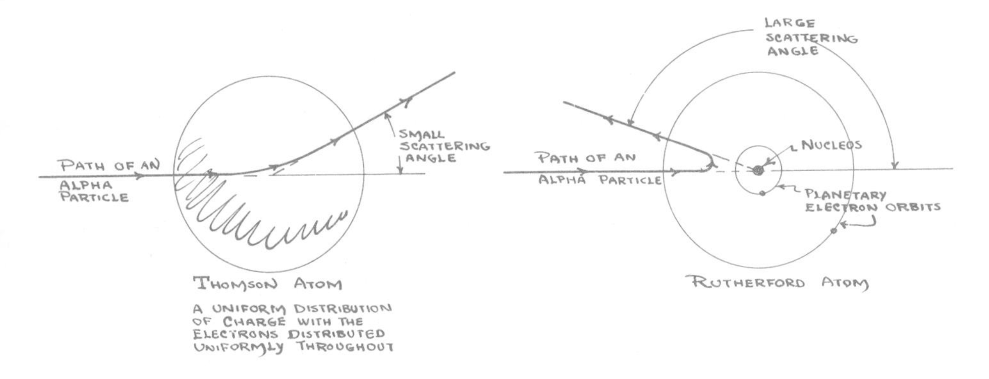

empirical formulas for the series of lines resulting from the excitation of hydrogen and the alkali elements. In 1908 W. Ritz discovered that the natural lines could be accounted for as the differences in pairs of terms involving a universal constant and the inverse of the squares of the integers. The first successful application of the quantum concept of Planck to the atom was achieved in 1912 by Niels Bjerrum, who quantized the rotational energy of a molecule. Bjerrum assumed that the two atoms making up the molecule were charged, one positively and the other negatively. Further, the atoms were assumed to vibrate along the line joining them with a natural frequency, ν_0. The rotational energy of the line of centers in a plane was then quantized in integral multiples of $h\nu_n$, providing a rotational spectrum with the frequencies $\nu_0 \pm \nu_n$. This assumption did not fit the experimental results available; however, one year later P. Ehrenfest improved this approach by assuming that the rotational energy was quantized in units of $(1/2)h\nu$. Ehrenfest's argument was that the energy is purely kinetic, without the usual oscillator potential energy contribution. Thus, if I was the moment of inertia of the molecule and ν the rotational frequency, the energy quantization could be written as

$$(1/2)I(2\pi\nu_n)^2 = n[(1/2)h\nu_n],$$

where n is an integer. The resulting frequency spectrum was then given by

$$\nu = \nu_0 \pm \nu_n = \nu_0 \pm \frac{nh}{4\pi^2 I}.$$

This assumption turned out to contain the major clue to the construction of a quantized atom. Here the angular momentum $\mathscr{L}$ could be obtained as

$$\mathscr{L}_n = I\omega_n = I(2\pi\nu_n) = n\frac{h}{2\pi},$$

suggesting the quantization of the angular momentum of the electrons in the Rutherford planetary atom. In 1913 Niels Bohr (1885–1962) constructed a Rutherford-type atom in which the electrons could only occupy certain quantized states which were characterized by the quantization of the angular momentum. Further, in accordance with some previous views that emission occurred only from excited atoms, he postulated that the emission of light resulted from the jump of an electron from one quantized state to a state of lower energy. In one model, therefore, Bohr accommodated the concept of the photon, the Rutherford planetary atom, and Ehrenfest's quantization of angular momentum. Classical theory predicts that an accelerating charge radiates, and in a bold stand Bohr postulated that this classical law did not apply to his quantized electron orbits. The orbits were assumed to be circular; thus the mass of the electron times the centripetal acceleration could be equated with the Coulomb force between the electron $-e$ and the nucleus of charge Ze; $mv^2/r = Ze^2/r^2$, where m is the mass of the electron, r is the radius of the circular orbit, and v is the tangential velocity. The angular momentum $\mathscr{L}$ of a point particle in a circular orbit has a particularly simple form, being the mass times the radius times the tangential velocity quantizing in units of $h/2\pi = \hbar$:

$$\mathscr{L}_n = mr_nv_n = \frac{nh}{2\pi} = n\hbar,$$

where n is an integer (modern notation employs the symbol $\hbar$ for $h/2\pi$). The two equations above allow one to solve for either v_n or r_n:

$$v_n = \frac{Ze^2}{n\hbar}$$

and

$$r_n = \frac{n^2\hbar^2}{Zme^2}.$$

The total energy of the bound electron is the kinetic energy $(1/2)mv_n^2$ plus the potential energy in the Coulomb field $-Ze^2/r_n$. Substituting for v_n and for r_n, the total energy E could be obtained in terms of Planck's constant and in terms of n, e, and m:

$$E_n = \frac{1}{2}mv_n^2 - \frac{Ze^2}{r_n} = -\frac{1}{2}\frac{mZ^2e^4}{n^2\hbar^2},$$

where $n = 1, 2, 3, \cdots$. Because the electron is bound, the total energy must be negative.

The energy of an emitted photon is, therefore, given by the energy difference between two quantized levels. Let the energy of the initial state (or level) be E_n, and the energy of the final state E_m; then the energy of the photon is

$$h\nu_{nm} = E_n - E_m = \frac{mZ^2e^4}{2\hbar^2}\left(\frac{1}{m^2} - \frac{1}{n^2}\right).$$

For hydrogen with $Z = 1$, the lowest state corresponds to $n = 1$; this gives

$$E_1 = -\frac{mZ^2e^4}{2\hbar^2} = -2.2(10)^{-11} \text{ ergs} = 13.6 \text{ electron volts.*}$$

* The electron volt is an hybrid unit from the rationalized mks system; 1 electron volt is equal to $1.6(10)^{-12}$ ergs.

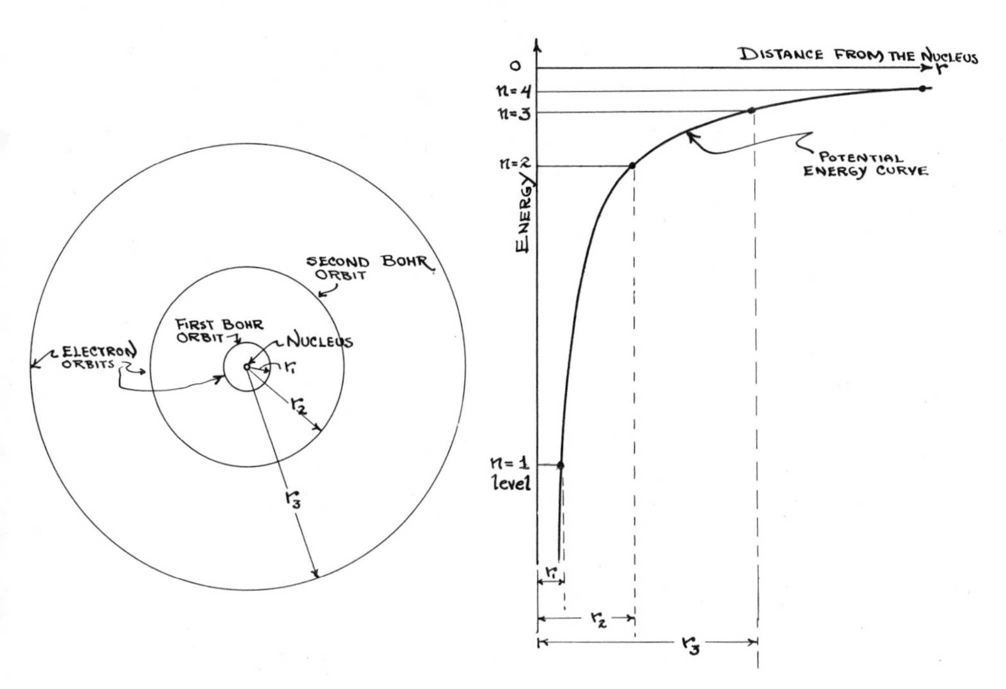

The success of this very simple model was immediate. Balmer's series terminated in the $m = 2$ level; Paschen's series terminated in $m = 3$. For large values of n the frequency of the radiation seemed to have a semi-classical connection with the frequency of revolution:

$$\nu \xrightarrow[\text{large } n]{} \frac{mZ^2e^4}{4\pi\hbar^3}\left\{\frac{1}{n^2} - \frac{1}{(n+1)^2}\right\} \simeq \frac{mZ^2e^4}{2\pi\hbar^3 n^3}.$$

This value corresponded to the frequency of electromagnetic radiation which would be emitted by an electron rotating about a nucleus at a distance given by the quantized radius. Although the initial success of this model was rather spectacular, it soon became apparent that it was only a step in the right direction rather than a dependable final result. Bohr made some minor improvements: as an example, he introduced the reduced mass of the electron relative to the nucleus, and because in many electron atoms there is some shielding by the innermost charges, a modified Z was instituted.

After a prediction by Woldemar Voigt in 1901, Johannes Stark, while investigating the emitted light from canal rays, discovered that the spectral emission lines were split into components grouped about the frequency of the radiation from unperturbed atoms. Zeeman magnetic splitting of spectral lines had been found in 1896. Here again was an effect which the Bohr quantized atom was incapable of representing. A classical analysis of Zeeman splitting had been furnished by Lorentz in 1897, in which he conceived of electron orbits within the atom. Because the motion produced a subsequent interaction with the external magnetic field, a shift in the rotational frequency of the electron in the orbit occurred.

Bohr's theory was obviously limited in view of the fact that there was not a sufficient number of available degrees of freedom to describe the magnetic splitting of the spectral lines that occurred in the simplest of systems. Sommerfeld in 1915 introduced elliptical orbits, sometimes called Keplerian orbits. His formalism was based upon the assumption that the action integral $\int p\,dq$ (where p is the momentum and q is the coordinate) was always an integral multiple of Planck's constant, h. By assuming elliptical orbits, Sommerfeld was able to introduce a second quantum number in linear combination with that connected with the Bohr model. However, his result was no different from the simple Bohr picture because the total number of possible energy levels remained the same. Realizing that the speed of the electron in an orbit is sufficiently high, Sommerfeld modified his theory to include the relativistic corrections to the motion of the electron. These calculations showed that the relativistic orbit is an ellipse with a moving perihelion (semi-major axis) and that the motion of the perihelion depended upon a new constant $e^2/c\hbar$, which was to be known as the fine structure constant. The Bohr theory gives E_n as $-mZ^2e^4/2\hbar^2n^2$, and on the average this is equal in magnitude to the kinetic energy of the electron. Using these approximations, the velocity of the electron is $v_n/c \simeq Ze^2/nc\hbar$; in other words, the fine structure constant is roughly the ratio of the speed of the electron in the first Bohr orbit of hydrogen ($Z = 1$) to the speed of light. Additional quantum numbers were required when Keplerian orbits were assumed. In 1916 Sommerfeld and Debye showed that the Zeeman effect could be accounted for by their quantized atom. In the presence of a magnetic field the plane of the orbit of the electron must precess about the direction of the magnetic field, $\vec{\mathbf{B}}$, with a precession frequency $eB/2mc$, called the Larmor frequency.

This is a classical effect because the instantaneous torque on the system arising from the magnetic force term $(-e\vec{\mathbf{v}} \times \vec{\mathbf{B}})$ causes the angular momentum vector, $\vec{\mathscr{L}} = \vec{\mathbf{r}} \times \vec{\mathbf{p}} = m\vec{\mathbf{r}} \times \vec{\mathbf{v}}$, to precess or change in the

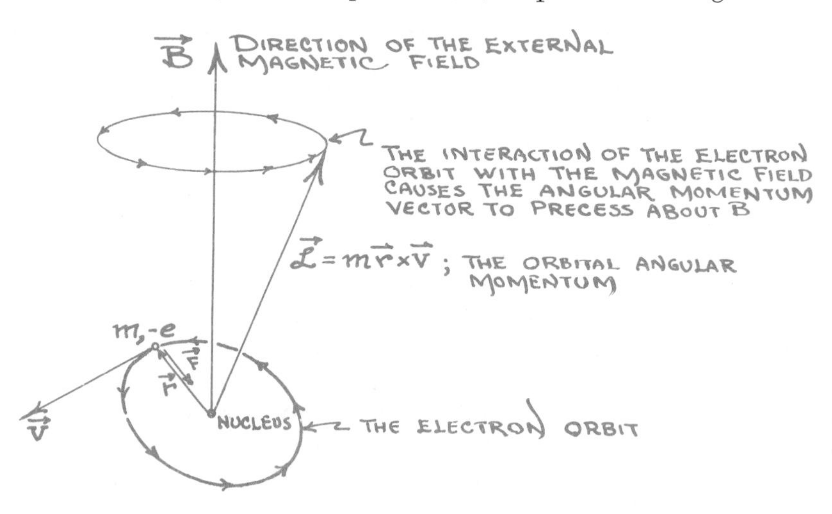

direction of the torque. The Sommerfeld and Debye condition on $\vec{\mathscr{L}}$, the angular momentum, stipulated that $\vec{\mathscr{L}}$ could only assume a discrete set of orientations relative to $\vec{\mathbf{B}}$ and would only have projections along $\vec{\mathbf{B}}$ which were whole integral multiples of $\hbar$.

As yet the quantum numbers were not represented in a consistent way, with the exception that the square of the angular momentum would be quantized in units of $\hbar^2$ and, further, that the projections of the angular momentum along a magnetic field $\vec{\mathbf{B}}$ could only appear in units of $\hbar$. Thus, if the maximum projection of the angular momentum was $l\hbar$, then the remaining projections could only take on the values

$$(l-1)\hbar,\ (l-2)\hbar, \cdots, \hbar,\ 0,\ -\hbar, \cdots,\ -(l-1)\hbar,\ -l\hbar.$$

Working with this hypothesis, Otto Stern and Walther Gerlach in 1921 demonstrated that, when a well-defined beam of silver atoms emerging from a furnace was deflected by an inhomogeneous magnetic field, the deflected portions of the beam were split into two distinct beams rather than into the continuous distribution, as would have been predicted by classical statistical mechanics. This single experiment confirmed the hypothesis of space quantization of the angular momentum.

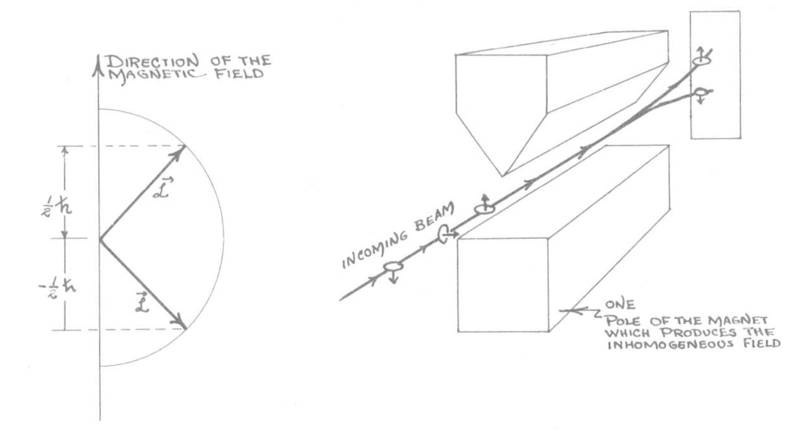

In spite of these successes, the atomic spectra still were not adequately described by the theory of Sommerfeld and Debye. The climax of the quantization of angular momentum came in 1925, when G. E. Uhlenbeck (1900–) and S. Goudsmit (1902–) proposed that the electron possessed an intrinsic spin of magnitude $(1/2)\hbar$; this implied that because of its charge the electron had an intrinsic magnetic moment. Both angular momentum and the magnetic moment have equivalent classical forms; therefore an angular momentum $(1/2)\hbar$ associated with a charged particle will give rise to a magnetic moment. Although the result is general, one can compute the relation from the parameters of an extended circular orbit. The magnitude of the magnetic moment of a charged particle in an orbit is equal to the current times the area of the orbit. A charged particle, $-e$, will move about a circular orbit of radius r once in the time interval

$2\pi r/v$. The equivalent current, then, is the charge divided by the time, or $-ev/2\pi r$.

Classically, the magnitude of the magnetic moment, μ, is given by

$$\mu = \text{area} \times \text{current} = \pi r^2\left(-\frac{ev}{2\pi r}\right) = -\frac{erv}{2}.$$

The angular momentum of the particle is $\vec{\mathscr{L}} = m\vec{r} \times \vec{v}$, giving

$$\vec{\mu}_{\text{classical}} = -\frac{e\vec{\mathscr{L}}}{2m};$$

and

$$\vec{\mu}_{\text{spin}} = -g\frac{e\vec{S}}{2m},$$

where $\vec{S}$ is the spin angular momentum with a projected magnitude $(1/2)\hbar$, and g, the gyromagnetic ratio, is a constant, which experimentally turns out to be about 2. Classically $\vec{\mu}$ is either parallel or anti-parallel to $\vec{\mathscr{L}}$, depending upon the sign of the charge. In the instance of the electron, the magnetic moment is only proportional to the classical term—the constant of proportionality is the g factor.

The Uhlenbeck-Goudsmit hypothesis indicated that the total interaction of an orbiting electron with an external field must be expressed in terms of the total vector sums of the orbital magnetic moment and the intrinsic magnetic moment. This, in turn, implies a vector addition of the orbital angular momentum and the spin angular momentum.

Wolfgang Pauli (1900–1958) postulated a principle known today as the Pauli principle, which restricts the total number of electrons in any single state to two. Spin completed this picture, in that the two electrons could be conceived of as filling a state in terms of their spin orientation, one with spin up and one with spin down. One could rephrase this and claim that each state split into two states. By now, the mechanics of the atom were complex because, even in the absence of an external field, the intrinsic magnetic moment of the electron would interact with the orbital magnetic moment, leading to the coupling or vector addition referred to above, and there would be different energies of interaction, called fine structure. Pauli also suggested, correctly, that there might well be an intrinsic spin and magnetic moment at the nucleus, giving rise to hyperfine structure through the interaction of the nuclear moment with the total electron moment.

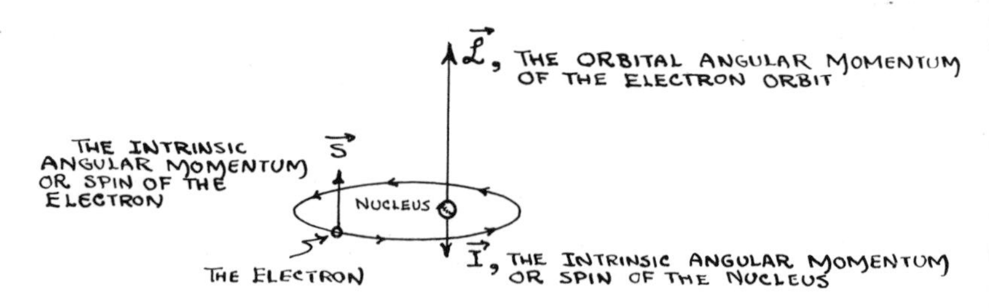

X

A Pair of Dice Regained

"You believe in the dice-playing God, and I in the perfect rule of law," wrote Albert Einstein to Max Born in 1944,* expressing his many reservations about the new physics. While it is true that unity, and even some consistency, had been achieved by the wave representation of material particles, not all physicists were convinced of an assured future for wave mechanics. However, although far from complete, the wave mechanical description of microscopic systems has proved to be remarkably versatile as well as successful.

From a broad point of view, the alteration of the concept of a microscopic particle to a wave packet is only startling if the history of physics is viewed in the short span of time from 1900 to 1920. Other theories had been modified in the past, often with much more reluctance. The corpuscular theories of light gave way to the wave theories of Young and Fresnel only after prolonged argument. As late as 1890 the vortex theory of matter still had many adherents; in fact, this theory may have had a beneficial effect, for the notions of scientists in 1923 were not very far removed from the older nonparticle views of matter. Einstein's photon, for instance, provided a clue: here a well-established wave theory was

* Quoted in Max Born, *Natural Philosophy of Cause and Chance* (New York: Dover, 1964), p. 122.

reinterpreted in terms of particles. Even more suggestive was A. H. Compton's discovery in 1922 that photons will scatter off of free electrons as if possessing an initial momentum $h\nu/c$. This implied a duality which might be applied to nuclei and electrons (i.e., particles).

Louis de Broglie (1892–) in 1923 proposed that particles in analogy with photons should display wave properties. In the case of a photon of energy $E = h\nu$, the momentum is given by E/c or $h\nu/c = h/\lambda$. By a similar argument, he adduced that if the energy of a free particle were assumed to have the dual representation, it must then have a frequency ν given by

$$E = h\nu = \frac{m_0 c^2}{\sqrt{1 - v^2/c^2}}.$$

A calculation of the group velocity of the wave $d\nu/d(1/\lambda)$ gave the momentum in terms of the wavelength λ, $p = h/\lambda$. Looking back, one can argue that this was not an unreasonable assumption, given the knowledge available at that time. In fact, faith in the concept of either point particles or rigid body particles was in no way sustained by observations—nor is it now.

It is a fact that every attempt to measure the numerical value of any dynamic quantity in the laboratory gives a distribution of numbers relative to some average value, rather than a set of numbers that are all identical. Questioning the importance of these distributions leads one immediately to wonder whether the distribution of measurements may not represent a fundamental characteristic of nature, rather than a limitation of laboratory technique. If observed distributions are an intrinsic aspect of our physical world, a wavelike description must have some validity, in that distributions can always be analyzed into a Fourier expansion, and such an expansion represents a wavelike description.

Any disturbance (such as a scattering process) of a system represented by a distribution results in filtering, with the result that individual Fourier (wave) components may be suppressed. A further interesting aspect of all Fourier series is the existence of conjugate variables. When one expands a function of space and time (i.e., x and t) in a Fourier series, terms occur of the type $\psi \sim \sin(kx - \omega t)$, where k = the variable conjugate to x = the wave number $= 2\pi/\lambda$, and $\omega = 2\pi\nu$ = the variable conjugate to t = the radial frequency $= 2\pi c/\lambda$.

It was quite natural, then, with de Broglie's matter waves, to associate functions of the form:

$$\sin\left(\frac{2\pi x}{\lambda} - 2\pi\nu t\right) = \sin\left(\frac{px}{\hbar} - \frac{Et}{\hbar}\right).$$

The postulate relating wave length to momentum and radial frequency to energy specifies the variables conjugate to x and t. Once the matter wave is indicated, there remains the basic question of interpretation. By analogy with electromagnetic theory, one would assume that the measurable qualities should involve the intensities of the waves, classically given by the squares of the magnitudes. De Broglie's description was incomplete, and therefore insufficient, when applied to complex atomic systems, although the electron diffraction experiments of C. J. Davisson and L. H. Germer in 1927 demonstrated that free particles, such as electrons, could be diffracted as waves.

The duality of microscopic systems seemed to present a fundamental paradox: electromagnetic waves were now conceived of as systems of photons, while electrons assumed a wavelike behavior in certain experiments. All of microscopic physics was thus plagued by the dual behavior of systems. Similarities between the dual behavior of electrons and photons disappear, however, in the limit of classical description. When large ensembles of photons are encountered, the description approaches that of classical electromagnetic theory; electrons, on the other hand, tend to the classical description in the limit where single particles are considered. As opposed to photon systems, large numbers of electrons do not conform to a classical representation but rather necessitate a quantum mechanical description.

The full quantum mechanics of atomic systems was initiated by the joint efforts of Werner Heisenberg (1901–) and Erwin Schrödinger (1887–1961). The early work of each of these men seems to have been done independently. Heisenberg analyzed the quantum states of harmonic oscillators and other dynamic systems in terms of matrices, which were similar in form to the older mathematical systems of matrices, but of infinite order. Matrix mechanics leads to the eigenvalue problem, wherein the solutions appear as sets of discrete states, which in turn can be interpreted as quantum states. Working with Max Born, Heisenberg succeeded in formulating his theory with the full power of the Hamilton-Jacobi classical dynamics built into it. By introducing Poisson brackets, or the commutators of dynamic operators, he was led to the uncertainty principle which bears his name—a principle limiting the accuracy of *simultaneous* measurements of two dynamic conjugate variables. From the Sommerfeld relation, Born found that $\mathbb{P} \cdot \mathbb{X} - \mathbb{X} \cdot \mathbb{P} = -i\hbar\mathbb{I}$. Here $\mathbb{P}$ is the momentum matrix, $\mathbb{X}$ is the matrix corresponding to the position operator, and $\mathbb{I}$ is the identity matrix. This equation led to the uncertainty relation between momentum and position. If Δp represents the accuracy of measurement of the momentum of a particle, and Δx the measurement of variation in the position, then the uncertainty principle appears as $\Delta p \, \Delta x \geqslant \hbar$. This particular form has an equivalent relation in the theory of electromagnetic waves. A fundamental consequence of Fourier analysis is the uncertainty relation between the width of a distribution function of one variable (say k) and that of its conjugate variable (in such a case, x). If a distribution has a mean width Δx in space and a mean width in terms of the distribution of wave number Δk, then classically $\Delta k \, \Delta x \geqslant 1$. From De Broglie's hypothesis, $k = p/\hbar$; therefore, the classical uncertainty is equivalent to the quantum mechanical uncertainty.

From an intuitive conviction that his matter waves must have an associated wave equation, De Broglie, as well as Schrödinger and Klein, carried the search for such a defining equation through several stages. The earliest attempt was a direct analogue of the classical wave equation, with the modification that a rest mass term was included. Formally, this defining relation for a free particle appeared as

$$c^2\nabla^2\psi - \frac{(m_0c^2)^2}{\hbar^2}\psi = \frac{\partial^2\psi}{\partial t^2}.$$

From the identification of momentum with wave number (or wave length) and energy with frequency, one observes that functions of the type $\sin(px/\hbar - Et/\hbar)$, when substituted in this differential equation, give the relativistic energy momentum equation: $E^2 = c^2p^2 + m_0c^4$.

This attempt was unsuccessful, for reasons only fully understood later. First, this form did not reduce to an equivalent nonrelativistic form. Second, the solutions of this differential equation (known as the Klein-Gordon equation) could not be interpreted as matter waves because the probability density was not conserved.

Starting with Hamiltonian of classical dynamics, in 1926 Schrödinger developed a nonrelativistic wave equation for matter waves. His first analysis led to a stationary state equation (no time dependence). Using a more general form of the Hamilton principle function, in the same year he published the nonrelativistic time-dependent differential equation for matter waves which bears his name:

$$-\frac{\hbar^2}{2m}\nabla^2\psi + V(x, y, z)\psi = i\hbar\frac{\partial\psi}{\partial t},$$

where $V(x, y, z)$ is the potential energy function.

To appreciate the physical significance of this equation, one must consider the product function $\psi^*\psi$ and its relation to the conjugate variables. Here the function $\psi(x, y, z; t)$ is in general a complex function, such as $e^{i(\vec{k}\cdot\vec{r} - \omega t)}$, and the function $\psi^*(x, y, z; t)$ is the complex conjugate of ψ. The magnitude squared of ψ, which is $\psi^*\psi$, gives the total probability of finding the particle at a time t between x, y, z, and $x + dx$, $y + dy$, and $z + dz$. To simplify the arguments, consider the motion of a particle or a system of particles in one dimension; for example, regard $\psi(x, t)$. The product function $P(x; t)\,dx = \psi^*\psi\,dx$ gives the probability that the particle is between x and $x + dx$ at t. The area under the curve, between x and $x + dx$, is $P(x, t)\,dx$, the probability of finding it in dx. If there is only one particle in the system, the probability of finding it between $x = -\infty$ and $x = +\infty$ must be unity. Mathematically, one asserts this constraint by the relation

$$\int_{-\infty}^{\infty} P(x; t)\,dx = \int_{-\infty}^{\infty} \psi^*(x; t)\psi(x; t)\,dx = 1.$$

By taking $P(x; t)$ as the product of the function, ψ, and its complex conjugate, one guarantees that $P(x; t)$ will be real and positive (measurable quantities result in real numbers). A positive probability is essential since

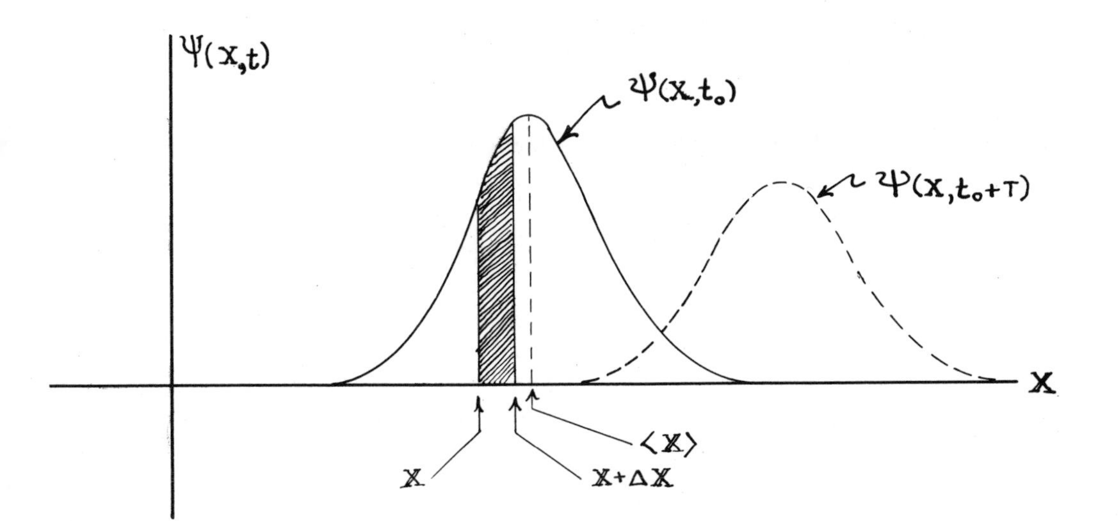

a negative probability is undefined. If one now wishes to know the average distance $\langle x \rangle$ of the particle, described by $\psi(x; t)$, from the origin, it is necessary to compute the average value of x in $P(x; t)$. This interpretation of length implies that x is an operator which operates on $\psi(x; t)$ or $\psi^*(x; t)$. With this type of description, the average distance from the origin is $\langle x \rangle$, where

$$\langle x \rangle = \int_{-\infty}^{\infty} xP(x; t)\, dx = \int_{-\infty}^{\infty} \psi^*(x; t) x \psi(x; t)\, dx.$$

To discover the average value of the conjugate variable (in this case the momentum) of the probability distribution, $P(x; t)$, one must introduce some fundamental properties of Fourier analysis. Here we employ the fact that the function $\psi(x; t)$ has a representation in the space $(k; \omega)$, which is conjugate to that formed by x and t. Fourier analysis has the property that for a well-behaved function $\psi(x; t)$ (in the mathematical sense) there is a transformed function $\phi(k; \omega)$, defined by

$$\psi(x, t) = \mathscr{F}^{-1}[\phi(k, \omega)] = \frac{1}{2\pi} \int_{-\infty}^{\infty} dk \int_{-\infty}^{\infty} d\omega \phi(k, \omega) e^{i(kx - \omega t)},$$

or, in the inverse form,

$$\phi(k, \omega) = \mathscr{F}[\psi(x, t)] = \frac{1}{2\pi} \int_{-\infty}^{\infty} dx \int_{-\infty}^{\infty} dt \psi(x, t) e^{-i(kx - \omega t)}.$$

De Broglie's hypothesis suggests that the average value of k computed in (k, ω) space will provide the average momentum of $P(x; t)$. Because $p = h/\lambda = \hbar k$, we can write

$$\text{the average momentum} = \langle p \rangle = \hbar \langle k \rangle$$

$$= \frac{\hbar}{2\pi} \int_{-\infty}^{\infty} \int_{-\infty}^{\infty} \phi^*(k, \omega) k \phi(k, \omega)\, dk\, d\omega.$$

The clue to the operators of wave mechanics is uncovered when this integral over k and ω is transformed to the space of their conjugate variables, x and t. Direct substitution of the Fourier integrals of ϕ^* and ϕ with integrations over k, ω, x', and t' gives the average momentum in (x, t) space as

$$\langle p \rangle = \frac{1}{2\pi} \int_{-\infty}^{\infty} \int_{-\infty}^{\infty} \psi^*(x, t) \left\{ \frac{\hbar}{i} \frac{\partial}{\partial x} \right\} \psi(x, t) \, dx \, dt.$$

Therefore, the term $\hbar k$ in (k, ω) space appears as the differential operator $(\hbar/i)(\partial/\partial x)$ in (x, t) space. Using the same arguments, one finds that the total energy $E = h\omega$ appears in (x, t) space as the operator $i\hbar(\partial/\partial t)$. Further, the kinetic energy in (k, ω) space (which has the form $p^2/2m = \hbar^2 k^2/2m$) appears in (x, t) space as the operator $-(\hbar^2/2m)(\partial^2/\partial x^2)$. With these connections, the Schrödinger equation may be interpreted as a representation of the average value of the operator equation, $p^2/2m + V(x, y, z) = E$, in the form

$$-\frac{\hbar^2}{2m} \frac{\partial^2 \psi}{\partial x^2} + V\psi = i\hbar \frac{\partial \psi}{\partial t}.$$

This equation implies that the total energy, E, is equal to the kinetic energy, $p^2/2m$, plus the potential energy, V, *on the average*, where the averages are spatial averages only.*

The triumph of Schrödinger's approach became apparent when this differential equation was applied to such simple systems as the hydrogen atom. In the stationary state form, where the time dependence of $\psi(x, t)$ takes on the simple functional behavior $e^{-iEt/\hbar}$, the operator $i\hbar(\partial/\partial t)$ is then replaced by the constant E. This has surprising consequences: to obtain well-behaved solutions to the differential equation, E can only take on discrete ordered values, called eigenvalues; these correspond to the primary quantum states of the atom and agree to a remarkable degree with observations gathered from experiments. This view of the Schrödinger equation illustrates the reason for writing it as a Hamiltonian operator H (H equals the kinetic plus potential energy operator), acting upon ψ to give the energy operator acting upon ψ:

$$H\psi = i\hbar \frac{\partial \psi}{\partial t}.$$

In 1927 Enrico Fermi (1901–1954) indicated that, in a stationary state, the time dependence of ψ^* and ψ cancel, suggesting that the electric charge distribution in the atom (the charge times $\psi^*\psi$) has no time dependence. As a result, the semiclassical theory of radiation predicts no radiation from the stationary state in conformity with Bohr's hypothesis.

* In the mathematical development of this argument, time averages are ultimately taken over arbitrary time intervals. Unless this is done, the dynamic quality of the result is lost.

A partial differential equation of the Schrödinger type is solved by separation of variables. Initially, with the variables x, y, z, and t, one assumes a product solution of four functions, each a function of a single variable:

$$\psi(x, y, z; t) = X(x)Y(y)Z(z)T(t),$$

or, in spherical coordinates with central forces,

$$\psi(r, \theta, \phi; t) = R(r)Y_l^m(\theta, \phi)T(t).$$

In many examples these separate functions are formed from the classical special functions of mathematics. Here it is important to observe that as each coordinate is included, another term is incorporated into the product function. When the angular momentum $\mathscr{L}$ is conserved, the magnitude and projection of $\mathscr{L}$ are given by the eigenvalues $l\hbar$ and $m\hbar$, respectively [l and m appear in the function $Y_l^m(\theta, \phi)$].

Spin is often called a fifth coordinate by physicists. In keeping with this terminology, a wave function with spin then contains a fifth term in the product function, say χ_s^m, where χ_s^m is a function of the magnitude of the spin, s, and the projection of this spin (i.e., m_s) along some preferred axis. Electron spin has a magnitude $\sqrt{(1/2)(1 + 1/2)}\hbar$ [often it is described as $(1/2)\hbar$, which is a misnomer], and we associate $(1/2)\hbar$ with the spin of the electron because the projections of spin along a preferred axis may only be plus or minus $(1/2)\hbar$. To incorporate the spin coordinate in the wave function, one adds to the product another term indicating whether the spin is up or down: when this is done, the total wave function has the form

$$\psi(r, \theta, \phi; s, m_s; t) = R_{nl}(r)P_l^m(\theta)\Phi_m(\phi)\chi_s^m T(t).$$

An appropriate form for the function χ_s^m in the case of spin 1/2 is the two-element column vector:

$$\chi_{1/2}^{1/2} = \begin{bmatrix}1\\0\end{bmatrix} \text{(spin up)},$$

and

$$\chi_{1/2}^{-1/2} = \begin{bmatrix}0\\1\end{bmatrix} \text{(spin down)}.$$

Often one observes two-element vectors (called spinors) with non-vanishing functions in both positions, such as

$$\psi = \begin{bmatrix}F(r, \theta, \phi; t)\\G(r, \theta, \phi; t)\end{bmatrix}.$$

Such a representation implies only that a finite vector space expansion has been performed, with F as the coefficient of the base corresponding to spin "up" and with G as the coefficient of the base associated with spin "down." Using the spinor bases, $\begin{bmatrix}1\\0\end{bmatrix}$ and $\begin{bmatrix}0\\1\end{bmatrix}$, we observe that the form employed above can be written in the conventional vector fashion as a linear superposition:

$$\psi = F(r, \theta, \phi; t)\begin{bmatrix}1\\0\end{bmatrix} + G(r, \theta, \phi; t)\begin{bmatrix}0\\1\end{bmatrix}.$$

Sommerfeld's early criticism of the original Bohr theory was valid for the nonrelativistic Schrödinger equation. In many ways the nonrelativistic result was satisfactory in that it led directly to a matter conservation relation, which took the form

$$\frac{\hbar}{2mi} \operatorname{div} (\psi^* \operatorname{grad} \psi - \psi \operatorname{grad} \psi^*) + \frac{\partial(\psi^*\psi)}{\partial t} = 0,$$

where the term

$$\frac{\hbar}{2mi} (\psi^* \operatorname{grad} \psi - \psi \operatorname{grad} \psi^*)$$

is the matter current. This type of conservation relation had been an important invariant of electromagnetism, leading to the conception of a displacement current. Here again, the faith in invariance (or conservation) provided a constraint by which any new formulation of the wave mechanics could be judged. A logical extension of a nonrelativistic theory of wave mechanics to a relativistic theory should have taken the form:

$$(E - V(\mathbf{r}))^2 = c^2p^2 + (m_0c^2)^2.$$

The free particle solution ($V = 0$) of this equation (the Klein-Gordon equation) demonstrated clearly that any attempt to develop a conservation relation led to negative probabilities. Furthermore, the equation involved the operator E^2 or $-\hbar^2(\partial^2/\partial t^2)$, requiring specification of the first-time derivative of ψ at $t = 0$. In the most general sense, the solutions of the Klein-Gordon equation are scalar or one-component functions, whereas, as has been stated, even the nonrelativistic electron wave function requires a two-component solution. This is not to imply that it was impossible to form a linear combination of two independent scalar solutions; however, it would have been expected that an appropriate theory would supply a two-component solution with dependent components.

By an ingenious reduction of the second-degree Klein-Gordon equation, Paul A. M. Dirac (1902–) avoided this dilemma and in 1928 achieved the desired results with a very significant addition. He constructed a first-degree linear form incorporating four-by-four matrices in such a way as to provide the Klein-Gordon equation when the first-order form is squared. Without going into details, it is sufficient to note that the Dirac first-degree equation for a free particle ($V = 0$) appeared as

$$\left(\gamma_\mu \frac{\partial}{\partial x_\mu} + \mathbb{I} \frac{m_0c}{\hbar}\right)\psi(x_\nu) = 0.$$

The indices μ and ν range from 1 to 4; the four operators γ_μ are 4×4 matrices, each having 16 elements. To produce the appropriate Klein-Gordon equation when operating from the left with a similar first-degree operator, certain constraints are forced upon the γ_μ matrices, and these restrictions essentially define these matrices.

A problem such as that for a free particle has four independent solutions, two of the solutions being associated with the positive energy states of the electron and two associated with a new dynamic system, the negative energy states of the electron. In other words, for free particles, two solutions (one for each spin state) correspond to an energy $+\sqrt{c^2p^2 + m_0^2c^4}$ and two correspond to an energy $-\sqrt{c^2p^2 + m_0^2c^4}$. Thus, in the limit of zero momentum, one set is associated with a rest mass energy $+m_0c^2$, while the other is associated with a rest mass energy of $-m_0c^2$. The existence of two sets of energy states led Dirac to suggest that electrons exist in positive and negative energy states. A vacuum with no electrons present was to be described, then, by an infinite sea of electrons, all in negative energy states, which would not interact with a charged particle. When an electron was excited from a negative state to a positive state, however, a hole would be left in the infinite sea; this hole would exhibit the properties of an anti-electron, being of positive charge with the mass of the electron and having two spin states, up and down. Dirac's original conjecture was that the hole in the distribution of negative energy electrons would be a proton—a conclusion which he soon abandoned.

When in 1932 C. D. Anderson discovered the existence of positively charged particles having a mass equal to that of the electron, the results of the Dirac equation appeared more plausible. Earlier, a seventeen-inch vertical Wilson cloud chamber had been constructed at the California Institute of Technology to investigate cosmic rays originating outside and from the upper atmosphere. This chamber had equipment capable of producing a 24,000-gauss field perpendicular to the path of the cosmic ray showers. In a series of photographs Anderson found particles of positive charge which produced an ionization equivalent to that of an electron. The subsequent supporting evidence established the "positron" as a fundamental particle and as the anti-particle of the electron. Dirac's initial concept of the sea of electrons in negative energy states was not an essential interpretation, and current formalisms which are not concerned with parity regard the two particles, the electron and positron, in a symmetric configuration, each described by a two-component spinor. In 1932 B. L. van der Waerden developed the two component solutions. The vacuum, instead of consisting of a positive and negative energy sea, is viewed as the state of lowest energy in which neither particles nor anti-particles are present. In this representation, both the electron and the positron have positive energy.

From 1926 to 1933, the studies of statistical mechanics assumed a new and broader scope. Planck's law of black body radiation was re-derived in 1916 by Albert Einstein: using Gibbs's canonical distribution as modified for discrete states (the probability of the state n being $e^{-E_n/kT}$), he created a detailed balance relation between the processes of spontaneous emission, stimulated emission, and the absorption of photons. If emission occurs from states at an energy E_n, and if absorption occurs in states having an energy E_m, Einstein showed that the energy density of the radiation was given by the Planck relation. As a result, S. N. Bose and Einstein postulated independently in 1924 a statistical formalism for photon fields. Unlike classical theories, which implied an identity for each particle and allowed an arbitrary number of particles in any one state, the Bose-Einstein statistics removed the uniqueness of particles, so that the interchange of two particles would not produce a new complexion of a distribution. This type of analysis gave the Planck radiation law directly, and as expected, it approached a Maxwellian form as Planck's constant, h, was allowed to approach zero. The significance of this result lay in its application to many problems of photon fields, and later to phonon fields, arising from the vibrational modes of crystals.

Pauli's exclusion principle for the allowed number of electrons per state in an atom suggested further that, in addition to the classical statistics and the photon statistics of Bose and Einstein, there should be also a special statistics for electrons, in fact, for all particles of spin $(1/2)\hbar$. In 1926 Fermi constructed a system of statistics in which only one particle was allowed to occupy a given state. Further, assuming that spin up and spin down split any energy state into two independent states, the particles were indistinguishable, and interchanges which maintained the same spin orientation did not result in a new complexion of the distribution.

All three types of statistics can be derived from arguments similar to those employed in the calculation of the general binomial coefficients. The total number of configurations for N things taken n_i at a time is set equal to the product of the number of distinguishable configurations and the number of indistinguishable configurations. From this, the probability per state of finding a particle with an energy between E and $E + dE$ is obtained as the number of distinguishable configurations in that energy interval, subject to the constraints that the total number of particles in the system and the total energy is constant. The total energy is the sum of the individual energies of the particles. These three theories differ only in the manner by which the number of indistinguishable configurations are calculated.

The result of Fermi's analysis for electrons gave a distribution function (probability per state of finding an electron between E and $E + dE$) of the form

$$f(E, T) = \frac{1}{e^{(E-E_F)/kT} + 1},$$

where T is the absolute temperature and E_F is a constant known as the Fermi energy. Immediate success accompanied this result, particularly in the theory of solids. To account for the paramagnetism of the alkali metals, Pauli in 1927 assumed that the conducting electrons in a metal can be viewed as a degenerate electron gas, described by Fermi statistics. Following this, Sommerfeld developed a general approach to the electron theory of metals based upon the same statistics. Sommerfeld's electron theory of metals was then employed by R. H. Fowler, J. A. Becker, E. O. Lawrence, and others to describe the photoelectric sensitivity of metals and to derive the Richardson equation for thermionic emission.

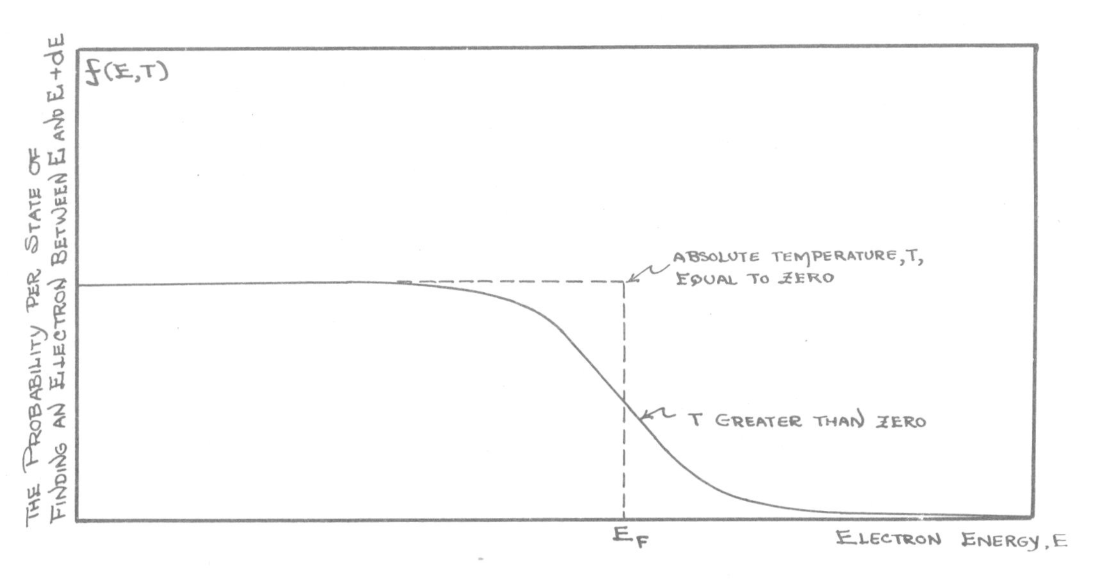

All of these calculations involved a work function, which was a measure of the energy difference between the top of the Fermi sea at E_F and the zero of potential outside the metallic surface.

Convincing evidence for the nuclear atom had been uncovered in the Rutherford alpha scattering experiments and their subsequent analysis in 1911. Barely eight years later Rutherford discovered the artificial disintegration of nitrogen. After the famous scattering experiments of 1911, the investigation of the ranges of scattered nuclei became a common interest, and all scatterings seemed to obey classical elastic laws, conserving energy and momentum, until *RaC′* alphas were scattered off of nitrogen. These experiments showed a number of long-range protons being ejected as a result of the alpha bombardment. The number of protons observed was far and above what could be accounted for by any impurity, and as a result Rutherford suggested that the heavier nuclei were being broken up into lighter components, yielding knocked out protons, in some cases. These experiments were immediately followed by a series of studies of the disintegration of nuclei by alphas.

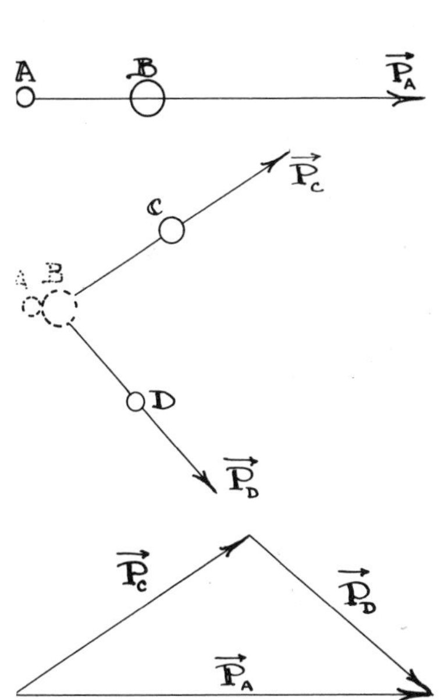

In light of the mass-energy relationship, the energetics of these disintegrations are quite simple. Assume that a nuclear system A is incident upon a nuclear system B at rest; after collision systems C and D emerge. Momentum is conserved; therefore, if the initial momentum of the projectile is $\vec{\mathbf{P}}_A$, the final total vector momentum $\vec{\mathbf{P}}_C + \vec{\mathbf{P}}_D$ is equal to $\vec{\mathbf{P}}_A$: $\vec{\mathbf{P}}_A = \vec{\mathbf{P}}_C + \vec{\mathbf{P}}_D$.

The total initial energy must equal the total final energy (total meaning kinetic plus rest mass energy). Thus, if we assume that B is initially at rest,

$$\frac{M_A c^2}{\sqrt{1 - V_A^2/c^2}} + M_B c^2 = \frac{M_C c^2}{\sqrt{1 - V_C^2/c^2}} + \frac{M_D c^2}{\sqrt{1 - V_D^2/c^2}},$$

where M_N is the rest mass of the Nth particle.

In the nonrelativistic limit, where the velocities are much less than c, the velocity of light, this expression can be written in terms of the nonrelativistic kinetic energy* as:

$$M_A c^2\left(1 + \frac{1}{2}\frac{V_A^2}{c^2}\right) + M_B c^2 = M_C c^2\left(1 + \frac{1}{2}\frac{V_C^2}{c^2}\right) + M_D c^2\left(1 + \frac{1}{2}\frac{V_D^2}{c^2}\right),$$

or

$$\frac{1}{2} M_A V_A^2 + (M_A c^2 + M_B c^2 - M_C c^2 - M_D c^2) = \frac{1}{2} M_C V_C^2 + \frac{1}{2} M_D V_D^2.$$

The term in brackets is the difference in the total rest mass energy before and after the collision. This quantity is known as the Q value (an energy) of the reaction, and if the Q value is positive, the final total kinetic energy will be greater than the initial kinetic energy, denoting a loss in mass during the reaction.

From the initial discovery until 1932, investigations of artificial disintegrations of nuclear reaction studies were limited by the fact that the only sources of projectiles were naturally occurring alpha emitters. Thus the available currents of bombarding particles were low, and there was little opportunity to exert control over the projectiles. By 1930 several methods of producing high-energy charged particles had been discovered, the most notable being the Cockcroft and Walton cascaded rectifiers. Cockcroft and Walton were the first to succeed in producing nuclear reactions with high-energy ions generated in the laboratory. Their original experiments employed positive hydrogen ions. The invention of the cyclotron in 1930–1931 by E. O. Lawrence made possible the production of beams of ions of much higher energy and of much higher intensity. After these developments, the control and study of nuclear reactions became a routine matter.

One of the major puzzles in the theory of nuclear structure at that time was the nature of the neutral participant. It was recognized from the initial postulate of the planetary atom, with its heavy positively charged nucleus, that the charge of the nucleus could be accounted for by incorporating into it a set of protons equal in number to the number of electrons in the atomic orbits. The mass of the nucleus, however, was on the average twice, or more than twice, what it would be if only the protons were present. For example, it was known that the helium nucleus had a charge equivalent to two protons but a mass equivalent to four. At first it was assumed that the additional mass was made up of protons whose charge was masked by very tightly bound (within 10^{-12} cm) electrons; the presence of electrons in radioactive decay seemed to support such a conjecture until the uncertainty relation of quantum theory was considered. To confine an electron in a space of dimension Δx would require momentum

* Here we expand $1/\sqrt{1 - V^2/c^2}$ as $1 + (1/2)(V^2/c^2)$ + higher-order terms.

variations of the order of $\hbar/\Delta x$, which in turn would correspond to hundreds of millions of electron volts of energy. Radioactive beta decays were on the order of one million electron volts, and therefore the electron confinement was not borne out by the energies observed in nuclear electron decays.

While investigating the interaction of the alpha rays from polonium incident upon the light nuclei, beryllium, boron, and lithium, Bothe and Becker in 1930 had found that a very penetrating radiation was emitted from these reactions. By 1932 Irène Curie and Frédéric Joliot showed that this radiation seemed to increase its ionization potential after passing through paraffin or some other hydrogen-bearing material. Their conjecture was that the unknown radiation was a very high energy photon of the order of 50 mev which interacted with hydrogenous material by a Compton-type scattering in which neither energy nor moment was conserved. Clearly, the suppression of the conservation laws of dynamics was most unsatisfactory, and physicists viewed this hypothesis with justified suspicion.

A solution to this perplexing situation was provided by Sir James Chadwick in 1932. He emphasized the fallacy of the assumption of a photon-proton collision. By measuring the recoil energies of both protons and nitrogen nuclei when subjected to the penetrating radiation formed when polonium alphas were incident upon beryllium, Chadwick showed that this highly penetrating radiation was an uncharged particle with a mass essentially equal to that of the proton. This was the discovery of the neutron, the second massive building block of the nucleus. In a remarkable paper, Chadwick analyzed the experimental recoil data to arrive at a mass for the neutron. He found that its mass was slightly larger than that of the proton, a fact that was to suggest that the free neutron would decay into a negative electron and a proton. Beta decays were to bring to light still another new particle, the neutrino.

Beta radiation from radioactive sources had been investigated since the early part of the twentieth century by various means; absorption, ionization chambers, cloud chambers with and without magnetic fields superimposed, and, ultimately, vacuum magnetic spectrographs employing either photographic plates or geiger counters for recording the particles. The remarkable attribute of the beta rays was that they possessed a range of energies extending from zero to some fixed maximum, rather than a unique single energy—which might have been expected if the nuclear states were quantized, with the beta decay represented as a transition between states. Gamma rays (photons) often accompany a beta decay, and it had been shown by Rutherford and his collaborators that the gamma rays preceded the electron emission. Thus the presence of a continuous range of electron decay energies could not be explained by some gamma transitions, and some doubts lingered concerning the appropriate theory of these decays.

Pauli proposed that beta decays were actually two-particle emissions involving an electron and a second particle, the anti-neutrino, which

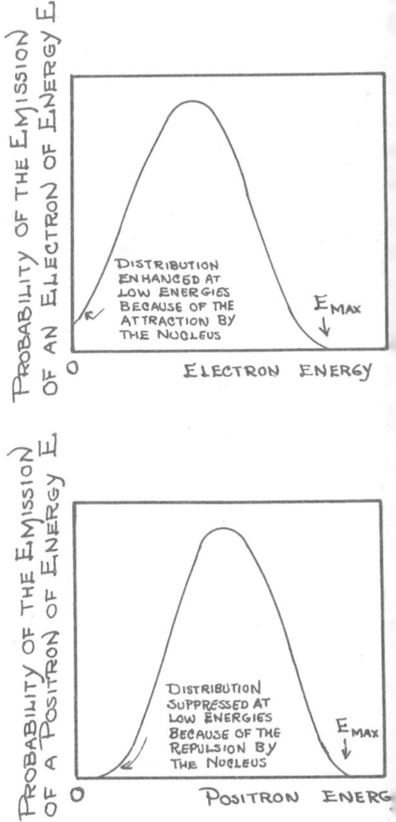

has zero charge, zero mass, and an intrinsic spin of $1/2\ \hbar$. Acting upon this, Enrico Fermi in 1934 constructed a theory which accounted for the continuous energy spectrum in a highly satisfactory fashion. He assumed that the nuclear transition was indeed a transition between quantized states. This nuclear transition was represented by a quantum mechanical integral or matrix element, and the probability for the decay was further modulated by the density of states in phase space for the electron and the density of states for the neutrino. By imposing the constraint of energy conservation between the initial and final quantum states, the total product was a function of the electron energy, and the major aspects of the spectral function were taken into account by the statistical phase space factors. It is convenient that the same analysis can be applied to the positron decays, with the exception that the positron is accompanied by the anti-anti-neutrino, or neutrino.

Positron decays do not occur from naturally occurring radioactive nuclei. In 1934 Irène Curie and Frédéric Joliot discovered, however, that after certain light elements are bombarded with alpha particles, they become radioactive and, in some instances, emit positrons; one of their early experiments involved the bombardment of boron 10, written ${}_5B^{10}$, with alphas. Symbolically, they pictured the resulting chain of events as leading to a radioactive nitrogen 13, meaning 7 protons and 6 neutrons for the nitrogen core: ${}_5B^{10} + {}_2He^4 \rightarrow {}_7N^{13} + n$, and ${}_7N^{13} \rightarrow {}_6C^{13} + \beta^+ + \nu$.

In this symbolic equation, ${}_7N^{13}$ decays to a ${}_6C^{13}$, a carbon isotope (6 protons and 7 neutrons), accompanied by the emission of a positron, β^+, and a neutrino, ν. One can think of the emission as the transformation of a proton in the nitrogen 13 into a neutron producing a carbon 13. We observe that the total number of nuclei in beta decay remains constant and that the beta decays, or positron decays, represent transitions of neutrons to protons or protons to neutrons, respectively.

With this discovery of artificial radioactivity, the attention of physicists was concentrated on the nucleus for some twenty years. Shortly after the discovery by Joliot and Curie, Fermi and his coworkers in Rome were able to slow down the relatively fast neutrons from nuclear reactions to essentially thermal velocities. These thermalized or slow neutrons proved to be many orders of magnitude more effective in nuclear reactions —they led first to the neutron pile and later to the nuclear bomb.

As one might surmise, in this period of almost constant discovery of new particles, there were no broad and encompassing theories. The major question of the age centered about the nature of the nuclear force binding the protons and neutrons together. There was a need for the equivalent of Newton's universal law of gravitation, a force law which was very strong and of very short range. With the discovery of the photon, theorists had conceived of the electric field, and even the gravitational field, as composed of field particles which give rise to the appropriate potentials through exchanges between interacting bodies. To borrow from a more recent diagrammatic representation, one can plot the interaction of two nucleons in a space-time diagram. Here the interaction appears as a scattering in space time accompanied by the exchange of a field particle.

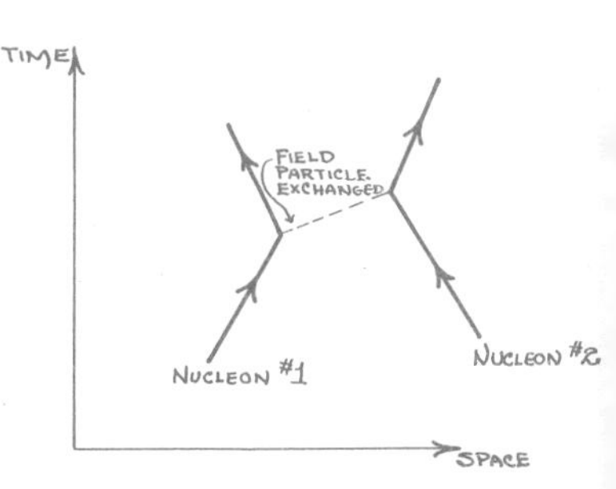

Modeling his approach to the nuclear interaction upon these photon field theories, in 1935 H. Yukawa proposed that the field particles governing the nuclear potential were derived from the scalar Klein-Gordon equation and had a rest mass determined by the range of the nuclear force:

$$\{-c^2\hbar^2\nabla^2 + \mu_0^2c^4\}\phi(x, y, z; t) = -\hbar^2 \frac{\partial^2\phi}{\partial t^2}\cdot$$

The solutions of this equation gave rise to a potential function of the form

$$\frac{e^{-r/R}}{r},$$

where $R = \hbar/\mu_0 c$ is the characteristic length of the interaction. If one takes R as $1.3(10)^{-13}$ cm, the corresponding particle mass, μ_0, is of the order of 230 electron masses. G. C. Wick showed in 1938 that Yukawa's mass formula was equivalent to the distance associated with the uncertainty in time if an energy of the order of $\mu_0 c^2$ is absorbed during the exchange: $\Delta t = \hbar/(\mu_0 c^2)$. Thus the lifetime of the field particle must be at least Δt, and in this time it will travel a characteristic distance R, given by: $R = c\,\Delta t = \hbar/(\mu_0 c)$, giving Yukawa's result: $\hbar/(Rc) = \mu_0$.

Within the same year, C. D. Anderson and S. H. Neddermeyer discovered in the multitude of charged particles in cosmic rays a particle with a mass intermediate between that of the electron and the nucleon. The natural assumption was to link the Yukawa field particle and this new particle, called a meson. What Anderson and Neddermeyer actually observed was the mu-meson, which is the decay product of the pi-meson: the pi-meson was to become the field particle to which Yukawa gave substance.

The half-life of this first intermediate particle, the mu-meson, was measured first by inference, using the rate of disappearance in the atmosphere. F. Rasetti (1902–) in 1941 made a direct measurement of the half-life of what is now called the mu-meson by stopping these particles in a carbon block and measuring the subsequent decay into positrons. His measurement led to the value $2.2(10)^{-6}$ sec. Conversi, Pancini, and Piccioni, after analyzing all available data, concluded in 1947 that the cosmic ray mu-meson did not interact strongly with nucleons and, therefore, could not be the field particles involved in nucleon-nucleon forces. C. F. Powell and G. P. S. Occhialini shed light on the confusion of mesons in 1947, when they detected the track of a pi-meson stopping in a thick photographic emulsion: this event was followed by the emission of a mu-meson plus a neutral particle, later found to be a neutrino. The masses of the pi-meson and mu-meson turned out be 273 and 207 electron masses, respectively. The pi-meson was indeed Yukawa's field particle, and it decays into a mu-meson plus a mu-neutrino with a half-life of $2.55(10)^{-8}$ sec. Mu decays proceed by electron or positron emission, depending upon the sign of the charge of the μ, and are accompanied by two neutrinos, one an electron neutrino and the other a mu-neutrino: $\pi^+ \rightarrow \mu^+ + \nu_\mu$, and $\pi^- \rightarrow \mu^- + \bar{\nu}_\mu$. Here ν_μ and $\bar{\nu}_\mu$ are the mu-neutrino and anti-neutrino, respectively. Finally, $\mu^+ \rightarrow e^+ + \nu_e + \bar{\nu}_\mu$, and $\mu^- \rightarrow e^- + \bar{\nu}_e + \nu_\mu$, where ν_e and $\bar{\nu}_e$ are, respectively, the electron neutrino and the anti-neutrino.

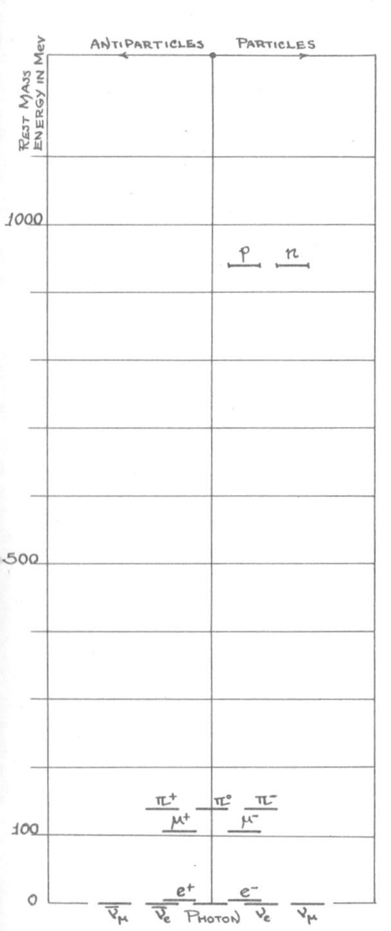

This collection of events does not quite clear up the pi-meson puzzle. With the completion of the large accelerators in the early 1950's, it became possible to study the scattering of pi-mesons from various substances. In other words, a high-energy proton beam was stopped in a target material, producing, as an example, a secondary beam of pi-mesons. Upon interaction with protons, these mesons were found to lead a fraction of the time to another pi-meson, the neutral pi, π^0, which, after slowing down, decays into two gamma rays (Panofsky, Steller, and Steinberger, 1951).

Symbolically, the production and decay of the π^0 can be diagrammed as follows: $\pi^- + p \rightarrow \pi^0 + n$, and $\pi^0 \rightarrow 2\gamma$.

This hasty summary allows us to list the elementary particles known by 1951 in a sort of preparatory table. After this period, the flood of newly discovered particles plunged the world of theoretical physics into a frantic upheaval. In this table some particles which were actually postulated later have been identified, namely, the mu-neutrinos and anti-neutrinos.

Starting with the discovery of the electron in 1898, it took but half a century to erect a structure containing the nucleons, the positron, the neutrinos, and finally, the mu- and pi-mesons. Progress was not steady and continuous, however, for in the period 1951–1968 the list of known particles was altered dramatically by the discovery of a multitude of new and strange particles.

Cloud chamber pictures and emulsion photographs had shown, on two occasions before 1951, the existence of a new species of particles. In the early days (1947–1949) these anomalous events were labeled "V particles," and with the introduction of the new high-energy accelerators at Brookhaven and Berkeley, the anomalous particles became regularly observed events; the subsequent array of new particles is still mounting.

All attempts to describe the new particles are based upon quantum theory—a quantum theory, however, with improvements and additions, making it a powerful analytic tool. To handle the calculations of quantum electrodynamics, R. Feynman introduced rather elementary space-time diagrams, each representing some basic integral in a quantum mechanical calculation of an interaction. As an example, we would, in modern theory, represent the interaction of two electrons as the exchange of a photon, rather than invoke "action at a distance," such as is done in Coulomb's law. This exchange would be represented as the space-time scattering with an intervening photon.

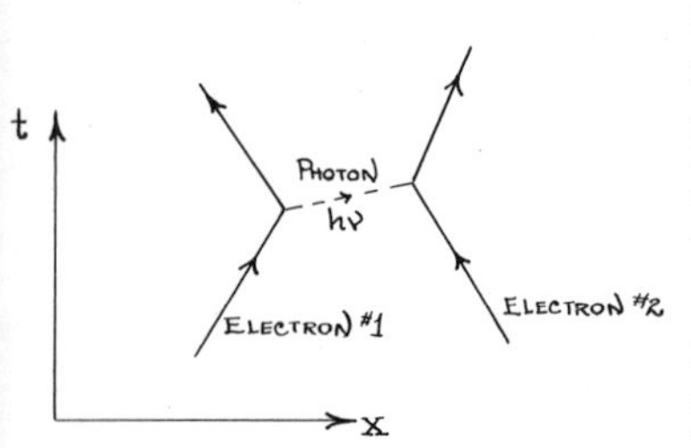

These are known as local field theories, in that the fundamental interactions occur only at specific points in space and time. An interesting picture in the original diagrammatics arose in positron motion: the anti-particles of all spin, 1/2 particles can be represented as the original particles in a negative energy state moving backward in time. The diagram illustrates a particularly simple example of pair production by a photon interacting with a heavy nucleon.

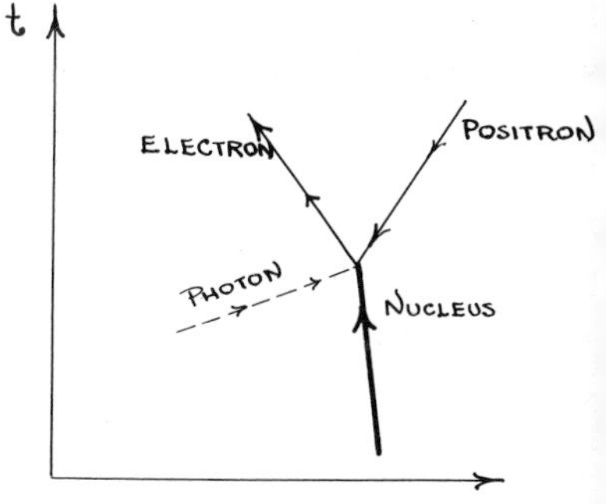

Quantum electrodynamics is the modern quantum method for performing calculations regarding the interaction of electromagnetic fields with electrons. The interaction of the Maxwell fields with various particles has a very unique and significant representation in terms of these space-time graphs. One characterizes a single process by the paths entering and leaving the region of local field interactions. At best, the calculation of this process is a series expansion, known as a perturbation expansion. Such a series can be represented diagrammatically by the sum of all of the single processes which can be portrayed by graphs of increasing complexity.

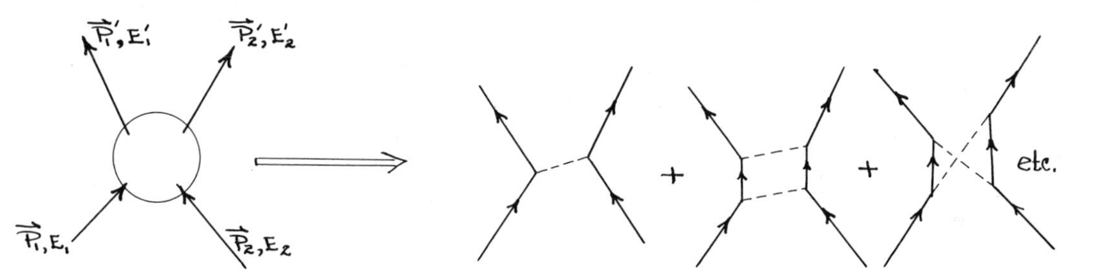

Each vertex has a strength in a quantum electrodynamic graph of the order of the square root of the fine structure constant; i.e., $\sqrt{e^2/hc} \simeq \sqrt{1/137}$. Thus, the first graph above is weighted by 1/137, compared with a weight of $(1/137)^2$ for the second diagram having four vertices.

In 1935, in the earliest days of quantum electrodynamics, the lowest or first-order calculations of electron interactions provided excellent agreement with observations. When higher-order corrections were attempted, however, the results tended to become undefined or infinite. Even when one considers the motion of a single electron in the absence of an external field, the possibilities of emitting and reabsorbing a photon are present. If calculated, these possible virtual photon emissions and absorptions lead to an infinite self-energy for the electron.

By the late 1940's, largely through the work of Kramers, Bethe, Feynman, Schwinger, Tomonaga, and Dyson, a consistent method was developed to separate the finite parts of these calculations from the infinite parts. To achieve this goal, the mass and charge of the electron were

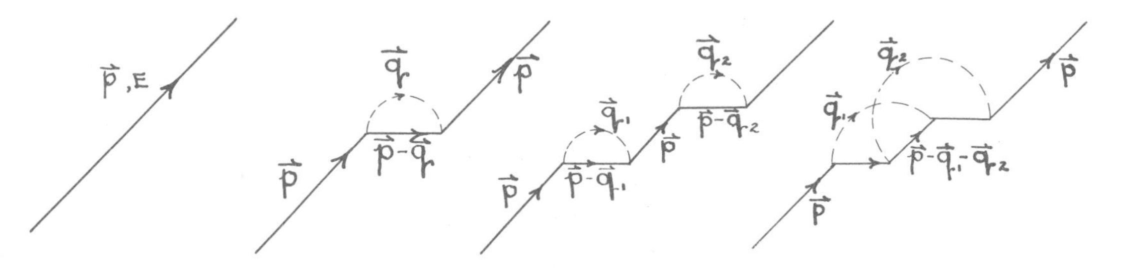

defined in terms of a bare mass and a bare charge, both of which were composed of a finite observed and an infinite component. This class of calculations is known as "re-normalization." The procedure is surprisingly effective, and with it a number of extraordinary higher-order corrections have been successfully carried out; for instance, the anomalous magnetic moment of the electron, corrections to Coulomb's law because of vacuum polarization, etc.

The whole of particle physics faced a very rapidly changing situation in the early 1950's. With the big accelerators (proton energies from 500 mev to 1 bev), a multitude of new particles were observed. The striking aspect of these events was the observation of lifetimes which were sufficiently long to be inconsistent with the strength of the interactions. Strong interactions, such as those between nucleons, must occur within times of the order of 10^{-22} to 10^{-23} seconds.

The discovery of the Λ^0 hyperon presented a puzzle: produced by the interaction of a pi-minus meson and a proton, the Λ^0 lives for $2.62(10)^{-10}$ sec and then decays into a negative pion and a proton. A. Pais and Y. Nambu suggested that such a particle must be produced in combination with a second particle, on the assumption that the two together proceed via a strong interaction; however, when separated, each decays by a weak interaction. This concept was named "associated production." In rapid order the K^0, $\overline{K^0}$, the Σ family, and the Ξ mesons were discovered.

To achieve some order in this confusion of particles, physicists have attempted to incorporate additional coordinates into their descriptions. The splitting of the atomic spectral lines had been explained by the introduction of internal spin (an additional coordinate). C. N. Yang had developed the analogue of spatial parity by the creation of intrinsic parity—intrinsic in much the same fashion that spin angular momentum is intrinsic, as compared with orbital angular momentum. It was quickly shown that high-energy and low-energy nuclear reactions proceeded in such a manner as to favor the conservation of overall parity, intrinsic plus normal. The concept of the parity of systems at the spatial level involves the reflection of each point of the system through the origin to produce a kind of mirror image of the system.

Consider the righthanded coordinate system reflected through the origin. From the above diagram we observe that after reflection the right-handed system becomes a lefthanded system, and any system or function which is unchanged upon reflection is said to have even parity, while a system which changes sign or changes from right- to lefthanded is said to have odd parity. For instance, the orbital angular momenta of quantum mechanics have parity assignments; the even angular momenta have even parity, while the odd have odd.

Parity was assumed to be conserved in nuclear decays or in nuclear reactions; that is to say, the evenness or oddness of the parity of a system before must be preserved after. Because of certain pi-meson reactions, it was concluded that particles themselves must have intrinsic parity in just the same manner as they possess intrinsic spin or angular momentum.

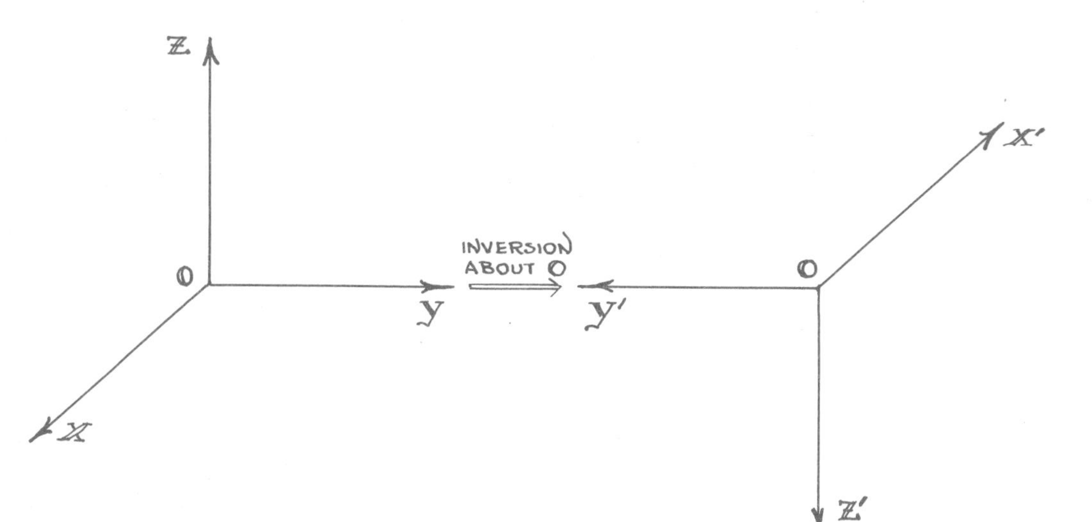

In 1956, a study of K-mesons led C. N. Yang and T. D. Lee to postulate that although parity must be conserved in strong interactions, weak interactions, involving beta decays, need not conserve parity. This revolutionary hypothesis was verified in the experimental work of C. S. Wu and R. Hayward, and also in studies of mu-mesonic decays by Garwin and Lederman.

Another quantum attribute, isotopic spin, had been employed by theoretical physicists since the early 1940's, although this attribute has nothing to do with spin; its mathematical representation takes the same form as the column vector representation of spinors. Here one regards the proton and neutron as two different isospin states, and in this way very useful representations can be constructed not only with the operators of the two-nucleon system but with those of other systems as well. For example, the various pi-mesons are viewed as the three projections of a spin 1 vector.

All of these quantum attributes are essentially empirical and without a unifying theoretical structure at this time. To complete the picture of artificial quantum numbers, in 1953 M. Gell-Mann and K. Nishijima postulated a "strangeness" quantum number, which became an enormous aid in the cataloging of the "whys" and "why nots" of associated production. In high-energy reactions and decays some processes seem to proceed favorably, while others do not. Ordinary quantum rules were insufficient to account for the observed cross sections, and it was postulated that in strong interactions the new quantum number "strangeness" must be conserved. The strangeness, S, of a particle was related to the charge Q, the isotopic spin component, T_3, and the baryon number N (the number of nucleons that ultimately appear as final products of the particle): $Q = T_3 + N/2 + S/2$.

By 1964 there were thought to be at least 82 strongly interacting particles and anti-particles, and the scheme of things was made to appear even more satisfactory by the discovery of the anti-proton and the antineutron by E. G. Segre and O. Chamberlain. During the last six years, several appealing attempts have been made to collect the known particles into families described by various group representations: the SU_3 group initiated by M. Gell-Mann and Y. Ne'eman; the SU_6 group by F. Gürsey,

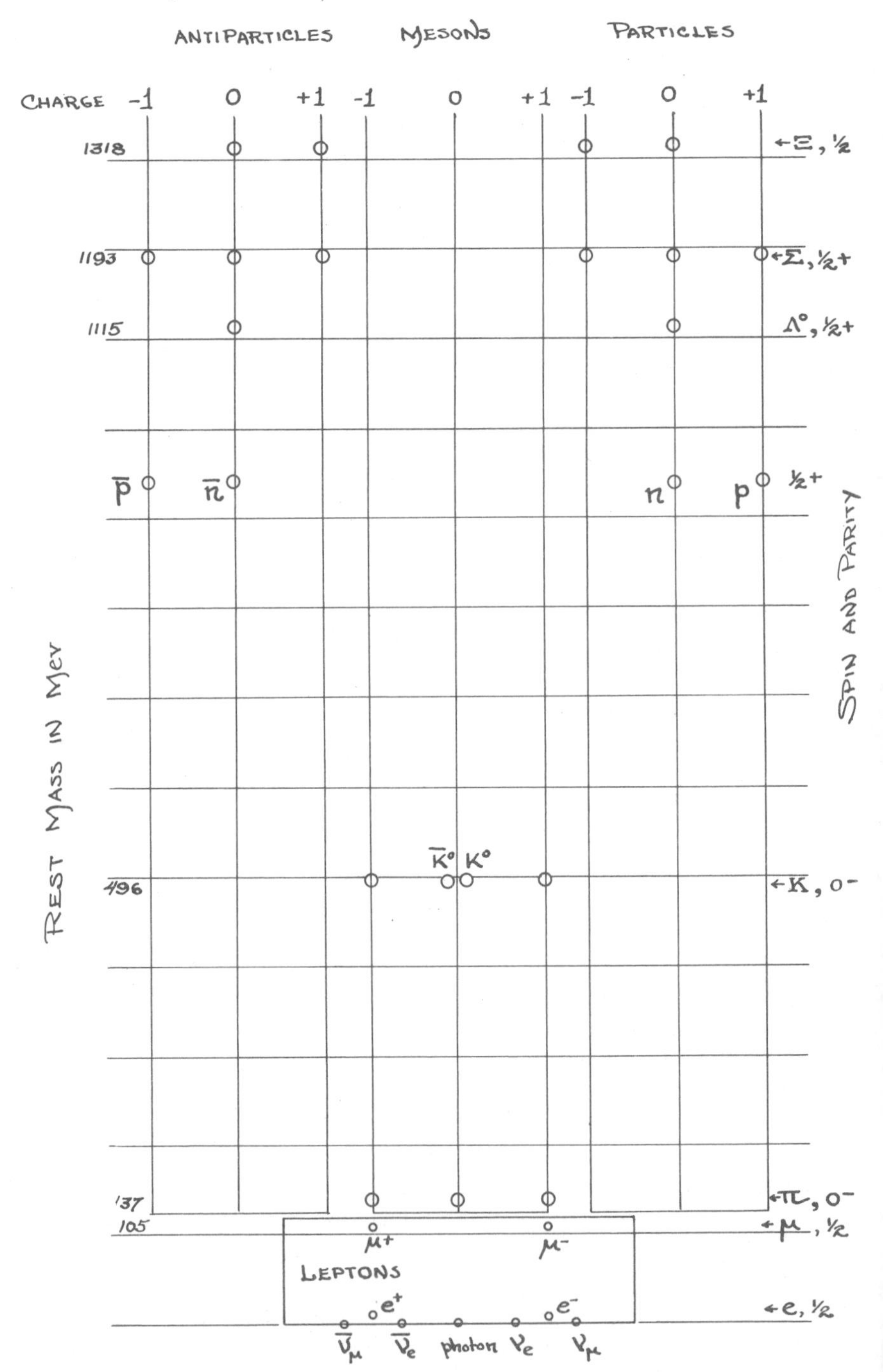

Radicatti, and A. Pais; and higher groups by A. Salam. A summary of the particles we now have to deal with is shown in the chart opposite.

The introduction of symmetry representations into the description of the elementary particles has as its analogue the type of description which might have been given for the levels of the hydrogen atom had the detailed model and Coulomb force been unknown. Because of the degeneracy in the various levels of different angular momenta in the hydrogen atom, one would have surmised that the surfaces of equipotential were spherical and the forces invariant under rotations. If one now views groups of baryons as merely energy levels of a single system, initially degenerate but split into the observed spacings by an interaction analogous to an external field on the hydrogen atom, the method has some merit. Gell-Mann proposed that all are formed from three basic building blocks, which he named "quarks." A consistent scheme then required that these three objects each have a baryon number $+1/3$, and charges $+2/3$, $+1/3$, and $-1/3$ of the charge of the proton. Quarks have supposedly been seen—however, the evidence is inconclusive. This inconclusiveness serves as an indication of the uncertain state of physics at present.

In our time we can observe the emphasis in physics gradually shifting. Rather profound alterations have occurred in the nature and training of physical scientists. Where breadth of interest was once the custom, one discovers now that extreme specialization within a subject is leading more and more to a community of mathematical technicians. Relations among the sciences, government, and society are creating an environment less and less tolerant of quiet contemplation. Modern physics has acquired an aura of political sophistication and showmanship: criticism of this state is futile because the importance of physics in a modern society creates and reinforces the condition.

All guesses as to the next generation of ideas are equally valid. The shift of interest from the microscopic to the cosmological is already taking place, and continued interest in astrophysics—in studies of gravitation, relativity, and the cosmos—is assured. Indeed, it is possible that the physical sciences have used up their legacy of mathematics: one should not expect this, however, for scientific curiosity is aroused, and the evidence of the past is that curiosity dies long before the art of mathematical expression expires.

Some might be more concerned with the state of mathematics. After centuries of remarkable progress, much of the interest in the field is now directed inward. All efforts to prove the consistency of the postulates of arithmetic have ended in strongly inconsistent results. In 1931 K. Gödel showed that in certain logical systems it is impossible to prove through the axioms of the system certain of its theorems which otherwise appear true. Gödel constructed a true theorem such that a formal proof within the system of arithmetic led to a contradiction: thus history was at a crossroads. In the past it had been assumed that every arithmetic theorem was provable by deduction, but this assumption seems to be invalid.

As geometry was the mathematics of the ancient scientists, analysis has become the tool of modern physics. Continuum analysis has already been discarded in some quantum theories and in the description of new particles: groups and group algebras have become of increasing importance. Quite possibly the future world of physics will lean more and more heavily upon qualitative mathematical techniques, such as those exhibited in topological research. The role of the computer in the evolution of mathematics has not yet been assessed. If one focuses upon changes in notation as signposts of new eras, the application of machines to new types of thinking and new types of representation may well contribute to progress.

The future of physics will continue to be governed largely by experimental evidence. Alternatives are always possible, and one must not disregard or minimize the activity in the more mystical regions of scientific philosophy. Future philosophers may not remain content merely to analyze what scientists have done—civilization could revert to a more ancient practice in which the philosophers lead more often than they follow.

XI

The Players in the Order of Their Appearance

B.C.E.

ca. 1800	*The Rhind papyrus* (*Ahmes the scribe*)
624–546	*Thales of Miletus* is credited with the introduction of mathematical proof into geometry; he was known to have traveled extensively in Egypt and Babylon
611–547	*Anaximander* introduced the gnomon and created models of the universe
569–500	*Pythagoras of Samos* according to legend created the Pythagorean school of geometry, number, and harmony; after studying in Babylon and Egypt, he returned to Samos, only to flee to southern Italy with his followers, where he is supposed to have established a mystical order based upon mathematics
500–428	*Anaxagoras of Clazomenae* described the planets and the moon and provided an explanation of eclipses
495–435	*Zeno the Eleatic* constructed famous paradoxes which, although not without merit, served to stifle the extension of the method of exhaustion
ca. 500	*Hippasus* split off from the Pythagorean sect and formed the "Mathematikoi"; one of the earliest to break the pledge of silence
b. 460	*Hippias of Elis* a Sophist, among other achievements invented the quadratrix to subdivide a given angle by a known fraction
ca. 450	*Oenopides of Chios* was concerned with astronomical periods of revolution and the calendar

ca. 430	*Hippocrates of Chios* lived in Athens and was concerned with the duplication of the cube; he is remembered for his analysis of lunes
ca. 430	*Democritus* both mathematician and scientist, wrote on perspective, astronomy, and geometry; according to Archimedes, he discovered that the volume of a cone or pyramid is one-third the volume of the cylinder or prism having the same base and altitude
ca. 430	*Theodorus of Cyrene* was called Plato's teacher; his main contributions were in the study of irrational line segments
429–348	*Plato* was the pupil and friend of Socrates; according to some accounts after traveling in Asia Minor he returned to Athens in 389 and founded his school of philosophy there
428–347	*Archytas of Tarentum* according to legend a contemporary of Plato and a member of the Academy, is supposed to have invented the pulley; the solution of the duplication of the cube in terms of mean proportionals is ascribed to him
415–369	*Theaetetus* an Athenian, was one of the great mathematicians of the time of Plato; his death is described by Euclid of Megara; assumed to be the author of the first part of Book X of the *Elements*, he is particularly known for his work on the theory of proportions and the theory of the five regular solids
384–322	*Aristotle* was one of the great members of the Academy; his works on metaphysics, physics, and politics were to have a lasting effect upon the course and development of science and philosophy
408–355	*Eudoxus of Cnidus* was one of the most famous mathematicians of antiquity; his contributions to the Elements represent many of the more important propositions; the theory of proportion and the use of the method of exhaustion are some of his most noteworthy achievements
375–325	*Menaechmus* was a pupil of Eudoxus; it is reported that he discovered the conic sections; by employing the parabola and hyperbola he effected a solution to the Delian problem
ca. 350	*Dinostratus* the brother of Menaechmus, rediscovered the quadratrix of Hippias and developed a fair approximation for the value of π
ca. 320	*Autolycus of Pitane* an ancient Athenian mathematician and astronomer, developed the kinematics of points and circles on a rotating sphere and gave a lengthy account of the rising and setting of stars
330–275	*Euclid* a teacher of mathematics in Alexandria, his compilation of all of Greek mathematics up to his time represents one of the most influential mathematical works of all time; the series of books represented by the Elements has dominated mathematics for over two thousand years
287–212	*Archimedes* born in Syracuse and a student at Alexandria, was with Eudoxus the greatest of ancient mathematicians; his writings cover the areas of physics and mathematics, and use of the method of exhaustion brought him near to the calculus of the seventeenth century
276–195	*Eratosthenes of Cyrene* a young contemporary of Archimedes, was head of the famous library at Alexandria and is remembered for his measurement of the circumference of the earth
b. 264	*Aristarchus of Samos* a philosopher and astronomer, is famed

for his attempt to estimate the distances of the sun and moon from the earth

262–200 *Apollonius of Perga* was the last great geometer of antiquity; conic sections were his major interest, and the names of the sections were created by him; his astronomical studies led to the creation of the epicycles and eccentric curves

ca. 240 *Nicomedes* is known for his work on the trisection of the angle and the duplication of the cube

ca. 160 *Hipparchus of Nicaea* an astronomer, compiled a star catalogue and provided an accurate determination of the length of a year. Ptolemy's trigonometry was initiated by Hipparchus

C.E.

ca. 100 *Menelaus of Alexandria* one of the last geometers of antiquity, wrote on the sphere and spherical triangles

ca. 80 *Nicomachus of Gerasa* was an arithmetician whose arithmetic is more of an introduction to the philosophy of the subject than a scholarly treatment

ca. 100 *Sun Tzu* wrote *Arithmetic Classic in Five Books*; although this is one of the best-known early Chinese works, it is obscure and inaccurate

100–168 *Claudius Ptolemy* the celebrated astronomer and mathematician of the late Alexandrian school; summarized in a single treatise, the *Almagest*, all of the work of his predecessors on trigonometry and astronomy; he was preeminent in the application of mathematics to astronomy

ca. 150 *Heron of Alexandria* a scientist and mathematician, ranged from the mensuration of plane and solid figures to mechanics and astronomy

ca. 250 *Diophantus of Alexandria* was an algebraist; usually his *Arithmetica* is regarded as the first treatise on algebra and presents the syncopated form; his studies of indeterminant equations have influenced mathematicians of all periods

ca. 263 *Liu Hui* was the best-known Chinese mathematician of the third century of the current era; much of his work was concerned with mensuration

ca. 350 *Pappus of Alexandria* a late Greek geometer and mathematical historian, compiled surveys of mathematics which provide a great deal of our knowledge of Greek mathematics

475–524 *Boethius* a Roman philosopher and mathematician, who, under the Ostrogoths, compiled works on arithmetic, geometry, logic, and music; he was mainly a translator of Greek works.

ca. 530 *Aryabhata* one of the earliest of the Hindu mathematicians of significance, was interested mainly in algebraic equations but also accurately computed the value of π to five places; his work was neglected by later generations

ca. 550 *Chang-Chiu-Chien* was an arithmetician; his work encompassed fractions and the modern rule of division by multiplying by the reciprocal of the divisor

b. 598 *Brahmagupta of Ujjayini* a Hindu authority on arithmetic and algebra, was most creative in the study of equations, where he gave rules for multiplying negative numbers and zero

673–735 *Venerable Bede of Jarrow* mainly a theologian, was significant in keeping the spirit of learning alive in Europe in the dark ages

d. 840 *Al-Khwarizmi of Bagdad* the greatest of the early Arab mathematicians, wrote the first work bearing the name "algebra," a manuscript based upon Greek models; he also wrote on arithmetic and coined the term "algorism"

ca. 850 — *Mahāvīra the Learned of Mysore* a Hindu algebraist and arithmetician, based much of his work upon that of Brahmagupta; his work is the most noteworthy of the Hindu contributions, with the exception of Bhaskara

ca. 929 — *Al-Battani of Batan, Syria* an Arab astronomer, compiled new astronomical tables and also made important advances in trigonometry

940–998 — *Abul-Wefa of Bagdad* an authority on trigonometry, introduced the tangent and compiled tables of sines and tangents in ten-minute intervals

ca. 991 — *Sridhara* a Hindu arithmetician, included operations with zero in his work; much of Bhaskara's work was anticipated by Sridhara

ca. 1010 — *Al-Karkhi of Bagdad* known for his works on arithmetic and algebra, was one of the last of the Bagdad mathematicians; his treatment of algebraic equations includes fractional solutions for indeterminant polynomials in three unknowns

1114–1185 — *Bhaskara of Biddur* the most accomplished of the Hindu mathematicians, ranged through astronomy, mensuration, arithmetic, and algebra; he clearly foresaw the dangers of multiplication and division by zero and also introduced directed numbers

ca. 1123 — *Omar Khayyám of Persia* poet, astronomer, and algebraist, wrote on Euclid, astronomy, and algebra; the known records of his interest in the cubic suggest that he may have anticipated the sixteenth-century Italians

ca. 1150 — *Robert of Chester* (or *Adelard of Bath*) basically a translator, translated the algebra of Al-Khowarizmi into Latin under the title *Algoritmi de Numero indorum*

1135–1204 — *Moses Maimonides of Córdoba* a Jewish scholar, philosopher, and physician to the Sultan Saladin, became an astronomer of prominence

ca. 1150 — *Johannes Hispalensis* (or *John of Seville*) a Jew who converted to Christianity, made his most significant contribution as a translator of the *Almagest* into Latin

1114–1187 — *Gerard* (or *Gherardo*) *of Cremona* a translator of Arabic works, in his translation of Ptolemy's *Almagest* was one of the earliest to use *sinus* for the half chord

1175–1250 — *Leonardo of Pisa* (or *Fibonacci*) a translator and mathematician, published his *Liber abaci* in 1202; a treatise on arithmetic and algebra, it bore many marks of originality and introduced Arabic-Hindu numerals into Europe

1201–1274 — *Nasir-Eddin of Tus* (*Khorasan*) accomplished in trigonometry, astronomy, and geometry, developed plane trigonometry as a science

1214–1294 — *Roger Bacon* (English) was one of the most famous scholars of the Middle Ages; perhaps his foremost contribution was his insistence that science should be founded upon experiment. He studied at the universities of Paris and Oxford and in his written work exhibited a knowledge of Euclid, Ptolemy, Hipparchus, Apollonius, and Archimedes

ca. 1230 — *Jordanus Nemorarius of Saxony* compiled works on algebra, arithmetic, and statics and, by using letters of the alphabet as symbols, contributed to the evolution of algebraic symbolism

1323–1382 *Nicole Oresme, Bishop of Normandy* (French), in his *Algorismus proportionum* conceived of fractional indices; he also developed a crude coordinate geometry

1401–1464 *Nicolas of Cusa, Cardinal at Rome* (German), is best remembered for his attempts at the quadrature of the circle; his values were inferior to earlier Arab calculations.

1423–1461 *Georg Puerbach of Vienna* is known for his translation of the *Almagest*

1436–1476 *Regiomontanus* (or *Johann Müller of Königsberg*) was accomplished in the study of trigonometry and algebra; a student of Puerbach, he extended the science of trigonometry, calculating a table of sines and tangents for every minute of arc; he introduced a semi-symbolic rhetorical algebra

1450–1520 *Luca Pacioli of Tuscany* was a mathematician whose *Summa de arithmetica* and *De divina proportione* are merely collections of the earlier works; he did, however, introduce symbols for addition, subtraction, and the square root

1452–1519 *Leonardo da Vinci* (Italian) was a painter and amateur scientist; in his drawings and writings one perceives clearly the emphasis upon accurate observation so essential to experimental science

1473–1543 *Nicolaus Copernicus* (Polish), an astronomer, publicized the doctrine of the heliocentric universe and also wrote a treatise on trigonometry, although he was not a mathematician

1530–1590 *Giovanni Battista Benedetti* (Italian) postulated the first law of mechanics; he fully understood the concept of undisturbed motion

1506–1557 *Nicola Fontana* (or *Niccolò Tartaglia*) (Venetian) was credited by Cardano with the discovery of the solution of the cubic and also wrote on mechanics; Fontana was limited by his inability to write in Latin, the accepted scientific language of his day

1501–1576 *Girolamo Cardano* (or *Jerome Cardan*) of Milan and Pavia, perhaps the most gifted scientist and mathematician of his period, presented in his *Ars magna* the solutions of the cubic and quartic equations; he wrote many popular works on science and among his many publications is found a thorough analysis of his own dreams

1540–1603 *François Viète* the greatest French algebraist of the sixteenth century, was originally trained as a lawyer and later turned to mathematics; his *De aequationum recognitione et emendatione* is a landmark in the development of the theory of equations

1540–1603 *William Gilbert* (English) performed a systematic study of magnetism and electrostatics; he exemplifies the beginning of experimental science in Europe

1548–1620 *Simon Stevin* (Dutch) was a mathematician and theoretical physicist who initiated the use of decimal fractions and also introduced modern trigonometry by considering chords in terms of half angles; his contributions to mechanics and hydrostatics include a study of the conditions for static equilibrium

1550–1617 *John Napier, Baron Merchiston* (Scottish), discovered logarithms; his method was tedious but in its fundamentals was a logarithmic method.

1564–1642 *Galileo Galilei* (Italian), famed for his fundamental work on mechanics, both experimental and theoretical, was led by his understanding of the laws of falling bodies to a correct description of the path of a projectile; in his writings one finds many ingenious mathematical arguments, particularly concerning the indefiniteness of infinity

1571–1630 *Johannes Kepler* (German), a mathematician and astronomer, was professor of mathematics at Graz and later imperial mathematician to Rudolph II; his three laws of planetary motion were essential for later tests of the universal law of gravitation, and his analysis of areas and volumes made an important contribution to the approximate integral calculus

1591–1626 *Willebrord Snell* (or *Snell van Royen*) (Dutch) discovered the law of refraction

1593–1662 *Gérard Desargues* (French), a mathematician and engineer, is credited as the founder of modern projective geometry

1595–1633 *Albert Girard* (Dutch) was one of the most distinguished mathematicians of his time; although little known, he was one of the first to discuss imaginary roots of polynomials; in his work the contractions sin, cos, and tan of the trigonometric functions first appear

1596–1605 *René Descartes* (French), philosopher, mathematician, and scientist, began his career as a soldier and eventually settled in Holland; his writings cover almost every branch of knowledge, and his fame rests in part on his invention of analytic geometry; so overwhelming was his influence that his point of view on the approach to science was to impose a constraint upon generations to follow

1598–1647 *Bonaventura Cavalieri* of Bologna, was a student of Galileo and an early contributor to the integral calculus; although his results were correct, his method was unsound

1601–1665 *Pierre de Fermat* (French), although an amateur, was one of the most influential mathematicians of all time; he independently discovered coordinate geometry and anticipated the calculus with rigorous works on slopes, areas, and extrema; in physics his "Least Time Principle" is still a fundamental concept; he is best remembered for his contributions to number theory, the theory of probability, and the theory of equations

1602–1675 *Gilles Personne de Roberval* (French), a philosopher and mathematician, became known for his discoveries in the field of higher plane curves and, in particular, for his method for constructing tangents, which involved the vector resolution of the tangent

1608–1647 *Evangelista Torricelli* (Italian), a pupil of Galileo, made significant contributions in hydrodynamics and invented the barometer; his contributions to geometry included the integration of a number of curves and squaring the cycloid

1616–1703 *John Wallis* (English), famed for his advances in the integral calculus; introduced fractional indices for powers and roots and was also the first to obtain a quadrature of the circle in terms of an infinite product

1623–1662 *Blaise Pascal* (French), philosopher and mathematician, in his brief life laid the foundations of probability theory and accomplished the derivatives of certain curves; his essay on the equilibrium of fluids is Pascal's law

1627–1691 *Robert Boyle* (English) was a scientist whose fame rests in large measure on his work on the physics of gases and on his many contributions to chemistry

1629–1695 *Christian Huygens* (Dutch), was a great physicist ranking with Newton and Descartes; in his work the beginnings of a theoretical physics can be found; he applied mathematics to the pendulum, rotary motion, the refraction of light, and the laws of falling bodies; he was an accomplished mathematician, on which subject no more need be said than that he was the teacher of Leibniz

1630–1677 *Isaac Barrow* (English), a mathematician and teacher of Newton, was mainly interested in optics and geometry; Newton was obviously influenced by Barrow's contributions to the methods of obtaining tangents to curves

1635–1703 *Robert Hooke* (English) would be regarded, had it not been for Newton, as one of the outstanding scientists of his day; he made fundamental contributions to optics and elasticity, but his mathematical exposition was weak

1642–1727 *Sir Isaac Newton* perhaps the greatest theoretician of all time, was born in England at the time of the Cromwell rebellion; he showed only average aptitude in his studies until he reached Cambridge in 1660; in the next twenty years, however, he proposed the fundamental theorem of the calculus, developed a general binomial expansion, and provided the universal law of gravitation and its application to planetary orbits; his contributions to optics still bear his name

1644–1710 *Olaus Roemer* (Danish), first accurately measured the velocity of light

1646–1716 *Baron Gottfried Wilhelm von Leibniz* (German), although an amateur, discovered the fundamental theorem of the calculus with Newton; his method and notation were far superior to the fluxions of Newton, and his pupils and followers dominated the development of the calculus for the next several hundred years

1654–1705 *Jacques* (*James*) *Bernoulli* a Belgian born in Switzerland, was the first of a famous family of mathematicians; after fleeing the reign of terror instituted by the Duke of Alba, his father settled in the free city of Basel; Jacques traveled widely before taking up the study of the calculus in 1682, as established by Leibniz; he contributed to the fields of differential equations, probability theory, infinite series, and the calculus of variations

1667–1748 *Jean* (*John*) *Bernoulli* the brother and equal of Jacques, was appointed to a professorship at Groningen and, on Jacques' death, assumed his brother's professorship of mathematics at Basel; both men contributed to the invention of partial differentiation along with the calculus of two and three dimensions, but the discovery of the exponential calculus is attributed to Jean

1667–1754 *Abraham de Moivre* born in England of French parentage, made contributions to the theory of probability and to the trigonometry of certain complex functions

1682–1716 *Roger Cotes* was the logical successor to Newton; his early death deprived England of the man who, to judge from all evidence, would have extended Newton's work

1685–1731 *Brook Taylor* (English) developed the power series representation of functions

1698–1746 *Colin Maclaurin* (Scottish) initiated the power series bearing his name; in 1724 he was awarded the prize of the Academie des Sciences for his essay on the impact of infinitely hard bodies; unfortunately, his solution was incorrect by modern standards, while that submitted by Jean Bernoulli, which was correct, was rejected; in his work Bernoulli anticipated the delta function

1698–1759 *Pierre Louis Moreau de Maupertuis* (French) worked on the calculus and mechanics; to support Descartes' incorrect solution for the law of refraction, he introduced the "Principle of Least Action," which, in spite of its motivation, became a fundamental part of dynamics

1700–1782 *Daniel Bernoulli* was the son of Jean and one of the nine great mathematicians in this family; like his father and his uncle Jacques, he expanded the boundaries of the calculus; his studies of mechanics included a solution of the partial differential equation for the vibrating string

1707–1783 *Leonhard Euler* (Swiss), one of the greatest intuitive mathematicians of all time, was educated at Basel where he came under the influence of the Bernoullis; his three monumental works on the calculus freed analysis of its geometric limitation; the remainder of his scholarly work on dynamics and hydrodynamics was of the most fundamental significance

1717–1783 *Jean le Rond d'Alembert* (French) generalized Newton's approach to dynamics and anticipated the later methods of Lagrange and Hamilton; he created the concept of virtual work

1731–1810 *Henry Cavendish* (English) was one of the discoverers of the inverse square law of electrostatics and also paved the way for the study of the capacity of condensers

1733–1804 *Joseph Priestley* (English) discovered oxygen and deduced the inverse square law of electrostatics from the absence of a force field inside a charged sphere

1736–1806 *Charles Augustin de Coulomb* (French) demonstrated the inverse square law of electrostatics experimentally, using a torsion balance

1736–1813 *Joseph Louis Lagrange* (Italian) was one of the most inventive mathematicians and theoretical physicists; his interest in mathematics began with astronomy; he is known particularly for his generalization of mechanics; as important, however, were his contributions to number theory, to the calculus of probability, and to the theory of algebraic polynomials

1737–1798 *Aloisio Galvani* (Italian) discovered dynamic electricity

1745–1818 *Caspar Wessel* (Norwegian), a surveyor, first proposed the geometric interpretation of complex numbers; his priority was not recognized until the twentieth century

1749–1827 *Pierre Simon, Marquis de Laplace* (French), was a mathematician and astronomer; after working under D'Alembert, Laplace made important contributions to the theory of astronomical stability; his great work was *Mécanique céleste*, and he made notable contributions to the theory of probability and to the theory of partial differential equations

1752–1833 *Adrien Marie Legendre* (French) began his career with the study of ballistics but soon settled upon mathematics and made contributions of the highest order to number theory, analysis, and geometry; his work on elliptic integrals demonstrates a mathematical ability of the first rank

1768–1830 *Baron Jean Baptiste Joseph Fourier* (French), invented the Fourier series; after accompanying Napoleon to Egypt, he returned to France in 1801; while studying the theory of heat flow, he evolved the expansion method which bears his name

1768–1822 *Jean Robert Argand* (French) discovered, independent of Wessel but some years later, the complex plane and the diagrams named for him

1773–1829 *Thomas Young* (English), an intuitive scientist of the first rank, described optical interference by pictorial analogy and was instrumental in establishing the transverse nature of light

1774–1862 *Jean Baptiste Biot* (French) with Savart was the first to describe the magnetic field of a long straight wire

1791–1841 *Félix Savart* (French) with Biot was the first to describe the magnetic field of a long straight wire

1775–1812 *Étienne Louis Malus* (French) was the first to identify the polarization of light after reflection from a dielectric surface

1775–1836 *André Marie Ampère* (French) is renowned for his discovery and description of the magnetic interaction between current-carrying wires

1777–1851 *Hans Christian Oersted* (Danish) discovered the magnetism of current-carrying conductors

1777–1855 *Karl Friedrich Gauss* (German) founded modern mathematics; his work ranged over the whole of mathematics; he initiated the study of complex numbers and following Legendre he completely developed the modular number systems; his early work contained a proof of the fundamental theorem of algebra

1778–1853 *Hoene Wroński* (Polish) made a major contribution to a philosophy of mathematics

1781–1840 *Siméon Denis Poisson* (French) contributed to the theory of probability, algebraic equations, differential equations, and the calculus of variations; physicists know him for the inhomogeneous differential equation which bears his name and for his separation of the Navier-Stokes equation into rotational and irrotational components

1785–1836 *Claude L. M. H. Navier* (French) first derived the equations of motion for an elastic solid, in 1821

1786–1853 *Dominique François Jean Arago* (French), an astronomer and physicist, made important advances in the study of magnetism and optics

1787–1854 *Georg Simon Ohm* (German), known for his studies of electrical resistance, introduced the notion of electromotive force and electrical resistance

1787–1826 *Joseph von Fraunhofer* (German) discovered the double line of sodium and was the first to observe spectral lines using gratings

1788–1827 *Augustin Jean Fresnel* (French), noted for his contributions to optics, used mathematical analysis to prove the undulatory theory and was instrumental in establishing the transverse nature of light and in describing light propagation in anisotropic crystals

1789–1857 *Augustin Louis Cauchy* (French) is generally regarded as the first to introduce rigor into mathematics; his investigation of the theory of complex variables led to rigorous treatments of convergence of series and analyticity of functions; he made major contributions to number theory, the algebra of polynomials, and higher spaces and also worked on problems of electricity and optics

1790–1868 *August Ferdinand Moebius* (or *Möbius*) (German) made important contributions to higher geometry and topology; he is

	known for the bariocentric calculus, an early attempt at a vector space representation
1791–1867	*Michael Faraday* (English), a physicist of profound intuitive genius, is known in large part for his experimental discoveries; his influence upon the course of theoretical physics was significant; his pictorial description of lines of force and polarization in a dielectric led to sophisticated mathematical representations
1793–1841	*George Green* (English), whose interests lay in the mathematical analysis of problems of electrostatics and optics, is remembered for the mathematical theorems bearing his name
1793–1856	*Nicolai Ivanovitch Lobachevski* (Russian) rejected the parallel postulate of Euclid and substituted the contrary postulate, thus creating a valid new geometry, the non-Euclidean geometry
1796–1832	*Nicolas Léonard Sadi Carnot* (French) introduced the concept of reversibility into thermodynamics
1798–1895	*Franz Ernst Neumann* (German) stressed the importance of energy conservation in the analysis of the magnetic fields of currents and created the magnetic vector potential
1802–1829	*Niels Henrik Abel* (Norwegian) was the first to show that a general solution in rational radicals to the algebraic equation of fifth degree is impossible; with Galois he aided in the creation of the theory of groups; his work on elliptic functions was unique and did much to clarify a difficult subject; he was the first to recognize integral equations as an independent subject for study
1802–1860	*János Bolyai* (Hungarian) was the son of Farkas Bolyai, another well-known mathematician; independent of Lobachevski he also created a non-Euclidean geometry
1803–1855	*Jacques Charles François Sturm* (Swiss) is known for his analysis of linear differential equations of second order
1804–1891	*Wilhelm Weber* (German) first correctly identified an electric current as streaming electric charges; he initiated the electron theories of matter, and his view of the phenomenon of magnetism was quite accurate
1805–1865	*William Rowan Hamilton* (Irish), known for his reformulation of analytic dynamics, also furthered the cause of astronomy and contributed to optics; in 1843 he invented the quaternian algebra, which had a decided influence on the development of modern algebra
1805–1859	*Peter Gustav Lejeune Dirichlet* (French), noted for his studies in the theory of numbers, studied under Gauss and later became professor of mathematics at Göttingen
1806–1871	*Augustus De Morgan* (Irish) contributed to the theory of probability and also came close to quaternians in his studies of double algebra
1809–1877	*Hermann Günther Grassmann* (German) created modern algebra; because he was a secondary school teacher, and because of the elaborate presentation of his ideas, his work was little regarded in his lifetime; when the significance of his multiple algebra was finally appreciated, a controversy developed with the followers of W. R. Hamilton, which conflict in no way aided the cause of mathematics
1809–1882	*Joseph Liouville* (French), whose fame lies in the field of second-degree linear differential equations, number theory, and

complex variables, was the first to prove the existence of transcendental numbers

1811–1832 *Évariste Galois* (French) in the space of four years before his death at the age of twenty-one took a giant step in mathematics; his work on algebraic equations of higher degree led him to invent the theory of groups; the concept of a field also appears for the first time in his work

1814–1897 *James Joseph Sylvester* (English) made important contributions to the theory of algebraic invariants and determinants; with Cayley he contributed to the invention of matrices; at The Johns Hopkins University he established the first graduate department in mathematics in the United States

1815–1864 *George Boole* (Irish) invented the algebra which bears his name; because his algebra and fundamentals of logic have been so prominent, Boole's advances in the areas of algebraic invariances, difference equations, etc., have often been overlooked

1815–1897 *Karl Weierstrass* (German) led in the movement to arithmetize analysis and made major contributions to the theory of numbers, elliptic, and Abelian functions; he was a pioneer in the attempt to base mathematics on logic

1819–1903 *Sir George Gabriel Stokes* (English) was primarily a theoretical physicist; from vibrations in crystals and dynamics to theories of light propagation, his genius left its mark; in 1896 he advanced the hypothesis that X-rays were, in fact, electromagnetic radiation of short wave length

1821–1895 *Arthur Cayley* (English) developed matrices and the theory of algebraic invariants; his initial work was on elliptic functions, but his main impact on physics has been his algebraic creation, the matrix, and the multiplication table representation of groups

1821–1894 *Hermann Ludwig Ferdinand von Helmholtz* (German) stressed energy conservation as a basis for physics, supported the view of heat as energy, and advanced the energy concept of the electromagnetic field

1822–1888 *Rudolf Julius Emanuel Clausius* (German) proposed with Lord Kelvin the second law of thermodynamics, and in the analysis of this law created the concept of entropy

1823–1891 *Leopold Kronecker* (German) created the Euclidean program whereby all mathematics was to be reduced to pure formalism; an authority on number theory, algebra, and higher arithmetic, he was widely known to the mathematical public for his philosophy of mathematics

1824–1887 *Gustav Robert Kirchhoff* (German), noted for his researches on atomic spectra, heat radiation, and electromagnetic theory, gave his name to the fundamental rules of circuit analysis

1824–1907 *William Thomson, Lord Kelvin* (English), created the modern nomenclature for the magnetic field vectors $\vec{\mathbf{B}}$ and $\vec{\mathbf{H}}$, was the first to obtain the energy storage in terms of the fields alone, is credited, with Clausius, with the second law of thermodynamics, and influenced the course of physics for fifty years

1826–1866 *Georg Friedrich Bernhard Riemann* (German) created the n-dimensional manifolds; he set forth the system of surfaces for complex functions which bear his name, redefined integration, and made many contributions to analysis spanning differential equations and functions

1831–1879 *James Clerk Maxwell* (Scottish) founded modern electro-

	magnetic theory through his systematization of the basic equations of electromagnetism; he was one of the originators of the statistics for many-bodied systems, which are known as the Maxwell-Boltzmann statistics
1831–1916	*Julius Wilhelm Richard Dedekind* (German) was one of the most prominent nineteenth-century contributors to the theory of algebraic numbers, modular and Abelian functions, and quadratic forms
1834–1907	*Dimitri Ivanovich Mendeleev* (Russian) proposed the periodic table of the elements
1838–1916	*Ernst Mach* (German) was an authority on mechanics and a philosopher of physics whose principle became an essential part of modern relativity theory
1839–1903	*Josiah Willard Gibbs* (American) with Boltzmann completed and refined the classical theory of statistical mechanics; he invented modern vector analysis
1842–1919	*J. W. Strutt, Baron Rayleigh* (English), was an authority on wave propagation, optics, and spectroscopy; his early analysis of black body radiation was done in cooperation with Sir James Jeans
1842–1899	*Marius Sophus Lie* (Swedish) is known for his work in differential equations, differential geometry, transformation groups, and, most of all, for his theory of infinite continuous groups
1844–1906	*Ludwig Boltzmann* (Austrian) with Gibbs and Maxwell founded classical statistical mechanics
1845–1879	*William Kingdon Clifford* (English) contributed works on Riemann's surfaces, biquaternions, and topological surfaces; in 1870 he expressed his belief that matter is only a manifestation of curvature in a space-time manifold
1845–1918	*Georg Cantor* (German), born of a Danish family in Russia, a mathematician of great originality who created set theory, contributed to the foundation of the modern approach to the mathematically infinite
1845–1923	*Wilhelm Konrad Roentgen* (German) discovered X-rays
1848–1912	*G. Frobenius* (German) contributed to the development of group algebras
1848–1901	*Henry Augustus Rowland* (American) established the identity of static and dynamic electrical charges and is also known for the construction of diffraction gratings
1850–1925	*Oliver Heaviside* (English) is known for his theories of electromagnetism and for several intuitive mathematical discoveries; his many contributions included an analysis of submarine cables and the formulation of the law governing the force on a charged particle moving in an external field
1852–1908	*Antoine Henri Becquerel* (French) discovered radioactivity
1853–1928	*Hendrik Antoon Lorentz* (Dutch) proposed the electron theories of electromagnetism and the space-time transformation named for him
1853–1925	*M. M. G. Ricci* (Italian) created the general tensor calculus
1854–1912	*Jules Henri Poincaré* (French) made some contribution to almost every branch of modern mathematics or physics; his researches are seen in the theory of Fuchsian functions, modular functions, and the theory of groups, to mention but a few, and in astronomy and in the theory of relativity he played a primary role

1856–1940 *Sir Joseph John Thomson* (English) discovered the electron, but his influence on the Cavendish Laboratory and his students —for example, Rutherford—was also very significant

1856–1894 *T. J. Stieltjes* (Dutch) with Poincaré laid the foundations of the theory of real variables

1857–1894 *Heinrich Rudolph Hertz* (German) is best known for the discovery of electromagnetic radiation; he was also one of the first to detect the photoelectric effect

858–1947 *Max Planck* (German), in deriving a complete black body radiation formula postulated the quantization of oscillators; he is the founder of the quantum theory; his contributions to the theory of relativity include the relativistic mechanics and the postulate of the gravitation of energy

1859–1906 *Pierre Curie* (French) with his wife discovered polonium and radium; he was also famous for a theoretical analysis of ferromagnetism

1862–1943 *David Hilbert* (German) is the modern giant of mathematical physics and of critical mathematics; he reintroduced the postulational basis of geometry and went on to a critical examination of the structure of common arithmetic; his famous work on mathematical analysis became the bible of modern theoretical physics

1864–1928 *Wilhelm Wien* (German) is noted for his preliminary theories of the black body radiation spectrum

1864–1909 *Hermann Minkowski* (Polish) formulated physics in terms of four-dimensional manifolds

1866–1927 *I. Fredholm* (Swedish) pioneered in the study and analysis of integral equations

1867–1934 *Marja Skłodowska Curie* (Polish) discovered with her husband the radioactive substances polonium and radium

1868–1951 *Arnold Sommerfeld* (German) refined the Bohr representation of the atom to include space quantization

1871–1937 *Ernest Rutherford* (English), a pupil of Thomson, discovered the planetary atom; his investigations of radioactivity were of the highest significance, and his leadership and that of J. J. Thomson made the Cavendish Laboratory

1873–1942 *Tullio Levi-Civita* (Italian) with Ricci extended the tensor calculus and applied it to relativity

1875–1941 *Henri Léon Lebesgue* (French) created a usable theory of divergent series and revolutionized the theory of integration; his additions to measure theory played an important role in the modern theory of real variables

1879–1955 *Albert Einstein* (German) was one of the greatest scientific minds of all time; his work on the photon, the quantum theory of specific heats, and relativity both general and special represented major advances in several fields of physics

1880–1933 *Paul Ehrenfest* (Dutch) is mainly noted for his extension of statistical mechanics into the realm of quantum theory; he was one of the first to propose the quantization of angular momentum

1882– *Max Born* (German) influenced several generations of younger scientists through his contributions to theoretical physics

1884– *Peter Debye* (Dutch) developed the theory of specific heats of solids

1885–1962 *Niels Bohr* (Danish) discovered the model for the quantized atom

1887–1961	*Erwin Schrödinger* (Austrian) created the nonrelativistic equation of quantum mechanics and also made contributions to special relativity theory and the theory of specific heats
1892–	*Louis Victor de Broglie* (French) proposed the theory that particles, like photons, behave as waves
1901–1954	*Enrico Fermi* (Italian) is noted for the Fermi statistics, the theory of beta decay, and the discovery of the slow neutron
1901–	*Werner Heisenberg* (German) developed a matrix mechanics representation of quantum mechanics equivalent to the Schrödinger method; the uncertainty principle bears his name
1902–	*Paul A. M. Dirac* (English) discovered the appropriate relativistic form for the defining equation of quantum mechanics

Index

Q

R

197

4418